P9-DEH-197

AMPL

A Modeling Language

for Mathematical Programming

About the Authors

Robert Fourer received his Ph.D. in operations research from Stanford University in 1980, and has been active as a researcher in mathematical programming and modeling language design. He joined the Department of Industrial Engineering and Management Sciences at Northwestern University in 1979, and was named chairman of the department in 1989.

David M. Gay received his Ph.D. in computer science from Cornell University in 1975, and has been a member of technical staff in the Computing Science Research Center at AT&T Bell Laboratories since 1981. His research interests include numerical analysis, optimization, and scientific computing.

Brian W. Kernighan is a member of technical staff in the Computing Science Research Center at AT&T Bell Laboratories. He received his Ph.D. in electrical engineering from Princeton in 1969. He is the co-author of several computer science books, including *The C Programming Language* and *The UNIX Programming Environment*.

AMPL

A Modeling Language

for Mathematical Programming

Robert Fourer

Northwestern University

David M. Gay

Brian W. Kernighan

AT&T Bell Laboratories

boyd & fraser publishing company

 The Scientific Press Series

Cover Design: Rogondino & Associates
Manufacturing Coordinator: Gordon Woodside

AMPL: A Modeling Language for Mathematical Programming
by Robert Fourer, David M. Gay, and Brian W. Kernighan

Originally published by the Scientific Press. ©1993 by American Telephone and Telegraph, Inc., and Robert Fourer

boyd & fraser publishing company
One Corporate Place • Ferncroft Village
Danvers, Massachusetts 01923

 The Scientific Press Series

I(T)P
International Thomson Publishing
boyd & fraser publishing company is an ITP company.
The ITP trademark is used under license.

All rights reserved. No part of this work may be reproduced or used in any form or by any means —graphic, electronic, or mechanical, including photocopying, recording, taping, or information storage and retrieval systems—without written permission from the publisher.

Names of all products mentioned herein are used for identification purposes only and may be trademarks and/or registered trademarks of their respective owners. boyd & fraser publishing company disclaims any affiliation, association, or connection with, or sponsorship or endorsement by such owners.

Manufactured in the United States of America.

ISBN: 0-78950-701-3

This book was typeset (`grap|pic|tbl|eqn|troff  -mpm`) in Times and Courier by the authors, using a Linotronic 200P phototypesetter and an SGI Challenge XL running the UNIX® operating system.

7 8 9 10   02 01 00 99 98

Contents

Introduction

As our title suggests, there are two aspects to the subject of this book. The first is mathematical programming, the optimization of a function of many variables subject to constraints. The second is the AMPL modeling language, which we designed and implemented to help people use computers to develop and apply mathematical programming models.

We intend this book as an introduction both to mathematical programming and to AMPL. For readers already familiar with mathematical programming, it can serve as a user's guide and reference manual for the AMPL software. We assume no previous knowledge of the subject, however, and hope that this book will also encourage the use of mathematical programming models by those who are new to the field.

Mathematical programming

The term "programming" was in use by 1940 to describe the planning or scheduling of activities within a large organization. "Programmers" found that they could represent the amount or level of each activity as a *variable* whose value was to be determined. Then they could mathematically describe the restrictions inherent in the planning or scheduling problem as a set of equations or inequalities involving the variables. A solution to all of these *constraints* would be considered an acceptable plan or schedule.

Experience soon showed that it was hard to model a complex operation simply by specifying constraints. If there were too few constraints, many inferior solutions could satisfy them; if there were too many constraints, desirable solutions were ruled out, or in the worst case no solutions were possible. The success of programming ultimately depended on a key insight that provided a way around this difficulty. One could specify, in addition to the constraints, an *objective*: a function of the variables, such as cost or profit, that could be used to decide whether one solution was better than another. Then it didn't matter that many different solutions satisfied the constraints — it was sufficient to find one such solution that minimized or maximized the objective. The term *mathematical programming* came to be used to describe the minimization or maximization of an objective function of many variables, subject to constraints on the variables.

In the development and application of mathematical programming, one special case stands out: that in which all the costs, requirements and other quantities of interest are terms strictly proportional to the levels of the activities, or sums of such terms. In mathematical terminology, the objective is a linear function, and the constraints are linear equations and inequalities. Such a problem is called a *linear program*, and the process of setting up such a problem and solving it is called *linear programming*. Linear programming is particularly important because a wide variety of problems can be modeled as linear programs, and because there are fast and reliable methods for solving linear programs even with thousands of variables and constraints. The ideas of linear programming are also important for analyzing and solving mathematical programming problems that are not linear.

All useful methods for solving linear programs require a computer. Thus most of the study of linear programming has taken place since the late 1940's, when it became clear that computers would be available for scientific computing. The first successful computational method for linear programming, the simplex algorithm, was proposed at this time, and was the subject of increasingly effective implementations over the next decade. Coincidentally, the development of computers gave rise to a now much more familiar meaning for the term "programming."

In spite of the broad applicability of linear programming, the linearity assumption is sometimes too unrealistic. If instead some smooth nonlinear functions of the variables are used in the objective or constraints, the problem is called a *nonlinear program*. Solving such a problem is harder, though in practice not impossibly so. Although the optimal values of nonlinear functions have been a subject of study for over two centuries, computational methods for solving nonlinear programs in many variables were developed only in recent decades, after the success of methods for linear programming. The field of mathematical programming is thus also known as *large scale optimization,* to distinguish it from the classical topics of optimization in mathematical analysis.

The assumptions of linear programming also break down if some variables must take on whole number, or integral, values. Then the problem is called *integer programming,* and in general becomes much harder. Nevertheless, a combination of faster computers and more sophisticated methods have made large integer programs increasingly tractable in recent years.

The AMPL modeling language

Practical mathematical programming is seldom as simple as running some algorithmic method on a computer and printing the optimal solution. The full sequence of events is more like this:

- Formulate a model — the abstract system of variables, objectives, and constraints that represent the general form of the problem to be solved.
- Collect data that define one or more specific problem instances.
- Generate a specific objective function and constraint equations from the model and data.

- Solve the problem — run a program to apply an algorithm that finds optimal values of the variables.
- Analyze the results.
- Refine the model and data as necessary, and repeat.

If people could deal with mathematical programs in the same way that algorithms do, the formulation and generation phases of modeling might be relatively straightforward. In reality, however, there are many differences between the form in which human modelers understand a problem and the form in which algorithms solve it. Conversion from the ''modeler's form'' to the ''algorithm's form'' is consequently a time-consuming, costly, and often error-prone procedure.

In the special case of linear programming, the largest part of the algorithm's form is the constraint coefficient matrix, which is the table of numbers that multiply all the variables in all the constraints. Typically this is a very sparse (mostly zero) matrix with hundreds or thousands of rows and columns, whose nonzero elements appear in intricate patterns. A computer program that produces a compact representation of the coefficients is called a matrix generator. Several programming languages have been designed specifically for writing matrix generators, and standard computer programming languages are also often used.

Although matrix generators can successfully automate some of the work of translation from modeler's form to algorithm's form, they remain difficult to debug and maintain. One way around much of this difficulty lies in the use of a *modeling language* for mathematical programming. A modeling language is designed to express the modeler's form in a way that can serve as direct input to a computer system. Then the translation to the algorithm's form can be performed entirely by computer, without the intermediate stage of computer programming. Modeling languages can help to make mathematical programming more economical and reliable; they are particularly advantageous for development of new models and for documentation of models that are subject to change.

Since there is more than one form that modelers use to express mathematical programs, there is more than one kind of modeling language. An *algebraic* modeling language is a popular variety based on the use of traditional mathematical notation to describe objective and constraint functions. An algebraic language provides computer-readable equivalents of notations such as $x_j + y_j$, $\Sigma_{j=1}^{n} a_{ij} x_j$, $x_j \geq 0$, and $j \in S$ that would be familiar to anyone who has studied algebra or calculus. Familiarity is one of the major advantages of algebraic modeling languages; another is their applicability to a particularly wide variety of linear, nonlinear and integer programming models. While successful algorithms for mathematical programming first came into use in the 1950's, the development and distribution of algebraic modeling languages only began in the 1970's. Since then, advances in computing and computer science have enabled such languages to become steadily more efficient and general.

This book describes AMPL, a relatively recent entry into the field of algebraic modeling languages for mathematical programming. AMPL is notable for the similarity of its arithmetic expressions to customary algebraic notation, and for the generality of its set and subscripting expressions. AMPL also extends algebraic notation to express common mathematical programming structures such as network flow constraints and piecewise linearities.

AMPL offers an interactive command environment for setting up and solving mathematical programming problems. A flexible interface enables several solvers to be available at once and allows a user to switch among solvers and to select options that may improve solver performance. Once optimal solutions have been found, they are automatically translated back to the modeler's form so that people can view and analyze them. All of the general set and arithmetic expressions of the AMPL modeling language can also be used for displaying data and results; a variety of options are available to format data for browsing on a screen, printing reports, or preparing input to other programs.

Through its emphasis on AMPL, this book differs considerably from the presentation of modeling in standard mathematical programming texts. The approach taken by a typical textbook is still strongly influenced by the circumstances of 10 or 20 years ago, when a student might be lucky to have the opportunity to solve a few small linear programs on any actual computer. As encountered in such textbooks, mathematical programming often appears to require only the conversion of a "word problem" into a small system of inequalities and an objective function, which are then presented to a simple optimization package that prints a short listing of answers. While this can be a good approach for introductory purposes, it is not workable for dealing with the hundreds or thousands of variables and constraints that are found in most real-world mathematical programs.

The availability of an algebraic modeling language makes it possible to emphasize the kinds of general models that can be used to describe large-scale optimization problems. Each AMPL model in this book describes a whole class of mathematical programming problems, whose members correspond to different choices of indexing sets and numerical data. Even though we use relatively small data sets for illustration, the resulting problems tend to be larger than those of the typical textbook. More important, the same approach, using still larger data sets, works just as well for mathematical programs of realistic size and practical value.

This book does not attempt to cover the optimization theory and algorithmic details that comprise the greatest part of most mathematical programming texts. Thus, for those readers who want to study the whole field in some depth, this book is a complement to existing textbooks, not a replacement. On the other hand, for those whose immediate concern is to apply mathematical programming to a particular problem, this book can provide a useful introduction on its own. It is accompanied by a version of AMPL and representative solvers, enough to easily handle problems of a few hundred variables and constraints on a personal computer. Versions that support much larger problems, additional solvers, and other operating systems are also available from the publisher.

Outline

The remainder of this book is organized conceptually into four parts. Chapters 1 through 4 are a tutorial introduction to models for linear programming:

1. Production Models: Maximizing Profits
2. Diets, Blending and Scheduling: Minimizing Costs
3. Transportation, Assignment and Minimum-Cost Flows
4. Building Larger Models

These chapters are intended to get you started using AMPL as quickly as possible. They include a brief review of linear programming and a discussion of a handful of simple modeling ideas that underlie most large-scale optimization problems. They also illustrate how to provide the data that convert a model into a specific problem instance, how to solve a problem, and how to display the answers.

The next four chapters describe the fundamental components of an AMPL linear programming model in detail, using more complex examples to examine major aspects of the language systematically:

5. Simple Sets and Indexing
6. Compound Sets and Indexing
7. Parameters and Expressions
8. Linear Programs: Variables, Objectives and Constraints

We have tried to cover the most important features, so that these chapters can serve as a general user's guide. Each feature is introduced by one or more examples, building on previous examples wherever possible.

The following two chapters describe how to use an AMPL model:

9. Specifying Data
10. Command Environment

Chapter 9 shows how to represent the data values that define a specific instance of a model. Chapter 10 explains the commands that read models and data, invoke solvers, and display or save results.

Finally, we turn to the rich variety of problems and applications beyond purely linear models. The remaining chapters deal with both special cases and generalizations:

11. Network Linear Programs
12. Columnwise Formulations
13. Nonlinear Programs
14. Piecewise-Linear Programs
15. Integer Linear Programs

Chapters 11 and 12 describe additional language features that help AMPL represent particular kinds of linear programs more naturally, and that may help to speed translation and solution. The last three chapters describe generalizations that can help models to be more realistic than linear programs, although they can also make the resulting optimization problems harder to solve.

Appendix A is the AMPL reference manual; it describes all language features, including some not mentioned elsewhere in the text.

Bibliography and exercises may be found in most of the chapters. All of the model and data files cited as examples are distributed with the AMPL software.

Acknowledgements

We are deeply grateful to Jon Bentley and Margaret Wright, who made extensive comments on several drafts of the manuscript. We also received many helpful suggestions on AMPL and the book from Collette Coullard, Gary Cramer, Arne Drud, Grace Emlin, Gus Gassmann, Eric Grosse, Paul Kelly, Mark Kernighan, Todd Lowe, Bob Richton, Michael Saunders, Robert Seip, Lakshman Sinha, Chris Van Wyk, Juliana Vignali, Thong Vukhac, and students in the mathematical programming classes at Northwestern University. Lorinda Cherry helped with indexing, and Jerome Shepheard with typesetting. Our sincere thanks to all of them.

Bibliography

E. M. L. Beale, ''Matrix Generators and Output Analyzers.'' In Harold W. Kuhn (ed.), *Proceedings of the Princeton Symposium on Mathematical Programming*, Princeton University Press (Princeton, NJ, 1970) pp. 25–36. A history and explanation of matrix generator software for linear programming.

Johannes Bisschop and Alexander Meeraus, ''On the Development of a General Algebraic Modeling System in a Strategic Planning Environment.'' Mathematical Programming Study 20 (1982) pp. 1–29. An introduction to GAMS, one of the first and most widely used algebraic modeling languages.

Anthony Brooke, David Kendrick and Alexander Meeraus, *GAMS: A User's Guide.* The Scientific Press (South San Francisco, CA, 1988). A complete description of the GAMS modeling language.

Robert Fourer, ''Modeling Languages versus Matrix Generators for Linear Programming.'' ACM Transactions on Mathematical Software **9** (1983) pp. 143–183. The case for modeling languages.

George B. Dantzig, ''Linear Programming: The Story About How It Began.'' In Jan Karel Lenstra, Alexander H. G. Rinnooy Kan and Alexander Schrijver, eds., *History of Mathematical Programming: A Collection of Personal Reminiscences.* North-Holland (Amsterdam, 1991) pp. 19–31. A source for our brief account of the history of linear programming. Dantzig was a pioneer of such key ideas as objective functions and the simplex algorithm.

1

Production Models: Maximizing Profits

As we saw in the Introduction, mathematical programming is a technique for solving certain kinds of optimization problems — maximizing profit, minimizing cost, etc. — subject to constraints on resources, capacities, supplies, demands, and the like. AMPL is a language for specifying such mathematical optimization problems. It provides a notation that is very close to the way that you might describe a problem mathematically, so that it is easy to convert from an algebraic description to AMPL.

We will concentrate initially on linear programming, which is the best known and easiest case; other kinds of mathematical programming are taken up toward the end of the book. This chapter addresses one of the most common applications of linear programming: maximizing the profit of some operation, subject to constraints that limit what can be produced. Chapters 2 and 3 are devoted to two other equally common kinds of linear programs, and Chapter 4 shows how linear programming models can be replicated and combined to produce truly large-scale problems. These chapters are written with the beginner in mind, but experienced practitioners of mathematical programming should find them useful as a quick introduction to AMPL.

We begin with a linear program (or LP for short) with only two variables, motivated by a mythical steelmaking operation. This will provide a quick review of linear programming to refresh your memory if you already have some experience, or to help you get started if you're just learning. We'll show how the same LP can be represented as a general algebraic model of production, together with specific data. Then we'll show how to express several linear programming problems in AMPL and how to run AMPL and a solver to produce a solution.

The separation of model and data is the key to describing more complex linear programs in a concise and understandable fashion. As an illustration, this chapter concludes by presenting several enhancements to the model.

1.1 A two-variable linear program

An (extremely simplified) steel company must decide how to allocate next week's time on a rolling mill. The mill takes unfinished slabs of steel as input, and can produce either of two semi-finished products, which we will call bands and coils. (The terminology is not entirely standard; see the bibliography at the end of the chapter for some accounts of realistic LP applications in steelmaking.) The mill's two products come off the rolling line at different rates:

Tons per hour:	Bands	200
	Coils	140

and they also have different profitabilities:

Profit per ton:	Bands	$25
	Coils	$30

To further complicate matters, the following weekly production amounts are the most that can be justified in light of the currently booked orders:

Maximum tons:	Bands	6,000
	Coils	4,000

The question facing the company is as follows: If 40 hours of production time are available this week, how many tons of bands and how many tons of coils should be produced to bring in the greatest total profit?

While we are given numeric values for production rates and per-unit profits, the tons of bands and of coils to be produced are as yet unknown. These quantities are the decision *variables* whose values we must determine so as to maximize profits. The purpose of the linear program is to specify the profits and production limitations as explicit formulas involving the variables, so that the desired values of the variables can be determined systematically.

In an algebraic statement of a linear program, it is customary to use a mathematical shorthand for the variables. Thus we will write X_B for the number of tons of bands to be produced, and X_C for tons of coils. The total hours to produce all these tons is then given by

$$(\text{hours to make a ton of bands}) \times X_B + (\text{hours to make a ton of coils}) \times X_C$$

This number cannot exceed the 40 hours available. Since hours per ton is the reciprocal of the tons per hour given above, we have a *constraint* on the variables:

$$(1/200)\, X_B + (1/140)\, X_C \le 40.$$

There are also production limits:

$$0 \le X_B \le 6000$$
$$0 \le X_C \le 4000$$

In the statement of the problem above, the upper limits were specified, but the lower limits were assumed — it was obvious that a negative production of bands or coils would be meaningless. Dealing with a computer, however, it is necessary to be quite explicit.

By analogy with the formula for total hours, the total profit must be

(profit per ton of bands) $\times X_B$ + (profit per ton of coils) $\times X_C$

That is, our objective is to maximize $25\,X_B + 30\,X_C$. Putting this all together, we have the following linear program:

Maximize $25\,X_B + 30\,X_C$

Subject to $(1/200)\,X_B + (1/140)\,X_C \le 40$
$0 \le X_B \le 6000$
$0 \le X_C \le 4000$

This is a very simple linear program, so we'll solve it by hand in a couple of ways, and then check the answer with AMPL.

First, by multiplying profit per ton times tons per hour, we can determine the profit per hour of mill time for each product:

Profit per hour: Bands $5,000
                   Coils $4,200

Bands are clearly a more profitable use of mill time, so to maximize profit we should produce as many bands as the production limit will allow — 6,000 tons, which takes 30 hours. Then we should use the remaining 10 hours to make coils — 1,400 tons in all. The profit is $25 times 6,000 tons plus $30 times 1,400 tons, for a total of $192,000.

Alternatively, since there are only two variables, we can show the possibilities graphically. If X_B values are plotted along the horizontal axis, and X_C values along the vertical axis, each point represents a choice of values, or solution, for the decision variables:

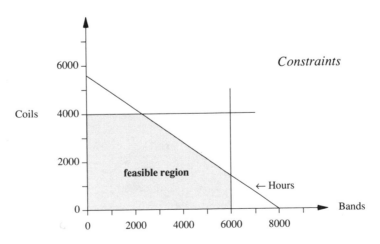

The horizontal line represents the constraint on coils, the vertical on bands. The diagonal line is the constraint on hours; each point on that line represents a combination of bands and coils that requires exactly 40 hours of production time, and any point downward and to the left requires less than 40 hours.

The shaded region bounded by the axes and these three lines corresponds exactly to the *feasible* solutions — those that satisfy all three constraints. Among all the feasible solutions represented in this region, we seek the one that maximizes the profit.

For this problem, a line of slope –25/30 represents combinations that produce the same profit; for example, in the figure below, the line from (4800, 0) to (0, 4000) represents combinations that yield $120,000 profit. Different profits give different but parallel lines, with higher profits giving lines that are higher and further to the right.

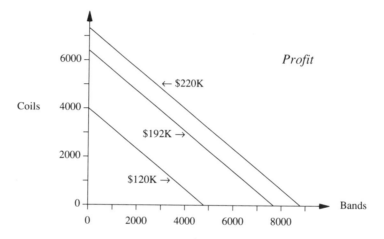

If we combine these two plots, we can see the profit-maximizing, or *optimal*, feasible solution:

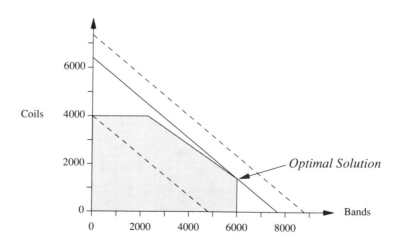

The line segment for profit equal to $120,000 is entirely within the feasible region; any point on this line corresponds to a solution that achieves a profit of $120,000. On the other hand, the line for $220,000 does not intersect the feasible region at all; this tells us that there is no way to achieve a profit as high as $220,000. Viewed in this way, solving the linear program reduces to answering the following question: Among all profit lines that intersect the feasible region, which is highest and furthest to the right? The answer is the middle line, which just touches the region at one of the corners. This point corresponds to 6,000 tons of bands and 1,400 tons of coils, and a profit of $192,000 — the same as we found before.

1.2 The two-variable linear program in AMPL

Solving this linear program with AMPL just requires entering it into the computer and typing *solve*:

```
> ampl
ampl: var XB;
ampl: var XC;
ampl: maximize profit: 25 * XB + 30 * XC;
ampl: subject to Time: (1/200) * XB + (1/140) * XC <= 40;
ampl: subject to B_limit: 0 <= XB <= 6000;
ampl: subject to C_limit: 0 <= XC <= 4000;

ampl: solve;
MINOS 5.4: optimal solution found.
2 iterations, objective 192000

ampl: display XB, XC;
XB = 6000
XC = 1400

ampl: quit;
>
```

You begin the session by typing *ampl* in response to your system's prompt (represented here by >). Then you type AMPL statements in response to the ampl: prompt, until you leave AMPL by typing *quit*. (Throughout the book, material you type is shown in *this slanted font*.)

The first part of the session above is devoted to entering the linear program. As in the algebraic form, the AMPL version specifies the decision variables, defines the objective, and lists the constraints. It differs mainly in being somewhat more formal and regular, to facilitate computer processing. Each variable is named in a var statement, and each constraint by a statement that begins with subject to and a name (Time, B_limit) for the constraint. Multiplication requires an explicit * operator, and the ≤ relation is written <=.

After you have typed in the linear program, type `solve` to have AMPL translate it, send it to a linear program solver, and return the answer. A final command, `display`, is used to show the optimal values of the variables.

The message `MINOS 5.4` directly following the `solve` command indicates that AMPL used version 5.4 of a solver called MINOS. We have used MINOS and several other solvers for the examples in this book. You may have a different set of solvers available on your machine, but any solver should give you the same optimal objective value for a linear program. Often there is more than one solution that achieves the optimal objective, however, in which case different solvers may report different optimal values for the variables. (Commands for choosing and controlling solvers will be explained in Chapter 10.)

Procedures for running AMPL can vary from one computer and operating system to another. Details are provided in supplementary instructions that come with your version of the AMPL software, rather than in this book. For subsequent examples, we will assume that AMPL has been started up, and that you have received the first `ampl:` prompt.

1.3 A linear programming model

The simple approach employed so far in this chapter is helpful for understanding the fundamentals of linear programming, but you can see that if our problem were only slightly more realistic — a few more products, a few more constraints — it would be a nuisance to write down and impossible to illustrate with pictures. And if the problem were subject to frequent change, either in form or merely in the data values, it would be hard to update as well.

If we are to progress beyond the very tiniest linear programs, we must adopt a more general and concise way of expressing them. This is where mathematical notation comes to the rescue. We can write a compact description of the general form of the problem, which we call a *model*, using algebraic notation for the objective function and the constraints. Figure 1-1 shows the production problem in algebraic notation.

Figure 1-1 is a symbolic linear programming model. Its components are fundamental to all models:

- **sets**, like the products
- **parameters**, like the production and profit rates
- **variables**, the values of which the solver is to determine
- an **objective function**, to be maximized or minimized
- **constraints** that the solution must satisfy.

The model describes an infinite number of related optimization problems, of which the previous example was one. If we provide specific values for data, however, the model becomes a specific problem, or instance of the model, that can be solved. Each different collection of data values defines a different instance.

Given: P, a set of products
                    a_j = tons per hour of product j, for each $j \in P$
                    b = hours available at the mill
                    c_j = profit per ton of product j, for each $j \in P$
                    u_j = maximum tons of product j, for each $j \in P$

Define variables: X_j = tons of product j to be made, for each $j \in P$

Maximize: $$\sum_{j \in P} c_j X_j$$

Subject to: $$\sum_{j \in P} (1/a_j) X_j \leq b$$

                    $$0 \leq X_j \leq u_j, \text{ for each } j \in P$$

Figure 1-1: Basic production model in algebraic form.

It might seem that we have made things less rather than more concise, since our model is longer than the original statement of the linear program in Section 1.1. Consider what would happen, however, if the set P had 42 products rather than 2. The linear program would have 120 more data values (40 each for a_j, c_j, and u_j); there would be 40 more variables, with new lower and upper limits for each; and there would be 40 more terms in the objective and the hours constraint. Yet the abstract model, as shown above, would be no different. Without this ability of a short model to describe a long linear program, larger and more complex instances of linear programming would become impossible to deal with.

A mathematical model like this is thus usually the best compromise between brevity and comprehension; and fortunately, it is easy to convert into a language that a computer can process. From now on, we'll assume models are given in the algebraic form. As always, reality is rarely so simple, so most models will have more sets, more parameters, more variables, more complicated objectives, and more constraints. In fact, in any real situation, formulating a correct model and providing accurate data are by far the hardest parts of the job; solving a specific problem only requires computing power.

1.4 The linear programming model in AMPL

Now we can talk about AMPL. The AMPL language is intentionally as close to the mathematical form as it can get while still being easy to type on an ordinary keyboard and process by a program. There are AMPL constructions for each of the basic components listed above — sets, parameters, variables, objectives, and constraints — and ways to write arithmetic expressions, sums over sets, and so on.

```
set P;

param a {j in P};
param b;
param c {j in P};
param u {j in P};

var X {j in P};

maximize profit: sum {j in P} c[j] * X[j];

subject to Time: sum {j in P} (1/a[j]) * X[j] <= b;

subject to Limit {j in P}: 0 <= X[j] <= u[j];
```

Figure 1-2: Basic production model in AMPL (file prod.mod).

We will first give an AMPL model that resembles our algebraic model as much as possible, and then present an improved version that takes better advantage of the language.

The basic model

For the basic production model of Figure 1-1, a direct transcription into AMPL would look like Figure 1-2.

The keyword set declares a set name, as in

```
set P;
```

The members of set P will be provided in separate data statements, which we'll show in a moment.

The keyword param declares a parameter, which may be a single scalar value, as in

```
param b;
```

or a collection of values indexed by a set. Where algebraic notation says that "there is an a_j for each j in P", one writes in AMPL

```
param a {j in P};
```

which means that a is a collection of parameter values, one for each member of the set P. Subscripts in algebraic notation are written with brackets in AMPL, so an individual value like a_j is written a[j].

The var declaration

```
var X {j in P};
```

names a collection of variables, one for each member of P, whose values the solver is to determine.

The objective function is given by the declaration

```
maximize profit: sum {j in P} c[j] * X[j];
```

The name `profit` is arbitrary; a name is required by the syntax, but any name will do. The precedence of the `sum` operator is lower than that of `*`, so the expression is indeed a sum of products, as intended.

Finally, the constraints are given by

```
subject to Time: sum {j in P} (1/a[j]) * X[j] <= b;
subject to Limit {j in P}: 0 <= X[j] <= u[j];
```

The `Time` constraint says that a certain sum over the set `P` may not exceed the value of parameter `b`. The `Limit` constraint is actually a family of constraints, one for each member `j` of `P`: each `X[j]` is bounded by zero and the corresponding `u[j]`.

The construct `{j in P}` is called an *indexing expression*. As you can see from our example, indexing expressions are used not only in declaring parameters and variables, but in any context where the algebraic model does something "for each j in P". Thus the `Limit` constraints are declared

```
subject to Limit {j in P}
```

because we want to impose a different restriction `0 <= X[j] <= u[j]` for each different product j in the set P. In the same way, the summation in the objective is written

```
sum {j in P} c[j] * X[j]
```

to indicate that the different terms `c[j] * X[j]`, for each `j` in the set `P`, are to be added together in computing the profit.

The layout of an AMPL model is quite free. Sets, parameters, and variables must be declared before they are used but can otherwise appear in any order. Statements end with semicolons. Upper and lower case letters are different, so `time`, `Time`, and `TIME` are three different names.

You have undoubtedly noticed several places where traditional mathematical notation has been adapted in AMPL to the limitations of normal keyboards and character sets. Instead of Σ, AMPL uses the word `sum` to express a summation, and `in` rather than $\in$ for set membership. Set specifications are enclosed in braces, as in `{P}` or `{j in P}`. Where mathematical notation uses adjacency to signify multiplication in $c_j X_j$, AMPL uses the `*` operator of most programming languages, and subscripts are replaced by square brackets, so $c_j X_j$ becomes `c[j]*X[j]`.

You will find that the rest of AMPL is similar — a few more arithmetic operators, a few more reserved words like `sum` and `in`, and many more ways to specify indexing expressions. Like any other computer language, AMPL has a precise grammar, but we won't stress the rules too much here; most will become clear as we go along, and full details are given in the reference manual, Appendix A.

Our original two-variable linear program is one of the many LPs that are instances of this model. To specify it or any other such instance, we need to supply the membership of P and the values of the various parameters. There is no standard way to describe these data values in algebraic notation; usually some kind of informal tables are used, such as the ones we showed earlier. In AMPL, there is a specific syntax for data tables, which is

```
set P := bands coils;

param:      a      c      u    :=
   bands   200     25    6000
   coils   140     30    4000 ;

param b := 40;
```

Figure 1-3: Production model data (file `prod.dat`).

sufficiently regular and unambiguous to be translated by a computer. Figure 1-3 gives data for the basic production model in that form. A `set` statement supplies the members (`bands` and `coils`) of set P, and a `param` table gives the corresponding values for `a`, `c`, and `u`. A simple `param` statement gives the value for `b`. These data statements, which are described in more detail in Chapter 9, have a variety of options that let you list or tabulate parameters in convenient ways; but essentially you must provide the name of each set member over which a parameter is indexed, together with the associated value.

An improved model

We could go on immediately to solve the linear program defined by Figures 1-2 and 1-3. Once we have written the model in AMPL, however, we need not feel constrained by all the conventions of algebra, and we can instead consider changes that might make the model easier to work with. Figures 1-4a and 1-4b show a possible "improved" version. The short "mathematical" names for the sets, parameters and variables have been replaced by longer, more meaningful ones. The indexing expressions have become p in PROD, or just PROD in those declarations that do not use the index p. The bounds on individual variables have been placed within their `var` declaration, rather than in a separate constraint; analogous bounds have been placed on the parameters, to indicate the ones that must be positive or nonnegative in any linear program derived from the model.

Finally, comments have been added to help explain the model to a reader. Comments begin with # and end at the end of the line. As in any programming language, judicious use of meaningful names, comments and formatting helps to make AMPL models more readable and understandable.

There are always many ways to describe a particular model in AMPL. It is left to the modeler to pick the way that seems clearest or most convenient. Our earlier, mathematical approach is often preferred for working quickly with a familiar model. On the other hand, the second version is more attractive for a model that will be maintained and modified by several people over months or years.

At this point, we could show how the model and data of Figures 1-4a and 1-4b would be typed interactively, as before. But clearly such an approach is unworkable for models of any size. Instead the statements of a model and its data are placed in one or more files, which are read as part of an AMPL session. Here is an example in which all of the model

```
set PROD;  # products

param rate {PROD} > 0;      # tons produced per hour
param avail >= 0;           # hours available in week

param profit {PROD};        # profit per ton
param market {PROD} >= 0;   # limit on tons sold in week

var Make {p in PROD} >= 0, <= market[p]; # tons produced

maximize total_profit: sum {p in PROD} profit[p] * Make[p];
            # Objective: total profits from all products

subject to Time: sum {p in PROD} (1/rate[p]) * Make[p] <= avail;
            # Constraint: total of hours used by all
            # products may not exceed hours available
```

Figure 1-4a: Steel production model (steel.mod).

```
set PROD := bands coils;

param:     rate  profit  market :=
  bands     200    25     6000
  coils     140    30     4000 ;

param avail := 40;
```

Figure 1-4b: Data for steel production model (steel.dat).

declarations have been put in a file called steel.mod, and the data specification has been put in steel.dat:

```
ampl: model steel.mod;
ampl: data steel.dat;
ampl: solve;
MINOS 5.4: optimal solution found.
2 iterations, objective 192000

ampl: display Make;
Make [*] :=
bands  6000
coils  1400
;
```

The model and data commands each specify a file whose contents are to be read immediately, as if all of its lines had been typed at that point. In this example, we first read the model from steel.mod, then read the data from steel.dat. The use of two separate commands encourages a clean separation of model from data.

Filenames can have any form recognized by your computer's operating system; AMPL doesn't check them for correctness. The filenames here refer to the files on the disk that accompanies the book.

Once the model has been solved, we can show all of the optimal values of the variables Make[p], by typing **display Make**. The output from display uses the same formats as AMPL data input, so that there is only one set of formats to learn. (The [*] indicates a variable or parameter with a single subscript. It is not strictly necessary for input, since Make is one-dimensional, but display prints it as a reminder.)

Error messages

You will inevitably make some mistakes as you develop a model. AMPL detects various kinds of incorrect statements, which are reported in error messages following the model, data or solve commands.

AMPL catches many errors as soon as the model is read. For example, if you use the wrong syntax for the bounds in the declaration of the variable Make, you will receive an error message like this, right after you enter the model command:

```
steel.mod, line 8 (offset 250):
        syntax error
context: var Make {p in PROD} >>> 0 <<< <= Make[p] <= market[p];
```

If you inadvertently use make instead of Make in an expression like profit[p] * make[p], you will receive this message:

```
steel.mod, line 11 (offset 353):
        make is not defined
context: maximize total_profit:
                sum {p in PROD} profit[p] *  >>> make[p] <<< ;
```

In each case, the offending line is printed, with the approximate location of the error surrounded by >>> and <<<.

Other sources of error messages include a variable or parameter with the wrong number of subscripts, a model component used before it is declared, a missing semicolon at the end of a command, or a reserved word like sum or in used in the wrong context. (Appendix A.17 contains a list of reserved words.) Syntax errors in data statements are similarly reported right after you enter a data command.

Errors in the data values are caught after you type solve. If the number of hours were given as –40, for instance, we would see:

```
ampl: model steel.mod; data steel.dat; solve;

error processing param avail:
        failed check: param avail is not >= 0;
currently the check reads...
        check -40 >= 0;
```

It is good practice to include as many validity checks as possible in the model. Errors that are not caught at an early stage often lead to a mystifyingly wrong or meaningless solution.

1.5 Adding lower bounds to the model

Once the model and data have been set up, it is a simple matter to change them and then re-solve. Indeed, we would not expect to find an LP application in which the model and data are prepared and solved just once, or even a few times. Most commonly, numerous refinements are introduced as the model is developed, and changes to the data continue for as long as the model is used.

Let's conclude this chapter with a few examples of changes and refinements. These examples also highlight some additional features of AMPL.

Suppose first that we add another product, steel plate. The model stays the same, but in the data we have to add `plate` to the list of members for the set `PROD`, and we have to add a line of parameter values for `plate`:

```
set PROD := bands coils plate;

param:    rate  profit  market :=
   bands   200    25     6000
   coils   140    30     4000
   plate   160    29     3500 ;

param avail := 40;
```

We put this version of the data in a file called `steel2.dat`, and use AMPL as before to get the solution:

```
ampl: model steel.mod; data steel2.dat; solve;
MINOS 5.4: optimal solution found.
2 iterations, objective 196400

ampl: display Make;
Make [*] :=
bands  6000
coils     0
plate  1600
;
```

Profits have increased compared to the two-variable version, but now it is best to produce no coils at all! On closer examination, this result is not so surprising. Plate yields a profit of $4640 per hour, which is less than for bands but more than for coils. Thus plate is produced to absorb the capacity not taken by bands; coils would be produced only if both bands and plate reached their market limits before the available hours were exhausted.

In reality, a whole product line cannot be shut down solely to increase weekly profits. The simplest way to reflect this in the model is to add lower bounds on the production amounts, as shown in Figures 1-5a and 1-5b. We have declared a new collection of parameters named `commit`, to represent the lower bounds on production that are imposed by sales commitments, and we have changed >= 0 to >= `commit[p]` in the declaration of the variables `Make[p]`.

After these changes are made, we can run AMPL again to get a more realistic solution:

```
set PROD;   # products

param rate {PROD} > 0;        # produced tons per hour
param avail >= 0;             # hours available in week
param profit {PROD};          # profit per ton

param commit {PROD} >= 0;     # lower limit on tons sold in week
param market {PROD} >= 0;     # upper limit on tons sold in week

var Make {p in PROD} >= commit[p], <= market[p]; # tons produced

maximize total_profit: sum {p in PROD} profit[p] * Make[p];
                # Objective: total profits from all products

subject to Time: sum {p in PROD} (1/rate[p]) * Make[p] <= avail;
                # Constraint: total of hours used by all
                # products may not exceed hours available
```

Figure 1-5a: Lower bounds on production (steel3.mod).

```
set PROD := bands coils plate;

param:      rate  profit  commit  market :=
   bands    200     25     1000    6000
   coils    140     30      500    4000
   plate    160     29      750    3500 ;

param avail := 40;
```

Figure 1-5b: Data for lower bounds on production (steel3.dat).

```
ampl: model steel3.mod; data steel3.dat; solve;
MINOS 5.4: optimal solution found.
2 iterations, objective 194828.5714

ampl: display commit,Make,market;
:        commit    Make    market      :=
bands     1000    6000       6000
coils      500     500       4000
plate      750    1028.57    3500
;
```

For comparison, we have displayed commit and market on either side of the actual production, Make. Evidently, after the commitments are met, it is most profitable to produce bands up to the market limit, and then to produce plate with the remaining available time.

1.6 Adding resource constraints to the model

Processing of steel slabs is not a single operation, but a series of steps that may proceed at different rates. To motivate a more general model, imagine that we divide pro-

```
set PROD;    # products
set STAGE;   # stages

param rate {PROD,STAGE} > 0; # tons per hour in each stage
param avail {STAGE} >= 0;      # hours available/week in each stage
param profit {PROD};           # profit per ton

param commit {PROD} >= 0;      # lower limit on tons sold in week
param market {PROD} >= 0;      # upper limit on tons sold in week

var Make {p in PROD} >= commit[p], <= market[p]; # tons produced

maximize total_profit: sum {p in PROD} profit[p] * Make[p];
                       # Objective: total profits from all products

subject to Time {s in STAGE}:
   sum {p in PROD} (1/rate[p,s]) * Make[p] <= avail[s];
                       # In each stage: total of hours used by all
                       # products may not exceed hours available
```

Figure 1-6a: Additional resource constraints (steel4.mod).

duction into a reheat stage that can process the incoming slabs at 200 tons per hour, and a rolling stage that makes bands, coils or plate at the rates previously given. Further imagine that there are only 35 hours of reheat time, even though there are 40 hours of rolling time.

To cover this kind of situation, we can add a set STAGE of production stages to our model. The parameter and constraint declarations are modified accordingly, as shown in Figure 1-6a. Since there is a potentially different number of hours available in each stage, the parameter avail is now indexed over STAGE. Since there is a potentially different production rate for each product in each stage, the parameter rate is indexed over both PROD and STAGE. In the Time constraint, the production rate for product p in stage s is referred to as rate[p,s]; this is AMPL's version of a doubly subscripted entity like a_{ps} in algebraic notation.

The only other change is to the constraint declaration, where we no longer have a single constraint, but a constraint for each stage imposed by limited time. In algebraic notation, this might have been written

$$\text{Subject to } \sum_{p \in P} (1/a_{ps}) X_p \le b_s, \text{ for each } s \in S.$$

Compare the AMPL version:

```
subject to Time {s in STAGE}:
   sum {p in PROD} (1/rate[p,s]) * Make[p] <= avail[s];
```

As in the other examples, this is a straightforward analogue, adapted to the requirements of a computer language. In complex models, most of the constraints are indexed collections like this one.

```
set PROD := bands coils plate;
set STAGE := reheat roll;

param rate:  reheat  roll :=
   bands        200    200
   coils        200    140
   plate        200    160 ;

param:      profit  commit  market :=
   bands       25     1000    6000
   coils       30      500    4000
   plate       29      750    3500 ;

param avail :=  reheat 35   roll   40 ;
```

Figure 1-6b: Data for additional resource constraints (`steel4.dat`).

Since `rate` is now indexed over combinations of two indices, it requires a data table all to itself, as in Figure 1-6b. This data also include the membership for the new set `STAGE`, and values of `avail` for both `reheat` and `roll`.

After these changes are made, we use AMPL to get another revised solution:

```
ampl: reset;
ampl: model steel4.mod; data steel4.dat; solve;
MINOS 5.4: optimal solution found.
4 iterations, objective 190071.4286

ampl: display Make.lb,Make,Make.ub,Make.rc;
:       Make.lb    Make    Make.ub    Make.rc      :=
bands    1000    3357.14    6000      5.32907e-15
coils     500     500       4000     -1.85714
plate     750    3142.86    3500      3.55271e-15
;

ampl: display Time;
Time [*] :=
reheat  1800
  roll  3200
  ;
```

If your computer system lets you work on files while AMPL remains active, you can use the `reset` command as shown above to erase the previous model and permit a new one to be read in. Otherwise, you'll need to type `quit` to leave AMPL, work on your files, then start up AMPL again.

At the end of the example above we have displayed the "dual values" or "shadow prices" associated with the `Time` constraints. The dual value of a constraint measures how much the value of the objective function would improve if the constraint were relaxed by a small amount. For example, here we would expect that up to some point, additional reheat time would produce another $1800 of extra profit per hour, and additional rolling time would produce $3200 per hour; decreasing these times would decrease the profit correspondingly. In output commands like `display`, AMPL interprets a constraint's name alone as referring to the associated dual values.

We also display several quantities associated with the variables `Make`. First there are lower bounds `Make.lb` and upper bounds `Make.ub`, which in this case are the same as `commit` and `market`. We also show the "reduced cost" `Make.rc`, which has the same meaning with respect to the bounds that the shadow prices have with respect to the constraints. Thus we see that, again up to some point, each increase of a ton in the lower bound (or commitment) for coil production should reduce profits by about $1.86; each one-ton decrease in the lower bound should improve profits by about $1.86. The production levels for bands and plates are between their bounds, so their reduced costs are essentially zero (recall that $e-15$ means 10^{-15}), and changing their levels will have no effect. Bounds, shadow prices, reduced costs and other quantities associated with variables and constraints are explored further in Section 10.7.

Comparing this session with our previous one, we see that the additional reheat time restriction reduces profits by about $4750, and forces a substantial change in the optimal solution: much higher production of plate and lower production of bands. Moreover, the logic underlying the optimum is no longer so obvious. It is the difficulty of solving LPs by logical reasoning alone that necessitates computer-based systems such as AMPL.

Bibliography

Julius S. Aronofsky, John M. Dutton and Michael T. Tayyabkhan, *Managerial Planning with Linear Programming: in Process Industry Operations.* John Wiley & Sons (New York, NY, 1978). A detailed account of a variety of profit-maximizing applications, with emphasis on the petroleum and petrochemical industries.

Vašek Chvátal, *Linear Programming*, W. H. Freeman (New York, NY, 1983). An introduction to theoretical and algorithmic topics in linear programming.

Tibor Fabian, "A Linear Programming Model of Integrated Iron and Steel Production." Management Science **4** (1958) pp. 415–449. An application to all stages of steelmaking — from coal and ore through finished products — from the early days of linear programming.

David A. Kendrick, Alexander Meeraus and Jaime Alatorre, *The Planning of Investment Programs in the Steel Industry.* The Johns Hopkins University Press (Baltimore, MD, 1984). Several detailed mathematical programming models, using the Mexican steel industry as an example.

Exercises

1-1. This exercise starts with a two-variable linear program similar in structure to the one of Sections 1.1 and 1.2, but with a quite different story behind it.

(a) You are in charge of an advertising campaign for a new product, with a budget of $1 million. You can advertise on TV or in magazines. One minute of TV time costs $20,000 and reaches 1.8 million potential customers; a magazine page costs $10,000 and reaches 1 million. You must sign up for at least 10 minutes of TV time. How should you spend your budget to maximize your audience? Formulate the problem in AMPL and solve it. Check the solution by hand using at least one of the approaches described in Section 1.1.

(b) It takes creative talent to create effective advertising; in your organization, it takes three person-weeks to create a magazine page, and one person-week to create a TV minute. You have only 100 person-weeks available. Add this constraint to the model and determine how you should now spend your budget.

(c) Radio advertising reaches a quarter million people per minute, costs $2,000 per minute, and requires only 1 person-day of time. How does this medium affect your solutions?

(d) How does the solution change if you have to sign up for at least two magazine pages? A maximum of 120 minutes of radio?

1-2. The steel model of this chapter can be further modified to reflect various changes in production requirements. For each part below, explain the modifications to Figures 1-6a and 1-6b that would be required to achieve the desired changes. (Make each change separately, rather than accumulating the changes from one part to the next.)

(a) How would you change the constraints so that total hours used by all products must *equal* the total hours available for each stage? Solve the linear program with this change, and verify that you get the same results. Explain why, in this case, there is no difference in the solution.

(b) How would you add to the model to restrict the total weight of all products to be less than a new parameter, max_weight? Solve the linear program for a weight limit of 6500 tons, and explain how this extra restriction changes the results.

(c) The incentive system for mill managers may tend to encourage them to produce as many tons as possible. How would you change the objective function to maximize total tons? For the data of our example, does this make a difference to the optimal solution?

(d) Suppose that instead of the lower bounds represented by commit[p] in our model, we want to require that each product represent a certain share of the total tons produced. In the algebraic notation of Figure 1-1, this new constraint might be represented as

$$X_j \geq s_j \sum_{k \in P} X_k, \text{ for each } j \in P$$

where s_j is the minimum share associated with project j. How would you change the AMPL model to use this constraint in place of the lower bounds commit[p]? If the minimum shares are 0.4 for bands and plate, and 0.1 for coils, what is the solution?

Verify that if you change the minimum shares to 0.5 for bands and plate, and 0.1 for coils, the linear program gives an optimal solution that produces nothing, at zero profit. Explain why this makes sense.

(e) Suppose there is an additional finishing stage for plates only, with a capacity of 20 hours and a rate of 150 tons per hour. Explain how you could modify the data, without changing the model, to incorporate this new stage.

1-3. This exercise deals with some issues of "sensitivity" in the steel models.

(a) For the linear program of Figures 1-5a and 1-5b, display Time and Make.rc. What do these values tell you about the solution? (You may wish to review the explanation of shadow prices and reduced costs in Section 1.6.)

(b) Explain why the reheat time constraints added in Figure 1-6a result in a higher production of plate and a lower production of bands.

(c) Use AMPL to verify the following statements: If the available reheat time is increased from 35 to 36 in the data of Figure 1-6b, then the profit goes up by $1800 as predicted in Section 1.6. (To change the reheat time without changing and reading the data file over again, type `let avail["reheat"] := 36`.) If the reheat time is further increased to 37, the profit goes up by another $1800. However, if the reheat time is increased to 38, there is a smaller increase in the profit, and further increases past 38 have no effect on the optimal profit at all.

By trying some other values of the reheat time, confirm that the profit increases by $1800 per extra hour for any number of hours between 35 and 37 9/14, but that any increase in the reheat time beyond 37 9/14 hours doesn't give any further profit.

Draw a plot of the profit versus the number of reheat hours available, for hours ≥ 35.

(d) To find the slope of the plot from (c) — profit versus reheat time available — at any particular reheat time value, you need only look at the shadow price of `Time["reheat"]`. Using this observation as an aid, extend your plot from (c) down to 25 hours of reheat time. Verify that the slope of the plot remains at $6000 per hour from 25 hours down to less than 12 hours of reheat time. Explain what happens when the available reheat time drops to 11 hours.

1-4. Here is a similar profit-maximizing model, but in a different context. An automobile manufacturer produces several kinds of cars. Each kind requires a certain amount of factory time per car to produce, and yields a certain profit per car. A certain amount of factory time has been scheduled for the next week, and it is desired to use all this time; but at least a certain number of each kind of car must be manufactured to meet dealer requirements.

(a) What are the data values that define this problem? How would you declare the sets and parameter values for this problem in AMPL? What are the decision variables, and how would you declare them in AMPL?

(b) Assuming that the objective is to maximize total profit, how would you declare an objective in AMPL for this problem? How would you declare the constraints?

(c) For purposes of experiment, suppose that there are three kinds of cars, known at the factory as T, C and L, that 120 hours are available, and that the time per car, profit per car and dealer orders for each kind of car are as follows:

Car	time	profit	orders
T	1	200	10
C	2	500	20
L	3	700	15

How much of each car should be produced, and what is the maximum profit? You should find that your solution specifies a fractional amount of one of the cars. As a practical matter, how could you make use of this solution?

(d) If you maximize the total number of cars produced instead of the total profit, how many more cars do you make? How much less profit?

(e) Each kind of car achieves a certain fuel efficiency, and the manufacturer is required by law to maintain a certain "fleet average" efficiency. The fleet average is computed by multiplying the efficiency of each kind of car times the number of that car produced, and dividing by the total cars produced. Extend your AMPL model to contain a minimum fleet average efficiency constraint. Rearrange the constraint as necessary to make it linear — no variables divided into other variables.

(f) Find the optimal solution for the case where cars T, C and L achieve fuel efficiencies of 50, 30 and 20 miles/gallon, and the fleet average efficiency must be at least 35 miles/gallon. Explain how this changes the production amounts and the total profit. Dealing with the fractional amounts in the solution is not so easy in this case. What might you do?

If you had 10 more hours of production time, you could make more profit. Does the addition of the fleet average efficiency constraint make the extra 10 hours more or less valuable?

(g) Explain how you could further refine this model to account for different production stages that have different numbers of hours available per stage, much as in the steel model of Section 1.6.

1-5. A group of young entrepreneurs earns a (temporarily) steady living by acquiring discontinued items from electronics stores and re-selling them. Each item has a street value, a weight, and a volume; there are limits on the numbers of available items, and on the total weight and volume that can be managed at one time.

(a) Formulate an AMPL model that will help to determine how much of each item to pick up, to maximize one day's profit.

(b) Find a solution for the case given by the following table,

	Value	Weight	Volume	Available
TV	50	35	8	20
radio	15	5	1	50
camera	85	4	2	20
CD player	40	3	1	30
VCR	50	15	5	30
camcorder	120	20	4	15

and by limits of 500 pounds and 300 cubic feet.

(c) Suppose that it is desirable to acquire some of each item, so as to always have stock available for re-sale. Suppose in addition that there are upper bounds on how many of each item you can reasonably expect to sell. How would you add these conditions to the model?

(d) How could the group use the dual variables on the maximum-weight and maximum-volume constraints to evaluate potential new partners for their activities?

(e) Through adverse circumstances the group has been reduced to only one member, who can carry a mere 75 pounds and five cubic feet. What is the optimum strategy now? Given that this requires a non-integral number of acquisitions, what is the best all-integer solution? (The integrality constraint converts this from a standard linear programming problem into a much harder problem called a Knapsack Problem. See Chapter 15.)

1-6. Profit-maximizing models of oil refining were one of the first applications of linear programming. This exercise asks you to model a simplified version of the final stage of the refining process.

A refinery breaks crude oil into some collection of intermediate materials, then blends these materials back together into finished products. Given the volumes of intermediates that will be available, we want to determine how to blend the intermediates so that the resulting products are most profitable. The decision is made more complicated, however, by the existence of upper limits on certain attributes of the products, which must be respected in any feasible solution.

To formulate an algebraic linear programming model for this problem, we can start by defining sets I of intermediates, J of final products, and K of attributes. The relevant technological data may be represented by

a_i barrels of intermediate i available, for each $i \in I$

r_{ik} units of attribute k contributed per barrel of intermediate i, for each $i \in I$ and $k \in K$

u_{jk} maximum allowed units of attribute k per barrel of final product j,
           for each $j \in J$ and $k \in K$

δ_{ij} 1 if intermediate i is allowed in the blend for product j, or 0 otherwise,
           for each $i \in I$ and $j \in J$

and the economic data can be given by

c_j revenue per barrel of product j, for each $j \in J$

There are two collections of decision variables:

X_{ij} barrels of intermediate i used to make product j, for each $i \in I$ and $j \in J$

Y_j barrels of product j made, for each $j \in J$

The objective is to

$$\text{maximize} \ \sum_{j \in J} c_j Y_j,$$

which is the sum of the revenues from the various products.

It remains to specify the constraints. The amount of each intermediate used in blending must equal the amount available:

$$\sum_{j \in J} X_{ij} = a_i, \text{ for each } i \in I.$$

The amount of a product made must equal the sum of amounts of the components blended into it:

$$\sum_{i \in I} \delta_{ij} X_{ij} = Y_j, \text{ for each } j \in J.$$

For each product, the total attributes contributed by all intermediates must not exceed the total allowed:

$$\sum_{i \in I} r_{ik} X_{ij} \le u_{jk} Y_j, \text{ for each } j \in J \text{ and } k \in K.$$

Finally, we bound the variables as follows:

$$0 \le X_{ij} \le \delta_{ij} a_i, \text{ for each } i \in I, j \in J,$$
$$0 \le Y_j, \text{ for each } j \in J.$$

The upper bound on X_{ij} assures that only the appropriate intermediates will be used in blending. If intermediate i is not allowed in the blend for product j, as indicated by δ_{ij} being zero, then the upper bound on X_{ij} is zero; this insures that X_{ij} cannot be positive in any solution. Otherwise, the upper bound on X_{ij} is just a_i, which has no effect since there are only a_i barrels of intermediate i available for blending in any case.

(a) Transcribe this model to AMPL, using the same names as in the algebraic form for the sets, parameters and variables as much as possible.

(b) Re-write the AMPL model using meaningful names and comments, in the style of Figure 1-4a.

(c) In a representative small-scale instance of this model, the intermediates are SRG (straight run gasoline), N (naphtha), RF (reformate), CG (cracked gasoline), B (butane), DI (distillate intermediate), GO (gas oil), and RS (residuum). The final products are PG (premium gasoline), RG (regular gasoline), D (distillate), and HF (heavy fuel oil). Finally, the attributes are vap (vapor pressure), oct (research octane), den (density), and sul (sulfur).

The following amounts of the intermediates are scheduled to be available:

SRG	N	RF	CG	B	DI	GO	RS
21170	500	16140	4610	370	250	11600	25210

The intermediates that can be blended into each product, and the amounts of the attributes that they possess, are as follows:

	Premium & regular gasoline		Distillate		Heavy fuel oil	
	vap	oct	den	sul	den	sul
SRG	18.4	−78.5				
N	6.54	−65.0	272	.283		
RF	2.57	−104.0				
CG	6.90	−93.7				
B	199.2	−91.8				
DI			292	.526		
GO			295	.353	295	.353
RS					343	4.70

The attribute limits and revenues/barrel for the products are:

	vap	oct	den	sul	revenue
PG	12.2	−90			10.50
RG	12.7	−86			9.10
D			306	0.5	7.70
HF			352	3.5	6.65

Limits left blank, such as density for gasoline, are irrelevant and may be set to some relatively large number.

Create a data file for your AMPL model and determine the optimal blend and production amounts.

(d) It looks a little strange that the attribute amounts for research octane are negative. What is the limit constraint for this attribute really saying?

2

Diets, Blending and Scheduling: Minimizing Costs

To complement the profit-maximizing models of Chapter 1, we now consider linear programming models in which the objective is to minimize costs. Where the constraints of maximization models tend to be upper limits on the availability of resources, the constraints in minimization models are more likely to be lower limits on the amounts of certain "qualities" in the solution.

As an intuitive example of a cost-minimizing model, this chapter uses the well-known "diet problem", which finds a mix of foods that satisfies requirements on the amounts of various vitamins. We will again construct a small, explicit linear program, and then show how a general model can be formulated for all linear programs of that kind. Since you are now more familiar with AMPL, however, we will spend more time on AMPL and less with algebraic notation.

After formulating the diet model, we will discuss a few changes that might make it more realistic. The full power of this model, however, derives from its applicability to many situations that have nothing to do with diets. Thus we conclude this chapter by rewriting the model in a more general way, and discussing its application to blending and scheduling.

2.1 A linear program for the diet problem

Consider the problem of choosing prepared foods to meet certain nutritional requirements. Suppose that precooked dinners of the following kinds are available for the following prices per package:

BEEF	beef	$3.19
CHK	chicken	2.59
FISH	fish	2.29
HAM	ham	2.89
MCH	macaroni & cheese	1.89
MTL	meat loaf	1.99
SPG	spaghetti	1.99
TUR	turkey	2.49

These dinners provide the following percentages, per package, of the minimum daily requirements for vitamins A, C, B1 and B2:

	A	C	B1	B2
BEEF	60%	20%	10%	15%
CHK	8	0	20	20
FISH	8	10	15	10
HAM	40	40	35	10
MCH	15	35	15	15
MTL	70	30	15	15
SPG	25	50	25	15
TUR	60	20	15	10

The problem is to find the cheapest combination of packages that will meet a week's requirements — that is, at least 700% of the daily requirement for each nutrient.

Let us write X_{BEEF} for the number of packages of beef dinner to be purchased, X_{CHK} for the number of packages of chicken dinner, and so forth. Then the total cost of the diet will be:

$$\text{total cost} =$$
$$3.19\,X_{BEEF} + 2.59\,X_{CHK} + 2.29\,X_{FISH} + 2.89\,X_{HAM} +$$
$$1.89\,X_{MCH} + 1.99\,X_{MTL} + 1.99\,X_{SPG} + 2.49\,X_{TUR}$$

The total percentage of the vitamin A requirement is given by a similar formula, except that X_{BEEF}, X_{CHK}, and so forth are multiplied by the percentage per package instead of the cost per package:

$$\text{total percentage of vitamin A daily requirement met} =$$
$$60\,X_{BEEF} + 8\,X_{CHK} + 8\,X_{FISH} + 40\,X_{HAM} +$$
$$15\,X_{MCH} + 70\,X_{MTL} + 25\,X_{SPG} + 60\,X_{TUR}$$

This amount needs to be greater than or equal to 700 percent. There is a similar formula for each of the other vitamins, and each of these also needs to be ≥ 700.

Putting these all together, we have the following linear program:

Minimize

$$3.19\,X_{BEEF} + 2.59\,X_{CHK} + 2.29\,X_{FISH} + 2.89\,X_{HAM} +$$
$$1.89\,X_{MCH} + 1.99\,X_{MTL} + 1.99\,X_{SPG} + 2.49\,X_{TUR}$$

Subject to

$$60\,X_{BEEF} + 8\,X_{CHK} + 8\,X_{FISH} + 40\,X_{HAM} +$$
$$15\,X_{MCH} + 70\,X_{MTL} + 25\,X_{SPG} + 60\,X_{TUR} \geq 700$$

$$20\,X_{BEEF} + 0\,X_{CHK} + 10\,X_{FISH} + 40\,X_{HAM} +$$
$$35\,X_{MCH} + 30\,X_{MTL} + 50\,X_{SPG} + 20\,X_{TUR} \geq 700$$

$$10\,X_{BEEF} + 20\,X_{CHK} + 15\,X_{FISH} + 35\,X_{HAM} +$$
$$15\,X_{MCH} + 15\,X_{MTL} + 25\,X_{SPG} + 15\,X_{TUR} \geq 700$$

$$15\,X_{BEEF} + 20\,X_{CHK} + 10\,X_{FISH} + 10\,X_{HAM} +$$
$$15\,X_{MCH} + 15\,X_{MTL} + 15\,X_{SPG} + 10\,X_{TUR} \geq 700$$

$$X_{BEEF} \geq 0, X_{CHK} \geq 0, X_{FISH} \geq 0, X_{HAM} \geq 0,$$
$$X_{MCH} \geq 0, X_{MTL} \geq 0, X_{SPG} \geq 0, X_{TUR} \geq 0$$

At the end we have added the common-sense requirement that no fewer than zero packages of a food can be purchased.

As in the production problem of Chapter 1, our explicit diet problem LP can be transcribed and typed directly into AMPL. For example, we could write it as follows:

```
ampl:  var Xbeef >= 0; var Xchk >= 0; var Xfish >= 0;
ampl:  var Xham >= 0;  var Xmch >= 0; var Xmtl >= 0;
ampl:  var Xspg >= 0;  var Xtur >= 0;

ampl: minimize cost:
ampl:  3.19*Xbeef + 2.59*Xchk + 2.29*Xfish + 2.89*Xham +
ampl:  1.89*Xmch  + 1.99*Xmtl + 1.99*Xspg  + 2.49*Xtur;

ampl: subject to A:
ampl:  60*Xbeef +  8*Xchk +  8*Xfish + 40*Xham +
ampl:  15*Xmch  + 70*Xmtl + 25*Xspg  + 60*Xtur >= 700;

ampl: subject to C:
ampl:  20*Xbeef +  0*Xchk + 10*Xfish + 40*Xham +
ampl:  35*Xmch  + 30*Xmtl + 50*Xspg  + 20*Xtur >= 700;

ampl: subject to B1:
ampl:  10*Xbeef + 20*Xchk + 15*Xfish + 35*Xham +
ampl:  15*Xmch  + 15*Xmtl + 25*Xspg  + 15*Xtur >= 700;

ampl: subject to B2:
ampl:  15*Xbeef + 20*Xchk + 10*Xfish + 10*Xham +
ampl:  15*Xmch  + 15*Xmtl + 15*Xspg  + 10*Xtur >= 700;

ampl: solve;
CPLEX 2.0: optimal solution; objective 88.2
2 iterations (1 in phase I)
```

```
ampl: display Xbeef,Xchk,Xfish,Xham,Xmch,Xmtl,Xspg,Xtur;
Xbeef = 0
Xchk = 0
Xfish = 0
Xham = 0
Xmch = 46.6667
Xmtl = 0
Xspg = 0
Xtur = 0
```

The optimal solution is found quickly, but it is hardly what we might have hoped for. The cost is minimized by a monotonous diet of $46^{2}/3$ packages of macaroni and cheese! You can check that this neatly provides $15\% \times 46^{2}/3 = 700\%$ of the requirement for vitamins A, B1 and B2, and a lot more vitamin C than necessary; the cost is only $\$1.89 \times 46^{2}/3 = \$88.20$.

You might guess that a better solution would be generated by requiring the amount of each vitamin to equal 700% exactly. Such a requirement can easily be imposed by changing each >= to = in the AMPL constraints. If you go ahead and solve the changed LP, you will find that the diet does indeed become more varied: approximately 19.5 packages of chicken, 16.3 of macaroni and cheese, and 4.3 of meat loaf. But, since equalities are more restrictive than inequalities, the cost goes up to $89.99.

2.2 An AMPL model for the diet problem

Clearly we will have to consider more extensive modifications to our linear program in order to produce a diet that is even remotely acceptable. We will probably want to change the sets of food and nutrients, as well as the nature of the constraints and bounds. As in the production example of the previous chapter, this will be much easier to do if we rely on a general model that can be coupled with a variety of specific data files.

This model deals with two things: nutrients and foods. Thus we begin an AMPL model by declaring sets of each:

```
set NUTR;
set FOOD;
```

Next we need to specify the numbers required by the model. Certainly a positive cost should be given for each food:

```
param cost {FOOD} > 0;
```

The cost of a particular food is written as, say, cost["BEEF"], although references to specific values in an LP model are usually in terms of indices like j, rather than specific set members like "BEEF".

To make the model somewhat more general than the LPs so far, we will also specify that for each food there are lower and upper limits on the number of packages in the diet:

```
param f_min {FOOD} >= 0;
param f_max {j in FOOD} >= f_min[j];
```

Notice that we need a dummy index j to run over FOOD in the declaration of f_max, in order to say that the maximum for each food must be greater than or equal to the corresponding minimum.

We will also find it convenient to specify similar lower and upper limits on the amount of each nutrient in the diet:

```
param n_min {NUTR} >= 0;
param n_max {i in NUTR} >= n_min[i];
```

Finally, for each combination of a nutrient and a food, we need a number that represents the amount of the nutrient in one package of the food. You may recall from Chapter 1 that such a "product" of two sets is written by listing them both:

```
param amt {NUTR,FOOD} >= 0;
```

References to this parameter require two indices. For example, amt[i,j] is the amount of nutrient i in a package of food j.

The decision variables for this model are the numbers of packages to buy of the different foods:

```
var Buy {j in FOOD} >= f_min[j], <= f_max[j];
```

The number of packages of some food j to be bought will be called Buy[j]; in any acceptable solution it will have to lie between f_min[j] and f_max[j].

The total cost of buying a food j is the cost per package, cost[j], times the number of packages, Buy[j]. The objective to be minimized is the sum of this product over all foods j:

```
minimize total_cost:  sum {j in FOOD} cost[j] * Buy[j];
```

This minimize declaration works the same as maximize did in Chapter 1.

Similarly, the amount of a nutrient i supplied by a food j is the nutrient per package, amt[i,j], times the number of packages Buy[j]. The total amount of nutrient i supplied is the sum of this product over all foods j:

```
sum {j in FOOD} amt[i,j] * Buy[j]
```

To complete the model, we need only specify that each such sum must lie between the appropriate bounds. Our constraint declaration begins

```
subject to diet {i in NUTR}:
```

to say that a constraint named diet[i] must be imposed for each member i of NUTR. The rest of the declaration gives the algebraic statement of the constraint for nutrient i: the variables must satisfy

```
n_min[i] <= sum {j in FOOD} amt[i,j] * Buy[j] <= n_max[i]
```

```
set NUTR;
set FOOD;

param cost {FOOD} > 0;
param f_min {FOOD} >= 0;
param f_max {j in FOOD} >= f_min[j];

param n_min {NUTR} >= 0;
param n_max {i in NUTR} >= n_min[i];

param amt {NUTR,FOOD} >= 0;

var Buy {j in FOOD} >= f_min[j], <= f_max[j];

minimize total_cost:  sum {j in FOOD} cost[j] * Buy[j];

subject to diet {i in NUTR}:
   n_min[i] <= sum {j in FOOD} amt[i,j] * Buy[j] <= n_max[i];
```

Figure 2-1: Diet model in AMPL (diet.mod).

A ''double inequality'' like this is interpreted in the obvious way: the value of the sum in the middle must lie between n_min[i] and n_max[i]. The complete model is shown in Figure 2-1.

2.3 Using the AMPL diet model

By specifying appropriate data, we can solve any of the linear programs that correspond to the above model. Let's begin by using the data from the beginning of this chapter, which is shown in AMPL format in Figure 2-2.

The values of f_min and n_min are as given originally, while f_max and n_max are set, for the time being, to large values that won't affect the optimal solution. In the table for amt, the notation (tr) indicates that we have ''transposed'' the table so the columns correspond to the first index (nutrients), and the rows to the second (foods). Alternatively, we could have changed the model to say

```
param amt {FOOD,NUTR}
```

in which case we would have had to write amt[j,i] in the constraint.

Suppose that model and data are stored in the files diet.mod and diet.dat, respectively. Then AMPL is used as follows to read these files and to solve the resulting linear program:

```
ampl: model diet.mod;
ampl: data diet.dat;

ampl: solve;
CPLEX 2.0: optimal solution; objective 88.2
2 iterations (1 in phase I)
```

```
set NUTR := A B1 B2 C ;
set FOOD := BEEF CHK FISH HAM MCH MTL SPG TUR ;

param:   cost   f_min   f_max :=
  BEEF   3.19     0      100
  CHK    2.59     0      100
  FISH   2.29     0      100
  HAM    2.89     0      100
  MCH    1.89     0      100
  MTL    1.99     0      100
  SPG    1.99     0      100
  TUR    2.49     0      100 ;

param:   n_min   n_max :=
  A       700    10000
  C       700    10000
  B1      700    10000
  B2      700    10000 ;

param amt (tr):
           A     C    B1    B2 :=
  BEEF    60    20    10    15
  CHK      8     0    20    20
  FISH     8    10    15    10
  HAM     40    40    35    10
  MCH     15    35    15    15
  MTL     70    30    15    15
  SPG     25    50    25    15
  TUR     60    20    15    10 ;
```

Figure 2-2: Data for diet model (diet.dat).

```
ampl: display Buy;
Buy [*] :=
BEEF    0
 CHK    0
FISH    0
 HAM    0
 MCH   46.6667
 MTL    0
 SPG    0
 TUR    0
;
```

Naturally, the result is the same as before.

Now suppose that we want to make the following enhancements. To promote variety, the weekly diet must contain between 2 and 10 packages of each food. The amount of sodium and calories in each package is also given; total sodium must not exceed 40,000 mg, and total calories must be between 16,000 and 24,000. All of these changes can be made through a few modifications to the data, as shown in Figure 2-3. Putting this new data in file diet2.dat, we can run AMPL again:

```
set NUTR := A B1 B2 C NA CAL ;
set FOOD := BEEF CHK FISH HAM MCH MTL SPG TUR ;

param:    cost   f_min   f_max :=
   BEEF   3.19    2       10
   CHK    2.59    2       10
   FISH   2.29    2       10
   HAM    2.89    2       10
   MCH    1.89    2       10
   MTL    1.99    2       10
   SPG    1.99    2       10
   TUR    2.49    2       10  ;

param:    n_min   n_max :=
   A       700    20000
   C       700    20000
   B1      700    20000
   B2      700    20000
   NA        0    40000
   CAL   16000    24000 ;

param amt (tr):
           A    C    B1   B2    NA    CAL :=
   BEEF   60   20    10   15    938   295
   CHK     8    0    20   20   2180   770
   FISH    8   10    15   10    945   440
   HAM    40   40    35   10    278   430
   MCH    15   35    15   15   1182   315
   MTL    70   30    15   15    896   400
   SPG    25   50    25   15   1329   370
   TUR    60   20    15   10   1397   450 ;
```

Figure 2-3: Data for enhanced diet model (diet2.dat).

```
ampl: model diet.mod;
ampl: data diet2.dat;

ampl: solve;
CPLEX 2.0: infeasible problem.
7 iterations (7 in phase I)
```

The message infeasible problem tells us that we have constrained the diet too tightly; there is no way that all of the restrictions can be satisfied.

AMPL lets us examine a variety of values produced by a solver as it attempts to find a solution. In Chapter 1, we used dual values to investigate the sensitivity of an optimum solution to changes in the constraints. Here there is no optimum, but the solver does return the last solution that it found while attempting to satisfy the constraints. We can look for the source of the infeasibility by displaying some values associated with this solution:

```
ampl: display diet.lb, diet.body, diet.ub;
:    diet.lb    diet.body diet.ub    :=
A         700    1993.09    20000
B1        700     841.091   20000
B2        700     601.091   20000
C         700    1272.55    20000
CAL     16000   17222.9     24000
NA          0   40000       40000
;
```

The values for diet.lb and diet.ub are the "lower bounds" and "upper bounds"
on the sum of the variables in the constraints diet[i] — in this case, just the values
n_min[i] and n_max[i]; diet.body is the current sum of the variables. We can
see that the diet does not supply enough vitamin B2, while the amount of sodium (NA)
has reached its upper bound. If we relax the sodium limit to 50,000 mg, a feasible solution becomes possible:

```
ampl: let n_max["NA"] := 50000; solve;
CPLEX 2.0: optimal solution; objective 118.0594032
11 iterations (7 in phase I)

ampl: display Buy;
Buy [*] :=
BEEF   5.36061
 CHK   2
FISH   2
 HAM  10
 MCH  10
 MTL  10
 SPG   9.30605
 TUR   2
;
```

This is at least a start toward a palatable diet, although we have to spend $118.06, com-
pared to $88.20 for the original, less restricted case. Clearly it would be easy, now that
the model is set up, to try many other possibilities. (The let statement, which permits
modifications of data, is described in Chapter 10.)

One still disappointing aspect of the solution is the need to buy 5.36061 packages of
beef, and 9.30605 of spaghetti. What if we can only buy whole packages? You might
think that we could just round the optimal values to whole numbers, but it is not so easy
to do so in a feasible way. Displaying the constraint bounds at the optimum,

```
ampl: display diet.lb, diet.body, diet.ub;
:    diet.lb  diet.body diet.ub    :=
A         700   1956.29    20000
B1        700   1036.26    20000
B2        700    700       20000
C         700   1682.51    20000
CAL     16000  19794.6     24000
NA          0  50000       50000
;
```

we see that, for example, 6 packages of beef and 10 of spaghetti will violate the sodium limit, while 5 of beef and 9 of spaghetti will provide insufficient vitamin B2. Even if we could find a nearby all-integer solution that satisfies the constraints, there would be no guarantee that it would be the least-cost all-integer solution. AMPL does provide for putting the integrality restriction directly into the declaration of the variables:

```
var Buy {j in FOOD} integer >= f_min[j], <= f_max[j];
```

This will only help, however, if you use a solver that can handle so-called integer programs. In general, integrality and other ''discrete'' restrictions make a model much harder to solve. This is discussed at length in Chapter 15.

2.4 Generalizations for blending and scheduling

Your personal experience probably suggests that diet models are not widely used by people to choose their dinners. These models would be much better suited to situations in which packaging and personal preferences don't play such a prominent role — for example, the blending of animal feed or perhaps food for college dining halls.

The diet model is a convenient, intuitive example of a linear programming formulation that appears in many contexts. Suppose that we rewrite the model in a more general way, as shown in Figure 2-4. The objects that were called foods and nutrients in the diet model are now referred to more generically as ''inputs'' and ''outputs''. For each input j, we must decide to use a quantity X[j] that lies between in_min[j] and in_max[j]; as a result we incur a cost equal to cost[j] * X[j], and we create io[i,j] * X[j] units of each output i. Our goal is to find the lowest cost combination of inputs that yields, for each output i, an amount between out_min[i] and out_max[i].

In one common class of applications for this model, the inputs are raw materials to be mixed together. The outputs are qualities of the resulting blend. The raw materials could be the components of an animal feed, but they could equally well be the crude oil derivatives that are blended to make gasoline, or the different kinds of coal that are mixed as input to a coke oven. The qualities can be amounts of something (sodium or calories for animal feed), or more complex measures (vapor pressure or octane rating for gasoline), or even physical properties such as weight and volume.

In another, less obvious application, the inputs are work schedules, and the outputs correspond to hours worked on certain days of a month. For a particular work schedule j, io[i,j] is the number of hours that a person following schedule j will work on day i (zero if none), cost[j] is the monthly salary for a person following schedule j, and X[j] is the number of workers assigned that schedule. Under this interpretation, the objective becomes the total cost of the monthly payroll, while the constraints say that for each day i, the total number of workers assigned to work that day must lie between the limits out_min[i] and out_max[i]. The same approach can be used in a variety of

```
set IN;     # inputs
set OUT;    # outputs

param cost {IN} > 0;
param in_min {IN} >= 0;
param in_max {j in IN} >= in_min[j];

param out_min {OUT} >= 0;
param out_max {i in OUT} >= out_min[i];

param io {OUT,IN} >= 0;

var X {j in IN} >= in_min[j], <= in_max[j];

minimize total_cost:   sum {j in IN} cost[j] * X[j];

subject to outputs {i in OUT}:
    out_min[i] <= sum {j in IN} io[i,j] * X[j] <= out_max[i];
```

Figure 2-4: Blending model (`blend.mod`).

other scheduling contexts, where the hours, days or months are replaced by other periods of time.

Although linear programming can be very useful in applications like these, we need to keep in mind the assumptions that underlie the LP model. We have already mentioned the "continuity" assumption whereby `X[j]` is allowed to take on any value between `in_min[j]` and `in_max[j]`. This may be a lot more reasonable for blending than for scheduling.

As another example, in writing the objective as

```
sum {j in IN} cost[j] * X[j]
```

we are assuming "linearity of costs", that is, that the cost of an input is proportional to the amount of the input used, and that the total cost is the sum of the inputs' individual costs.

In writing the constraints as

```
out_min[i] <= sum {j in IN} io[i,j] * X[j] <= out_max[i]
```

we are also assuming that the yield of an output i from a particular input is proportional to the amount of the input used, and that the total yield of an output i is the sum of the yields from the individual inputs. This "linearity of yield" assumption poses no problem when the inputs are schedules, and the outputs are hours worked. But in the blending example, linearity is a physical assumption about the nature of the raw materials and the qualities, which may or may not hold. In early applications to refineries, for example, it was recognized that the addition of lead as an input had a nonlinear effect on the quality known as octane rating in the resulting blend.

AMPL makes it easy to express discrete or nonlinear models, but any departure from continuity or linearity is likely to make an optimal solution much harder to obtain. At the least, it takes a more powerful solver to optimize the resulting mathematical programs. Chapters 13, 14 and 15 discuss these issues in more detail.

Bibliography

George B. Dantzig, ''The Diet Problem.'' Interfaces **20**, 4 (1990) pp. 43–47. An entertaining account of the origins of the diet problem.

Said S. Hilal and Warren Erikson, ''Matching Supplies to Save Lives: Linear Programming the Production of Heart Valves.'' Interfaces **11**, 6 (1981) pp. 48–56. A less appetizing equivalent of the diet problem, involving the choice of pig heart suppliers.

Exercises

2-1. Suppose the foods listed below have calories, protein, calcium, vitamin A, and costs per pound as shown. In what amounts should these food be purchased to meet at least the daily requirements listed while minimizing the total cost? (This problem comes from George B. Dantzig's classic book, *Linear Programming and Extensions*, page 118. We will take his word on nutritional values, and for nostalgic reasons have left the prices as they were when the book was published in 1963.)

	bread	meat	potatoes	cabbage	milk	gelatin	required
calories	1254	1457	318	46	309	1725	3000
protein	39	73	8	4	16	43	70 g.
calcium	418	41	42	141	536	0	800 mg.
vitamin A	0	0	70	860	720	0	500 I.U.
cost/pound	$0.30	$1.00	$0.05	$0.08	$0.23	$0.48	

2-2. (a) You have been advised by your doctor to get more exercise, specifically, to burn off at least 2000 extra calories per week by some combination of walking, jogging, swimming, exercise-machine, collaborative indoor recreation, and pushing yourself away from the table at mealtimes. You have a limited tolerance for each activity in hours/week; each expends a certain number of calories per hour, as shown below:

	walking	jogging	swimming	machine	indoor	pushback
Calories	100	200	300	150	300	500
Tolerance	5	2	3	3.5	3	0.5

How should you divide your exercising among these activities to minimize the amount of time you spend?

(b) Suppose that you should also have some variety in your exercise — you must do at least one hour of each of the first four exercises, but no more than four hours total of walking, jogging, and exercise-machine. Solve the problem in this form.

2-3. (a) A manufacturer of soft drinks wishes to blend three sugars in approximately equal quantities to ensure uniformity of taste in a product. Suppliers only provide combinations of the sugars, at varying costs/ton:

				SUPPLIER			
Sugar	A	B	C	D	E	F	G
Cane	10%	10	20	30	40	20	60
Corn	30%	40	40	20	60	70	10
Beet	60%	50	40	50	0	10	30
Cost/ton	$10	11	12	13	14	12	15

Formulate an AMPL model that minimizes the cost of supply while producing a blend that contains 52 tons of cane sugar, 56 tons of corn sugar, and 59 tons of beet sugar.

(b) The manufacturer feels that to ensure good relations with suppliers it is necessary to buy at least 10 tons from each. How does this change the model and the minimum-cost solution?

(c) Formulate an alternative to the model in (a) that finds the lowest-cost way to blend one ton of supplies so that the amount of each sugar is between 30 and 37 percent of the total.

2-4. At the end of Chapter 1, we indicated how to interpret the "shadow prices" of constraints and the "reduced costs" of variables in a production model. The same ideas can be applied to this chapter's diet model.

(a) Going back to the diet problem that was successfully solved in Section 2.3, we can display the shadow prices as follows:

```
ampl: display diet.lb,diet.body,diet.ub,diet;
:    diet.lb  diet.body diet.ub    diet          :=
A       700     1956.29   20000    0
B1      700     1036.26   20000    0
B2      700       700     20000    0.404585
C       700     1682.51   20000    0
CAL   16000     19794.6   24000    0
NA        0       50000   50000   -0.00306905
;
```

How can you interpret the two that are nonzero?

(b) For the same problem, this listing gives the reduced costs:

```
ampl: display Buy.lb,Buy,Buy.ub,Buy.rc;
:      Buy.lb    Buy    Buy.ub      Buy.rc      :=
BEEF     2      5.36061    10     -4.44089e-16
CHK      2      2          10      1.18884
FISH     2      2          10      1.14441
HAM      2     10          10     -0.302651
MCH      2     10          10     -0.551151
MTL      2     10          10     -1.3289
SPG      2      9.30605    10      0
TUR      2      2          10      2.73162
;
```

Based on this information, if you want to save money by eating more than 10 packages of some food, which one is likely to be your best choice?

2-5. A chain of fast-food restaurants operates 7 days a week, and requires the following minimum number of kitchen employees from Monday through Sunday: 45, 45, 40, 50, 65, 35, 35. Each employee is scheduled to work one weekend day (Saturday or Sunday) and four other days in a week. The management wants to know the minimum total number of employees needed to satisfy the requirements on every day.

(a) Set up and solve this problem as a linear program.

(b) In light of the discussion in Section 2.4, explain how this problem can be viewed as a special case of the blending model in Figure 2-4.

2-6. The output of a paper mill consists of standard rolls 110 inches (110") wide, which are cut into smaller rolls to meet orders. This week there are orders for rolls of the following widths:

Width	Orders
20"	48
45"	35
50"	24
55"	10
75"	8

The owner of the mill wants to know what cutting patterns to apply so as to fill the orders using the smallest number of 110" rolls.

(a) A cutting pattern consists of a certain number of rolls of each width, such as two of 45" and one of 20", or one of 50" and one of 55" (and 5" of waste). Suppose, to start with, that we consider only the following six patterns:

Width	1	2	3	4	5	6
20"	3	1	0	2	1	3
45"	0	2	0	0	0	1
50"	1	0	1	0	0	0
55"	0	0	1	1	0	0
75"	0	0	0	0	1	0

How many rolls should be cut according to each pattern, to minimize the number of 110" rolls used? Formulate and solve this problem as a linear program, assuming that the number of smaller rolls produced need only be greater than or equal to the number ordered.

(b) Re-solve the problem, with the restriction that the number of rolls produced in each size must be between 10% under and 40% over the number ordered.

(c) Find another pattern that, when added to those above, improves the optimal solution.

(d) All of the solutions above use fractional numbers of rolls. Can you find solutions that also satisfy the constraints, but that cut a whole number of rolls in each pattern? How much does your whole-number solution cause the objective function value to go up in each case? (See Chapter 15 for a discussion of how to find optimal whole-number, or integer, solutions.)

2-7. In the refinery model of Exercise 1-6, the amount of premium gasoline to be produced is a decision variable. Suppose instead that orders dictate a production of 42,000 barrels. The octane rating of the product is permitted to be in the range of 89 to 91, and the vapor pressure in a range of 11.7 to 12.7. The five feedstocks that are blended to make premium gasoline have the following production and/or purchase costs:

SRG	9.57
N	8.87
RF	11.69
CG	10.88
B	6.75

Other data are as in Exercise 1-6. Construct a blending model and data file to represent this problem. Run them through AMPL to determine the optimal composition of the blend.

2-8. Recall that Figure 2-4 generalizes the diet model as a minimum-cost input selection model, with constraints on the outputs.

(a) In the same way, generalize the production model of Figure 1-6a as a maximum-revenue output selection model, with constraints on the inputs.

(b) The concept of an "input-output" model was one of the first applications of linear programming in economic analysis. Such a model can be described in terms of a set A of activities and a set M of materials. The decision variables are the levels $X_j \geq 0$ at which the activities are run; they have lower limits u_j^- and upper limits u_j^+.

Each activity j has either a revenue per unit $c_j > 0$, or a cost per unit represented by $c_j < 0$. Thus total profit from all activities is $\sum_{j \in A} c_j X_j$, which is to be maximized.

Each unit of activity j produces an amount of material i given by $a_{ij} \geq 0$, or consumes an amount of material i represented by $a_{ij} < 0$. Thus if $\sum_{j \in A} a_{ij} X_j$ is > 0 it is the total production of material i by all activities; if < 0, it is the total consumption of material i by all activities.

For each material i, there is either an upper limit on the total production given by $b_i^+ > 0$, or a lower limit on the total consumption given by $b_i^+ < 0$. Similarly, there is either a lower limit on the total production given by $b_i^- > 0$, or an upper limit on the total consumption given by $b_i^- < 0$.

Write out a formulation of this model in AMPL.

(c) Explain how the minimum-cost input selection model and maximum-revenue output-selection model can be viewed as special cases of the input-output model.

3

Transportation, Assignment and Minimum-Cost Flows

The linear programs in Chapters 1 and 2 are all examples of classical "activity" models. In such models the variables and constraints deal with distinctly different kinds of activities — tons of steel produced versus hours of mill time used, or packages of food bought versus percentages of nutrients supplied. To use these models you must supply coefficients like tons per hour or percentages per package that convert a unit of activity in the variables to the corresponding amount of activity in the constraints.

This chapter addresses a significantly different but equally common kind of model, in which something is shipped or assigned, but not converted. The resulting constraints, which reflect both limitations on availability and requirements for delivery, have an especially simple form.

We begin by describing the so-called transportation problem, in which a single good is to be shipped from several origins to several destinations at minimum overall cost. This problem gives rise to the simplest kind of linear program for minimum-cost flows. We then generalize to a transportation model, an essential step if we are to manage all the data, variables and constraints effectively.

As with the diet model, the power of the transportation model lies in its adaptability. We continue by considering some other interpretations of the "flow" from origins to destinations, and work through one particular interpretation in which the variables represent assignments rather than shipments.

The transportation model is only the most elementary kind of minimum-cost flow model. More general models are often best expressed as networks, in which nodes — some of which may be origins or destinations — are connected by arcs that carry flows of some kind. AMPL offers convenient features for describing network flow models, including `node` and `arc` declarations that specify network structure directly. Network models and the relevant AMPL features are the topic of Chapter 11.

3.1 A linear program for the transportation problem

Suppose that we have decided (perhaps by the methods described in Chapter 1) to produce steel coils at three mill locations, in the following amounts:

GARY	Gary, Indiana	1400
CLEV	Cleveland, Ohio	2600
PITT	Pittsburgh, Pennsylvania	2900

The total of 6,900 tons must be shipped in various amounts to meet orders at seven locations of automobile factories:

FRA	Framingham, Massachusetts	900
DET	Detroit, Michigan	1200
LAN	Lansing, Michigan	600
WIN	Windsor, Ontario	400
STL	St. Louis, Missouri	1700
FRE	Fremont, California	1100
LAF	Lafayette, Indiana	1000

We now have an optimization problem: What is the least expensive plan for shipping the coils from mills to plants?

To answer the question, we need to compile a table of shipping costs per ton:

	GARY	CLEV	PITT
FRA	39	27	24
DET	14	9	14
LAN	11	12	17
WIN	14	9	13
STL	16	26	28
FRE	82	95	99
LAF	8	17	20

Let GARY:FRA be the number of tons shipped from GARY to FRA, and similarly for the other cities. Then the objective can be written as follows:

Minimize

    39 GARY:FRA + 27 CLEV:FRA + 24 PITT:FRA +
    14 GARY:DET + 9 CLEV:DET + 14 PITT:DET +
    11 GARY:LAN + 12 CLEV:LAN + 17 PITT:LAN +
    14 GARY:WIN + 9 CLEV:WIN + 13 PITT:WIN +
    16 GARY:STL + 26 CLEV:STL + 28 PITT:STL +
    82 GARY:FRE + 95 CLEV:FRE + 99 PITT:FRE +
     8 GARY:LAF + 17 CLEV:LAF + 20 PITT:LAF

There are 21 decision variables in all. Even small transportation problems like this one tend to have a lot of variables, because there is one for each possible pair of cities.

We could achieve the lowest possible shipping cost by supplying each factory from the closest mill. However, we would then end up shipping 900 tons from PITT, 1600 from CLEV, and all the rest from GARY — amounts quite inconsistent with the production levels previously decided upon. We need to add a constraint that the sum of the shipments from GARY to the seven factories is equal to the production level of 1400:

$$\text{GARY:FRA} + \text{GARY:DET} + \text{GARY:LAN} + \text{GARY:WIN} +$$
$$\text{GARY:STL} + \text{GARY:FRE} + \text{GARY:LAF} = 1400$$

There are analogous constraints for the other two mills:

$$\text{CLEV:FRA} + \text{CLEV:DET} + \text{CLEV:LAN} + \text{CLEV:WIN} +$$
$$\text{CLEV:STL} + \text{CLEV:FRE} + \text{CLEV:LAF} = 2600$$

$$\text{PITT:FRA} + \text{PITT:DET} + \text{PITT:LAN} + \text{PITT:WIN} +$$
$$\text{PITT:STL} + \text{PITT:FRE} + \text{PITT:LAF} = 2900$$

There also have to be constraints like these at the factories, to ensure that the amounts shipped equal the amounts ordered. At FRA, the sum of the shipments received from the three mills must equal the 900 tons ordered:

$$\text{GARY:FRA} + \text{CLEV:FRA} + \text{PITT:FRA} = 900$$

And similarly for the other six factories:

$$\text{GARY:DET} + \text{CLEV:DET} + \text{PITT:DET} = 1200$$
$$\text{GARY:LAN} + \text{CLEV:LAN} + \text{PITT:LAN} = 600$$
$$\text{GARY:WIN} + \text{CLEV:WIN} + \text{PITT:WIN} = 400$$
$$\text{GARY:STL} + \text{CLEV:STL} + \text{PITT:STL} = 1700$$
$$\text{GARY:FRE} + \text{CLEV:FRE} + \text{PITT:FRE} = 1100$$
$$\text{GARY:LAF} + \text{CLEV:LAF} + \text{PITT:LAF} = 1000$$

We have ten constraints in all, one for each mill and one for each factory. If we add the requirement that all variables be nonnegative, we have a complete linear program for the transportation problem.

We won't even try showing what it would be like to type these constraints one-by-one in response to AMPL's prompts. Clearly we want to set up a general model to deal with this problem.

3.2 An AMPL model for the transportation problem

Two fundamental sets of objects underlie the transportation problem: the sources or origins (mills, in our example) and the destinations (factories). Thus we begin the AMPL model with a declaration of these two sets:

```
set ORIG;
set DEST;
```

There is a supply of something at each origin (tons of steel coils produced, in our case), and a demand for the same thing at each destination (tons of coils ordered). AMPL defines nonnegative quantities like these with `param` statements indexed over a set; in this case we add one extra refinement, a `check` statement to test the data for validity:

```
param supply {ORIG} >= 0;
param demand {DEST} >= 0;

check: sum {i in ORIG} supply[i] = sum {j in DEST} demand[j];
```

The `check` statement says that the sum of the supplies has to equal the sum of the demands. The way that our model is to be set up, there can't possibly be any solutions unless this condition is satisfied. By putting it in a `check` statement, we tell AMPL to test this condition after reading the data, and to issue an error message if it is violated.

For each combination of an origin and a destination, there is a transportation cost and a variable representing the amount transported. Again, the ideas from previous chapters are easily adapted to produce the appropriate AMPL statements:

```
param cost {ORIG,DEST} >= 0;
var Trans {ORIG,DEST} >= 0;
```

For a particular origin `i` and destination `j`, we ship `Trans[i,j]` units from `i` to `j`, at a cost of `cost[i,j]` per unit; the total cost for this pair is

```
cost[i,j] * Trans[i,j]
```

Adding over all pairs, we have the objective function,

```
minimize total_cost:
    sum {i in ORIG, j in DEST} cost[i,j] * Trans[i,j];
```

which could also be written as either of

```
sum {j in DEST, i in ORIG} cost[i,j] * Trans[i,j];
```

or

```
sum {i in ORIG} sum {j in DEST} cost[i,j] * Trans[i,j];
```

As long as you express the objective in some mathematically correct way, AMPL will sort out the terms.

It remains to specify the two collections of constraints, those at the origins and those at the destinations. If we name these collections `Supply` and `Demand`, their declarations will start as follows:

```
subject to Supply {i in ORIG}:  ...
subject to Demand {j in DEST}:  ...
```

To complete the `Supply` constraint for origin `i`, we need to say that the sum of all shipments out of `i` is equal to the supply available. Since the amount shipped out of `i` to a particular destination `j` is `Trans[i,j]`, the amount shipped to all destinations must be

```
sum {j in DEST} Trans[i,j]
```

```
set ORIG;    # origins
set DEST;    # destinations

param supply {ORIG} >= 0;    # amounts available at origins
param demand {DEST} >= 0;    # amounts required at destinations

    check: sum {i in ORIG} supply[i] = sum {j in DEST} demand[j];

param cost {ORIG,DEST} >= 0;    # shipment costs per unit
var Trans {ORIG,DEST} >= 0;     # units to be shipped

minimize total_cost:
    sum {i in ORIG, j in DEST} cost[i,j] * Trans[i,j];

subject to Supply {i in ORIG}:
    sum {j in DEST} Trans[i,j] = supply[i];

subject to Demand {j in DEST}:
    sum {i in ORIG} Trans[i,j] = demand[j];
```

Figure 3-1a: Transportation model (transp.mod).

Since we have already defined a parameter supply indexed over origins, the amount available at i is supply[i]. Thus the constraint is

```
subject to Supply {i in ORIG}:
    sum {j in DEST} Trans[i,j] = supply[i];
```

(Note that the names supply and Supply are unrelated; AMPL distinguishes upper and lower case.) The other collection of constraints is much the same, except that the sum is over i in ORIG, and equals demand[j].

We can now present the complete transportation model, Figure 3-1a. As you might have noticed, we have been consistent in using the index i to run over the set ORIG, and the index j to run over DEST. This is not an AMPL requirement, but such a convention makes it easier to read a model. You may name your own indices whatever you like, but keep in mind that the scope of an index — the part of the model where it has the same meaning — is to the end of the expression that defines it. Thus in the Supply constraint

```
subject to Supply {i in ORIG}:
    sum {j in DEST} Trans[i,j] = supply[i];
```

the scope of i runs to the semicolon at the end of the declaration, while the scope of j extends only through the summand Trans[i,j]. Since the latter scope is inside the former, the two indices must have different names. Also an index may not have the same name as a set or other model component. Index scopes are discussed more fully, with further examples, in Section 5.5.

Data values for the transportation model are shown in Figure 3-1b. To define DEST and demand, we have used an input format that permits a set and one or more parameters indexed over it to be specified together. The set name is surrounded by colons. (We also show some comments, which can appear among data statements just as in a model.)

```
param: ORIG:   supply :=  # defines set "ORIG" and param "supply"
        GARY   1400
        CLEV   2600
        PITT   2900 ;
param: DEST:   demand :=  # defines "DEST" and "demand"
        FRA     900
        DET    1200
        LAN     600
        WIN     400
        STL    1700
        FRE    1100
        LAF    1000 ;

param cost:
        FRA  DET  LAN  WIN  STL  FRE  LAF :=
   GARY  39   14   11   14   16   82    8
   CLEV  27    9   12    9   26   95   17
   PITT  24   14   17   13   28   99   20 ;
```

Figure 3-1b: Data for transportation model (`transp.dat`).

If the model is stored in a file `transp.mod` and the data in `transp.dat`, we can solve the linear program and examine the output:

```
ampl: model transp.mod;
ampl: data transp.dat;
ampl: solve;
MINOS 5.4: optimal solution found.
13 iterations, objective 196200

ampl: display Trans;
Trans [*,*] (tr)
:     CLEV   GARY   PITT      :=
DET   1200      0      0
FRA      0      0    900
FRE      0   1100      0
LAF    400    300    300
LAN    600      0      0
STL      0      0   1700
WIN    400      0      0
;
```

By displaying the variable `Trans`, we see that most destinations are supplied from a single mill, but CLEV, GARY and PITT all ship to LAF.

It is instructive to compare this solution to one given by another solver:

```
ampl: option solver osl;
ampl: solve;
OSL 1.2: optimal solution; objective 196200
10 simplex iterations
```

```
ampl: display Trans;
Trans [*,*] (tr)
:      CLEV   GARY   PITT      :=
DET    1200      0      0
FRA       0      0    900
FRE       0   1100      0
LAF     400      0    600
LAN     600      0      0
STL       0    300   1400
WIN     400      0      0
;
```

The minimum cost is still 196200, but it is achieved in a different way. Alternative optimal solutions such as these are often exhibited by transportation problems, particularly when the coefficients in the objective function are round numbers.

Unfortunately, there is no easy way to characterize all the optimal solutions. You may be able to get a better choice of optimal solution by working with several objectives, however, as we will illustrate in Section 8.3.

3.3 Other interpretations of the transportation model

As the name suggests, a transportation model is applicable whenever some material is being shipped from a set of origins to a set of destinations. Given certain amounts available at the origins, and required at the destinations, the problem is to meet the requirements at a minimum shipping cost.

Viewed more broadly, transportation models do not have to be concerned with the shipping of "materials". They can be applied to the transportation of anything, provided that the quantities available and required can be measured in some units, and that the transportation cost per unit can be determined. They might be used to model the shipments of automobiles to dealers, for example, or the movement of military personnel to new assignments.

In an even broader view, transportation models need not deal with "shipping" at all. The quantities at the origins may be merely associated with various destinations, while the objective measures some value of the association that has nothing to do with actually moving anything. Often the result is referred to as an "assignment" model.

As one particularly well-known example, consider a department that needs to assign some number of people to an equal number of offices. The origins now represent individual people, and the destinations represent individual offices. Since each person is assigned one office, and each office is occupied by one person, all of the parameter values supply[i] and demand[j] are 1. We interpret Trans[i,j] as the "amount" of person i that is assigned to office j; that is, if Trans[i,j] is 1 then person i will occupy office j, while if Trans[i,j] is 0 then person i will not occupy office j.

What of the objective? One possibility is to ask people to rank the offices, giving their first choice, second choice, and so forth. Then we can let cost[i,j] be the rank

```
set ORIG := Coullard Hazen Hopp Hurter Jones Mehrotra
            Rieders Spearman Sun Tamhane Zazanis ;

set DEST := 1021 1049 1053 1055 1083 1087 2009 2019 2053 2083 2087;

param supply default 1 ;

param demand default 1 ;

param cost:
          1021 1049 1053 1055 1083 1087 2009 2019 2053 2083 2087 :=
Coullard    6    9    8    7   11   10    4    5    3    2    1
Hazen      11    8    7    6    9   10    1    5    4    2    3
Hopp        9   10   11    1    5    6    2    7    8    3    4
Hurter     11    9    8   10    6    5    1    7    4    2    3
Jones       3    2    8    9   10   11    1    5    4    6    7
Mehrotra   11    9   10    5    3    4    6    7    8    1    2
Rieders     6   11   10    9    8    7    1    2    5    4    3
Spearman   11    5    4    6    7    8    1    9   10    2    3
Sun        11    9   10    8    6    5    7    3    4    1    2
Tamhane     5    6    9    8    4    3    7   10   11    2    1
Zazanis    11    9    8    4    6    5    3   10    7    2    1 ;
```

Figure 3-2: Data for assignment problem (`assign.dat`).

that person i gives to office j. This convention lets us interpret each objective function term `cost[i,j] * Trans[i,j]` as follows: if `Trans[i,j]` is 1, the term equals the rank that person i gave the office to which that person is assigned; if `Trans[i,j]` equals 0, the term is zero. Since the objective is the sum of all these terms, it represents the sum of people's rankings for the offices to which they were assigned. By minimizing this sum, we can hope to find an assignment that will please a lot of people.

To use the transportation model for this purpose, we need only supply the appropriate data. Figure 3-2 is one example, with 11 people to be assigned to 11 offices. The `default` option has been used to set all the `supply` and `demand` values to 1 without typing all the 1's. If we store this data set in `assign.dat`, we can use it with the transportation model that we already have:

```
ampl: model transp.mod;
ampl: data assign.dat;
ampl: solve;
CPLEX 2.0: optimal solution; objective 28
8 iterations (2 in phase I)
```

By setting the option `omit_zero_rows` to 1, we can print just the nonzero terms in the objective. (Options for displaying results are presented in Chapter 10.) This listing tells us each person's assigned room and his or her preference for it:

```
ampl: option omit_zero_rows 1;
ampl: display {i in ORIG, j in DEST} cost[i,j] * Trans[i,j];
cost[i,j]*Trans[i,j] :=
Coullard 1021    6
Hazen    2053    4
Hopp     1055    1
Hurter   2009    1
Jones    1049    2
Mehrotra 1083    3
Rieders  2019    2
Spearman 1053    4
Sun      2083    1
Tamhane  1087    3
Zazanis  2087    1
;
```

The solution is reasonably successful, although it does assign two fourth choices and one sixth choice.

It is not hard to see that when all the supply[i] and demand[j] values are 1, any Trans[i,j] satisfying all the constraints must be between 0 and 1. But how did we know that every Trans[i,j] would equal either 0 or 1 in the optimal solution, rather than, say, ½? We were able to rely on a special property of transportation models, which guarantees that as long as all supply and demand values are integers, and all lower and upper bounds on the variables are integers, there will be an optimal solution that is entirely integral. Moreover, we used a solver that always finds one of these integral solutions. But don't let this favorable result mislead you into assuming that integrality can be assured in all other circumstances; even in examples that seem to be much like the transportation model, finding integral solutions can require a special solver, and a lot more work. Chapter 15 discusses issues of integrality at length.

A problem of assigning 100 people to 100 rooms has ten thousand variables; assigning 1000 people to 1000 rooms yields a million variables. In applications on this scale, however, most of the assignments can be ruled out in advance, so that the number of actual decision variables is not too large. After looking at an initial solution, you may want to rule out some more assignments — in our example, perhaps no assignment to lower than fifth choice should be allowed — or you may want to force some assignments to be made a certain way, in order to see how the rest could be done optimally. These situations require models that can deal with subsets of pairs (of people and offices, or origins and destinations) in a direct way. AMPL's features for describing pairs and other ''compound'' objects are the subject of Chapter 6.

Exercises

3-1. This transportation model, which deals with finding a least cost shipping schedule, comes from Dantzig's *Linear Programming and Extensions*. A company has plants in Seattle and San

Diego, with capacities 350 and 600 cases per week respectively. It has customers in New York, Chicago, and Topeka, which order 325, 300, and 275 cases per week. The distances involved are:

	New York	Chicago	Topeka
Seattle	2500	1700	1800
San Diego	2500	1800	1400

The shipping cost is $90 per case per thousand miles. Formulate this model in AMPL and solve it to determine the minimum cost and the amounts to be shipped.

3-2. A small manufacturing operation produces six kinds of parts, using three machines. For the coming month, a certain number of each part is needed, and a certain number of parts can be accommodated on each machine; to complicate matters, it does not cost the same amount to make the same part on different machines. Specifically, the costs and related values are as follows:

			Part				
Machine	1	2	3	4	5	6	Capacity
1	3	3	2	5	2	1	80
2	4	1	1	2	2	1	30
3	2	2	5	1	1	2	160
Required	10	40	60	20	20	30	

(a) Using the model in Figure 3-1a, create a file of data statements for this problem; treat the machines as the origins, and the parts as the destinations. How many of each part should be produced on each machine, so as to minimize total cost?

(b) If the capacity of machine 2 is increased to 50, the manufacturer may be able to reduce the total cost of production somewhat. What small change to the model is necessary to analyze this situation? How much is the total cost reduced, and in what respects does the production plan change?

(c) Now suppose that the capacities are given in hours, rather than in numbers of parts, and that it takes a somewhat different number of hours to make the same part on different machines:

			Part				
Machine	1	2	3	4	5	6	Capacity
1	1.3	1.3	1.2	1.5	1.2	1.1	50
2	1.4	1.1	1.1	1.2	1.2	1.1	90
3	1.2	1.2	1.5	1.1	1.1	1.2	175

Modify the supply constraint so that it limits total time of production at each "origin" rather than the total quantity of production. How is the new optimal solution different? On which machines is all available time used?

(d) Solve the preceding problem again, but with the objective function changed to minimize total machine-hours rather than total cost.

3-3. This exercise deals with generalizations of the transportation model and data of Figure 3-1.

(a) Add two parameters, supply_pct and demand_pct, to represent the maximum fraction of a mill's supply that may be sent to any one factory, and the maximum fraction of a factory's demand that may be satisfied by any one mill. Incorporate these parameters into the model of Figure 3-1a.

Solve for the case in which no more than 50% of a mill's supply may be sent to any one factory, and no more than 85% of a factory's demand may be satisfied by any one mill. How does this change the minimum cost and the optimal amounts shipped?

(b) Suppose that the rolling mills do not produce their own slabs, but instead obtain slabs from two other plants, where the following numbers of tons are to be made available:

```
MIDTWN   2700
HAMLTN   4200
```

The cost per ton of shipping a slab from a plant to a mill is as follows:

	GARY	CLEV	PITT
MIDTWN	12	8	17
HAMLTN	10	5	13

All other data values are the same as before, but with `supply_pct` reinterpreted as the maximum fraction of a plant's supply that may be sent to any one mill.

Formulate this situation as an AMPL model. You will need two indexed collections of variables, one for the shipments from plants to mills, and one for the shipments from mills to factories. Shipments from each mill will have to equal supply, and shipments to each factory will have to equal demand as before; also, shipments out of each mill will have to equal shipments in.

Solve the resulting linear program. What are the shipment amounts in the minimum-cost solution?

(c) In addition to the differences in shipping costs, there may be different costs of production at the plants and mills. Explain how production costs could be incorporated into the model.

(d) When slabs are rolled, some fraction of the steel is lost as scrap. Assuming that this fraction may be different at each mill, revise the model to take scrap loss into account.

(e) In reality, scrap is not really lost, but is sold for recycling. Make a further change to the model to account for the value of the scrap produced at each mill.

3-4. This exercise considers variations on the assignment problem introduced in Section 3.3.

(a) Try reordering the list of members of DEST in the data (Figure 3-2), and solving again. Find a reordering that causes your solver to report a different optimal assignment.

(b) An assignment that gives even one person a very low-ranked office may be unacceptable, even if the total of the rankings is optimized. In particular, our solution gives one individual her sixth choice; to rule this out, change all preferences of six or larger in the cost data to 99, so that they will become very unattractive. (You'll learn more convenient features for doing the same thing in later chapters, but this crude approach will work for now.) Solve the assignment problem again, and verify that the result is an equally good assignment in which no one gets worse than fifth choice.

Now apply the same approach to try to give everyone no worse than fourth choice. What do you find?

(c) Suppose now that offices 1021, 1083 and 1087 become unavailable, and you have to put two people each into 1049, 1053 and 1055. Add 20 to each ranking for these three offices, to reflect the fact that anyone would prefer a private office to a shared one. What other modifications to the model and data would be necessary to handle this situation? What optimal assignment do you get?

(d) Some people may have seniority that entitles them to greater consideration in their choice of office. Explain how you could enhance the model to use seniority level data for each person.

4

Building Larger Models

The linear programs that we have presented so far have been quite small, so their data and solutions could fit onto a page. Most of the LPs found in practical applications, however, have hundreds or thousands of variables and constraints, and some are even larger.

How do linear programs get to be so large? They might be like the ones we have shown, but with larger indexing sets and more data. A steel mill could be considered to make hundreds of different products, for example, if every variation of width, thickness, and finish is treated separately. Or a large organization could have thousands of people involved in one assignment problem. Nevertheless, these kinds of applications are not as common as one might expect. As a model is refined to greater levels of detail, its data values become harder to maintain and its solutions harder to understand; past a certain point, additional detail offers no benefit. Thus to plan production for next week, considerable detail may be justifiable; but to plan for next year, it may be better to have a small, aggregated model that can be easily run many times with different scenarios.

A more common source of large linear programs is the linking together of smaller ones. It is not unusual for an application to give rise to many simple LPs of the kinds we have discussed before; here are three possibilities:

- Many products are to be shipped, and there is a transportation problem (as in Chapter 3) for each product.
- Manufacturing is to be planned over many weeks, and there is a production problem (as in Chapter 1) for each week.
- Several products are made at several mills, and shipped to several factories; there is a production problem for each mill, and a transportation problem for each product.

When variables or constraints are added to tie these LPs together, the result can be one very large LP. No individual part need be particularly detailed; the size is more due to the large number of combinations of origins, destinations, products and weeks.

This chapter shows how AMPL models might be formulated for the three situations outlined above. The resulting models are necessarily more complicated than our previous ones, and require the use of a few more features from the AMPL language. Since they build on the terminology and logic of smaller models that have already been introduced, however, these larger models are still manageable.

4.1 A multicommodity transportation model

The transportation model of the previous chapter was concerned with shipping a single commodity from origins to destinations. Suppose now that we are shipping several different products. We can define a new set, PROD, whose members represent the different products, and we can add PROD to the indexing of every component in the model; the result can be seen in Figure 4-1. Because supply, demand, cost, and Trans are indexed over one more set in this version, they take one more subscript: supply[i,p] for the amount of product p shipped from origin i, Trans[i,j,p] for the amount of p shipped from i to j, and so forth. Even the check statement is now indexed over PROD, so that it verifies that supply equals demand for each separate product.

We now have (origins + destinations) × (products) constraints in (origins) × (destinations) × (products) variables. The result could be quite a large linear program, even if the individual sets do not have many members. For example, 5 origins, 20 destinations and 10 products give 250 constraints in 1000 variables. The size of this LP is misleading, however, because the shipments of the products are independent. That is, the amounts we ship of one product do not affect the amounts we can ship of any other product, or the costs of shipping any other product. We would do better in this case to solve a smaller transportation problem for each individual product. In AMPL terms, we would use the simple transportation model from the previous chapter, together with a different data file for each product.

The situation would be different if some additional circumstances had the effect of tying together the different products. As an example, imagine that there are restrictions on the total shipments of products from an origin to a destination, perhaps because of limited shipping capacity. To accommodate such restrictions in our model, we declare a new parameter limit indexed over the combinations of origins and destinations:

```
param limit {ORIG,DEST} >= 0;
```

Then we have a new collection of constraints, one for each origin i and destination j, which say that the sum of shipments from i to j of all products p may not exceed limit[i,j]:

```
subject to Multi {i in ORIG, j in DEST}:
    sum {p in PROD} Trans[i,j,p] <= limit[i,j];
```

Subject to these constraints (also shown in Figure 4-1), we can no longer set the amount of one product shipped from i to j without considering the amounts of other products also shipped from i to j, since it is the sum of all products that is limited. Thus we have no choice but to solve the one large linear program.

For the steel mill in Chapter 1, the products were bands, coils, and plate. Thus the data for the multicommodity model could look like Figure 4-2. We invoke AMPL in the usual way to get the following solution:

```
ampl: model multi.mod; data multi.dat; solve;
CPLEX 2.0: optimal solution; objective 199500
31 iterations (15 in phase I)
```

```
set ORIG;   # origins
set DEST;   # destinations
set PROD;   # products

param supply {ORIG,PROD} >= 0;  # amounts available at origins
param demand {DEST,PROD} >= 0;  # amounts required at destinations

    check {p in PROD}:
        sum {i in ORIG} supply[i,p] = sum {j in DEST} demand[j,p];

param limit {ORIG,DEST} >= 0;

param cost {ORIG,DEST,PROD} >= 0;   # shipment costs per unit
var Trans {ORIG,DEST,PROD} >= 0;    # units to be shipped

minimize total_cost:
   sum {i in ORIG, j in DEST, p in PROD}
      cost[i,j,p] * Trans[i,j,p];

subject to Supply {i in ORIG, p in PROD}:
   sum {j in DEST} Trans[i,j,p] = supply[i,p];

subject to Demand {j in DEST, p in PROD}:
   sum {i in ORIG} Trans[i,j,p] = demand[j,p];

subject to Multi {i in ORIG, j in DEST}:
   sum {p in PROD} Trans[i,j,p] <= limit[i,j];
```

Figure 4-1: Multicommodity transportation model (`multi.mod`).

```
ampl: display {p in PROD}: {i in ORIG, j in DEST} Trans[i,j,p];
Trans[i,j,'bands'] [*,*] (tr)
:    CLEV  GARY  PITT     :=
DET     0     0   300
FRA   225     0    75
FRE     0     0   225
LAF   150     0   100
LAN     0     0   100
STL   250   400     0
WIN    75     0     0
;

Trans[i,j,'coils'] [*,*] (tr)
:    CLEV  GARY  PITT     :=
DET   525     0   225
FRA     0     0   500
FRE    50   625   175
LAF    75   150   275
LAN   400     0     0
STL   300    25   625
WIN   250     0     0
;
```

```
set ORIG := GARY CLEV PITT ;
set DEST := FRA DET LAN WIN STL FRE LAF ;
set PROD := bands coils plate ;

param supply (tr):    GARY    CLEV    PITT :=
                bands  400     700     800
                coils  800    1600    1800
                plate  200     300     300 ;

param demand (tr):
              FRA    DET    LAN    WIN    STL    FRE    LAF :=
     bands    300    300    100     75    650    225    250
     coils    500    750    400    250    950    850    500
     plate    100    100      0     50    200    100    250 ;

param limit default 625 ;

param cost :=

  [*,*,bands]:  FRA   DET   LAN   WIN   STL   FRE   LAF :=
        GARY     30    10     8    10    11    71     6
        CLEV     22     7    10     7    21    82    13
        PITT     19    11    12    10    25    83    15

  [*,*,coils]:  FRA   DET   LAN   WIN   STL   FRE   LAF :=
        GARY     39    14    11    14    16    82     8
        CLEV     27     9    12     9    26    95    17
        PITT     24    14    17    13    28    99    20

  [*,*,plate]:  FRA   DET   LAN   WIN   STL   FRE   LAF :=
        GARY     41    15    12    16    17    86     8
        CLEV     29     9    13     9    28    99    18
        PITT     26    14    17    13    31   104    20 ;
```

Figure 4-2: Multicommodity transportation problem data (`multi.dat`).

```
Trans[i,j,'plate'] [*,*] (tr)
:    CLEV  GARY  PITT      :=
DET  100     0     0
FRA   50     0    50
FRE  100     0     0
LAF    0     0   250
LAN    0     0     0
STL    0   200     0
WIN   50     0     0
;
```

In both our specification of the shipping costs and AMPL's display of the solution, a three-dimensional collection of data (that is, indexed over three sets) must be represented on a two-dimensional screen or page. We accomplish this by "slicing" the data along one index, so that it appears as a collection of two-dimensional tables. The display command will make a guess as to the best index on which to slice, but by use of an

```
set PROD;      # products
param T > 0;  # number of weeks

param rate {PROD} > 0;          # tons per hour produced
param avail {1..T} >= 0;         # hours available in week
param profit {PROD,1..T};        # profit per ton
param market {PROD,1..T} >= 0;  # limit on tons sold in week

var Make {p in PROD, t in 1..T} >= 0, <= market[p,t];
                                # tons produced

maximize total_profit:
   sum {p in PROD, t in 1..T} profit[p,t] * Make[p,t];

        # total profits from all products in all weeks

subject to time {t in 1..T}:
   sum {p in PROD} (1/rate[p]) * Make[p,t] <= avail[t];

        # total of hours used by all products
        # may not exceed hours available, in each week
```

Figure 4-3: Production model replicated over periods (`steelT0.mod`).

explicit indexing expression as shown above, we can tell it to display a table for each product.

The optimal solution above ships only 25 tons of coils from GARY to STL. It might be reasonable to require that, if any amount at all is shipped from an origin to a destination, it must be at least, say, 50 tons. In terms of our model, either `Trans[i,j,p]` = 0 or `Trans[i,j,p]` >= 50. Unfortunately, although it is possible to write such an ''either/or'' constraint in AMPL, it is not a linear constraint, and so there is no way that an LP solver can handle it. Chapter 15 explains how more powerful (but costlier) integer programming techniques are required for this and related kinds of discrete restrictions.

4.2 A multiperiod production model

Another common way in which models are expanded is by replicating them over time. To illustrate, we consider how the model of Figure 1-4a might be used to plan production for the next T weeks, rather than for a single week.

We begin by adding another index set to most of the quantities of interest. The added set represents weeks numbered 1 through T, as shown in Figure 4-3. The expression `1..T` is AMPL's shorthand for the set of integers from 1 through `T`. We have replicated all the parameters and variables over this set, except for `rate`, which is regarded as fixed over time. As a result there is a constraint for each week, and the `profit` terms are summed over weeks as well as products.

So far this is merely a separate LP for each week, unless something is added to tie the weeks together. Just as we were able to find constraints that involved all the products, we could look for constraints that involve production in all of the weeks. Most multiperiod models take a different approach, however, in which constraints relate each week's production to that of the following week only.

Suppose that we allow some of a week's production to be placed in inventory, for sale in any later week. We thus add new decision variables to represent the amounts inventoried and sold in each week. The variables Make[j,t] are retained, but they represent only the amounts produced, which are now not necessarily the same as the amounts sold. Our new variable declarations look like this:

```
var Make {PROD,1..T} >= 0;
var Inv {PROD,0..T} >= 0;

var Sell {p in PROD, t in 1..T} >= 0, <= market[p,t];
```

The bounds market[p,t], which represent the maximum amounts that can be sold in a week, are naturally transferred to Sell[p,t].

The variable Inv[p,t] will represent the inventory of product p at the end of period t. Thus the quantities Inv[p,0] will be the inventories at the end of week zero, or equivalently at the beginning of the first week — in other words, now. Our model assumes that these initial inventories are provided as part of the data:

```
param inv0 {PROD} >= 0;
```

A simple constraint guarantees that the variables Inv[p,0] take these values:

```
subject to initial {p in P}:  Inv[p,0] = inv0[p];
```

It may seem "inefficient" to devote a constraint like this to saying that a variable equals a constant, but when it comes time to send the linear program to a solver, AMPL will automatically substitute the value of inv0[p] for any occurrence of Inv[p,0]. In most cases, we can concentrate on writing the model in the clearest or easiest way, and leave matters of efficiency to the computer.

Now that we are distinguishing sales, production, and inventory, we can explicitly model the contribution of each to the profit, by defining three parameters:

```
param revenue {PROD,1..T} >= 0;
param prodcost {PROD} >= 0;
param invcost {PROD} >= 0;
```

These are incorporated into the objective as follows:

```
maximize total_profit:
    sum {p in P, t in 1..T} (revenue[p,t]*Sell[p,t] -
        prodcost[p]*Make[p,t] - invcost[p]*Inv[p,t]);
```

As you can see, revenue[p,t] is the amount received per ton of product p sold in week t; prodcost[p] and invcost[p] are the production and inventory carrying cost per ton of product p in any week.

Finally, with the sales and inventories fully incorporated into our model, we can add the key constraints that tie the weeks together: the amount of a product made available in

a week, through production or from inventory, must equal the amount disposed of in that
week, through sale or to inventory:

```
subject to balance {p in P, t in 1..T}:
    Make[p,t] + Inv[p,t-1] = Sell[p,t] + Inv[p,t];
```

Because the index t is from a set of numbers, the period previous to t can be written as
t-1. In fact, t can be used in any arithmetic expression; conversely, an AMPL expression such as t-1 may be used in any context where it makes sense. Notice also that for a
first-period constraint (t equal to 1), the inventory term on the left is Inv[p,0], the initial inventory.

We now have a complete model, as shown in Figure 4-4. To illustrate a solution, we
use the small sample data file shown in Figure 4-5; it represents a four-week expansion of
the data from Figure 1-4b.

If we put the model and data into files steelT.mod and steelT.dat, then AMPL
can be invoked to find a solution:

```
ampl: model steelT.mod;
ampl: data steelT.dat;
ampl: solve;
MINOS 5.4: optimal solution found.
20 iterations, objective 515033

ampl: option display_1col 0;

ampl: display Make;
Make [*,*] (tr)
:  bands  coils     :=
1   5990   1407
2   6000   1400
3   1400   3500
4   2000   4200
;

ampl: display Inv;
Inv [*,*] (tr)
: bands   coils     :=
0   10        0
1    0     1100
2    0        0
3    0        0
4    0        0
;

ampl: display Sell;
Sell [*,*] (tr)
:  bands  coils     :=
1   6000    307
2   6000   2500
3   1400   3500
4   2000   4200
;
```

```
set PROD;      # products
param T > 0;   # number of weeks

param rate {PROD} > 0;         # tons per hour produced
param inv0 {PROD} >= 0;        # initial inventory
param avail {1..T} >= 0;       # hours available in week
param market {PROD,1..T} >= 0; # limit on tons sold in week

param prodcost {PROD} >= 0;      # cost per ton produced
param invcost {PROD} >= 0;       # carrying cost/ton of inventory
param revenue {PROD,1..T} >= 0;  # revenue per ton sold

var Make {PROD,1..T} >= 0;       # tons produced
var Inv {PROD,0..T} >= 0;        # tons inventoried
var Sell {p in PROD, t in 1..T} >= 0, <= market[p,t]; # tons sold

maximize total_profit:
   sum {p in PROD, t in 1..T} (revenue[p,t]*Sell[p,t] -
      prodcost[p]*Make[p,t] - invcost[p]*Inv[p,t]);

                 # Total revenue less costs in all weeks

subject to time {t in 1..T}:
   sum {p in PROD} (1/rate[p]) * Make[p,t] <= avail[t];

                 # Total of hours used by all products
                 # may not exceed hours available, in each week

subject to initial {p in PROD}:  Inv[p,0] = inv0[p];

                 # Initial inventory must equal given value

subject to balance {p in PROD, t in 1..T}:
   Make[p,t] + Inv[p,t-1] = Sell[p,t] + Inv[p,t];

                 # Tons produced and taken from inventory
                 # must equal tons sold and put into inventory
```

Figure 4-4: Multiperiod production model (steelT.mod).

```
param T := 4;
set PROD := bands coils;

param avail := 1 40  2 40  3 32  4 40 ;

param rate :=  bands 200   coils 140 ;
param inv0 :=  bands 10    coils  0 ;

param prodcost := bands 10    coils  11 ;
param invcost  := bands 2.5   coils  3 ;

param revenue:   1     2     3     4 :=
        bands   25    26    27    27
        coils   30    35    37    39 ;

param market:    1     2     3     4 :=
        bands  6000  6000  4000  6500
        coils  4000  2500  3500  4200 ;
```

Figure 4-5: Data for multiperiod production model (steelT.dat).

Production of coils in the first week is held over to be sold at a higher price in the second week. In the second through fourth weeks, coils are more profitable than bands, and so coils are sold up to the limit, with bands filling out the capacity. (Setting option `display_1col` to zero permits this output to appear in a nicer format, as explained in Section 10.4.)

4.3 A model of production and transportation

Large linear programs can be created not only by tying together small models of one kind, as in the two examples above, but by linking different kinds of models. We conclude this chapter with an example that combines features of both production and transportation models.

Suppose that the steel products are made at several mills, from which they are shipped to customers at the various factories. For each mill we can define a separate production model to optimize the amounts of each product to make. For each product we can define a separate transportation model, with mills as origins and factories as destinations, to optimize the amounts of the product to be shipped. We would like to link all these separate models into a single integrated model of production and transportation.

To begin, we replicate the production model of Figure 1-4a over mills — that is, origins — rather than over weeks as in the previous example:

```
set PROD;    # products
set ORIG;    # origins (steel mills)

param rate {ORIG,PROD} > 0;   # tons per hour at origins
param avail {ORIG} >= 0;      # hours available at origins

var Make {ORIG,PROD} >= 0;    # tons produced at origins

subject to Time {i in ORIG}:
    sum {p in PROD} (1/rate[i,p]) * Make[i,p] <= avail[i];
```

We have temporarily dropped the components pertaining to the objective, to which we will return later. We have also dropped the market demand parameters, since the demands are now properly associated with the destinations in the transportation models.

The next step is to replicate the transportation model, Figure 3-1a, over products, as we did in the multicommodity example at the beginning of this chapter:

```
set ORIG;    # origins (steel mills)
set DEST;    # destinations (factories)
set PROD;    # products

param supply {ORIG,PROD} >= 0; # tons available at origins
param demand {DEST,PROD} >= 0; # tons required at destinations

var Trans {ORIG,DEST,PROD} >= 0; # tons shipped
```

```
set ORIG;    # origins (steel mills)
set DEST;    # destinations (factories)
set PROD;    # products

param rate {ORIG,PROD} > 0;        # tons per hour at origins
param avail {ORIG} >= 0;           # hours available at origins
param demand {DEST,PROD} >= 0;   # tons required at destinations

param make_cost {ORIG,PROD} >= 0;          # manufacturing cost/ton
param trans_cost {ORIG,DEST,PROD} >= 0;  # shipping cost/ton

var Make {ORIG,PROD} >= 0;         # tons produced at origins
var Trans {ORIG,DEST,PROD} >= 0; # tons shipped

minimize total_cost:
   sum {i in ORIG, p in PROD} make_cost[i,p] * Make[i,p] +
   sum {i in ORIG, j in DEST, p in PROD}
                        trans_cost[i,j,p] * Trans[i,j,p];

subject to Time {i in ORIG}:
   sum {p in PROD} (1/rate[i,p]) * Make[i,p] <= avail[i];

subject to Supply {i in ORIG, p in PROD}:
   sum {j in DEST} Trans[i,j,p] = Make[i,p];

subject to Demand {j in DEST, p in PROD}:
   sum {i in ORIG} Trans[i,j,p] = demand[j,p];
```

Figure 4-6: Production/transportation model, 3rd version (`steelP.mod`).

```
subject to Supply {i in ORIG, p in PROD}:
   sum {j in DEST} Trans[i,j,p] = supply[i,p];

subject to Demand {j in DEST, p in PROD}:
   sum {i in ORIG} Trans[i,j,p] = demand[j,p];
```

Comparing the resulting production and transportation models, we see that the sets of origins (ORIG) and products (PROD) are the same in both models. Moreover, the "tons available at origins" (supply) in the transportation model are really the same thing as the "tons produced at origins" (Make) in the production model, since the steel available for shipping will be whatever is made at the mill.

We can thus merge the two models, dropping the definition of supply and substituting Make[i,p] for the occurrence of supply[i,p]:

```
subject to Supply {i in ORIG, p in PROD}:
   sum {j in DEST} Trans[i,j,p] = Make[i,p];
```

There are several ways in which we might add an objective to complete the model. Perhaps the simplest is to define a cost per ton corresponding to each variable. We define a parameter make_cost so that there is a term make_cost[i,p] * Make[i,p] in the objective for each origin i and product p; and we define trans_cost so that there is a term trans_cost[i,j,p] * Trans[i,j,p] in the objective for each origin i, destination j and product p. The full model is shown in Figure 4-6.

```
set ORIG := GARY CLEV PITT ;
set DEST := FRA DET LAN WIN STL FRE LAF ;
set PROD := bands coils plate ;

param avail :=  GARY 20  CLEV 15  PITT 20 ;

param demand (tr):
              FRA    DET    LAN    WIN    STL    FRE    LAF :=
      bands   300    300    100     75    650    225    250
      coils   500    750    400    250    950    850    500
      plate   100    100      0     50    200    100    250 ;

param rate (tr):   GARY   CLEV   PITT :=
           bands    200    190    230
           coils    140    130    160
           plate    160    160    170 ;

param make_cost (tr):
                  GARY   CLEV   PITT :=
          bands    180    190    190
          coils    170    170    180
          plate    180    185    185 ;

param trans_cost :=

 [*,*,bands]:  FRA   DET   LAN   WIN   STL   FRE   LAF :=
       GARY     30    10     8    10    11    71     6
       CLEV     22     7    10     7    21    82    13
       PITT     19    11    12    10    25    83    15

 [*,*,coils]:  FRA   DET   LAN   WIN   STL   FRE   LAF :=
       GARY     39    14    11    14    16    82     8
       CLEV     27     9    12     9    26    95    17
       PITT     24    14    17    13    28    99    20

 [*,*,plate]:  FRA   DET   LAN   WIN   STL   FRE   LAF :=
       GARY     41    15    12    16    17    86     8
       CLEV     29     9    13     9    28    99    18
       PITT     26    14    17    13    31   104    20 ;
```

Figure 4-7: Data for production/transportation model (`steelP.dat`).

Reviewing this formulation, we might observe that, according to the `Supply` declaration, the nonnegative expression

```
sum {j in DEST} Trans[i,j,p]
```

can be substituted for `Make[i,p]`. If we make this substitution for all occurrences of `Make[i,p]` in the objective and in the `Time` constraints, we no longer need to include the `Make` variables or the `Supply` constraints in our model, and our linear programs will be smaller as a result. Nevertheless, in most cases we will be better off leaving the model as it is shown above. By ''substituting out'' the `Make` variables we render the model harder to read, and not a great deal easier to solve.

As an instance of solving a linear program based on this model, we can adapt the data from Figure 4-2, as shown in Figure 4-7. Here are some representative result values:

```
ampl: model steelP.mod; data steelP.dat; solve;
OSL 1.2: optimal solution; objective 1392175
57 simplex iterations

ampl: option display_1col 5;
ampl: option omit_zero_rows 1, omit_zero_cols 1;

ampl: display Make;
Make [*,*]
:        bands   coils plate      :=
CLEV        0    1950      0
GARY     1125    1750    300
PITT      775     500    500
;

ampl: display Trans;
Trans [CLEV,*,*]
:    coils      :=
DET    750
LAF    500
LAN    400
STL     50
WIN    250

  [GARY,*,*]
:    bands coils plate      :=
FRE     225   850    100
LAF     250     0      0
STL     650   900    200

  [PITT,*,*]
:    bands coils plate      :=
DET     300     0    100
FRA     300   500    100
LAF       0     0    250
LAN     100     0      0
WIN      75     0     50
;

ampl: display Time;
Time [*] :=
CLEV  -1300
GARY  -2800
;
```

As one might expect, the optimal solution does not ship all products from all mills to all factories. We have used the options omit_zero_rows and omit_zero_cols to suppress the printing of table rows and columns that are all zeros. The dual values for Time show that additional capacity is likely to have the greatest impact on total cost if it is placed at GARY, and no impact if it is placed at PITT.

We can also investigate the relative costs of production and shipping, which are the two components of the objective:

```
ampl: display sum {i in ORIG, p in PROD}
            make_cost[i,p] * Make[i,p];
sum{i in ORIG, p in PROD} make_cost[i,p]*Make[i,p] = 1215250

ampl: display sum {i in ORIG, j in DEST, p in PROD}
            trans_cost[i,j,p] * Trans[i,j,p];
sum{i in ORIG, j in DEST, p in PROD}
    trans_cost[i,j,p]*Trans[i,j,p] = 176925
```

Clearly the production costs dominate in this case. These examples point up the ability of AMPL to evaluate and display any valid expression.

Bibliography

H. P. Williams, *Model Building in Mathematical Programming* (3rd edition). John Wiley & Sons (New York, 1990). An extended compilation of many kinds of models and combinations of them.

Exercises

4-1. Formulate a multi-period version of the transportation model, in which inventories are kept at the origins.

4-2. Formulate a combination of a transportation model for each of several foods, and a diet model at each destination.

4-3. The following questions pertain to the multiperiod production model and data of Section 4.2.

(a) Display the shadow prices associated with the constraints time[t]. In which periods does it appear that additional production capacity would be most valuable?

(b) By soliciting additional sales, you might be able to raise the upper bounds market[p,t]. Display the reduced costs Sell[p,t].rc, and use them to suggest whether you would prefer to go after more orders of bands or of coils in each week.

(c) If the inventory costs are all positive, any optimal solution will have zero inventories after the last week. Why is this so?

This phenomenon is an example of an "end effect". Because the model comes to an end after period T, the solution tends to behave as if production is to be shut down after that point. One way of dealing with end effects is to increase the number of weeks modeled; then the end effects should have little influence on the solution for the earlier weeks. Another approach is to modify the model to better reflect the realities of inventories. Describe some modifications you might make to the constraints, and to the objective.

4-4. A producer of packaged cookies and crackers runs several shifts each month at its large bakery. This exercise is concerned with a multiperiod planning model for deciding how many crews to employ each month. In the algebraic description of the model, there are sets S of shifts and P of products, and the planning horizon is T four-week periods. The relevant operational data are as follows:

l number of production lines: maximum number of crews that can work in any shift
r_p production rate for product p, in crew-hours per 1000 boxes
h_t number of hours that a crew works in planning period t

The following data are determined by market or managerial considerations:

w_s total wages for a crew on shift s in one period
d_{pt} demand for product p that must be met in period t
M maximum change in number of crews employed from one period to the next

The decision variables of the model are:

$X_{pt} \geq d_{pt}$ total boxes (in 1000s) of product p baked in period t
$0 \leq Y_{st} \leq l$ number of crews employed on shift s in period t

The objective is to minimize the total cost of all crews employed,

$$\sum_{s \in S} \sum_{t=1}^{T} w_s Y_{st}.$$

Total hours required for production in each period may not exceed total hours available from all shifts,

$$\sum_{p \in P} r_p X_{pt} \leq h_t \sum_{s \in S} Y_{st}, \text{ for each } t = 1, \dots, T.$$

The change in number of crews is restricted by

$$-M \leq \sum_{s \in S} (Y_{s,t+1} - Y_{st}) \leq M, \text{ for each } t = 1, \dots, T-1.$$

As required by the definition of M, this constraint restricts any change to lie between a reduction of M crews and an increase of M crews.

(a) Formulate this model in AMPL, and solve the following instance. There are $T = 13$ periods, $l = 8$ production lines, and a maximum change of $M = 3$ crews per period. The products are 18REG, 24REG, and 24PRO, with production rates r_p of 1.194, 1.509 and 1.509 respectively. Crews work either a day shift with wages w_s of \$44,900, or a night shift with wages \$123,100. The demands and working hours are given as follows by period:

Period t	$d_{18REG,t}$	$d_{24REG,t}$	$d_{24PRO,t}$	h_t
1	63.8	1212.0	0.0	156
2	76.0	306.2	0.0	152
3	88.4	319.0	0.0	160
4	913.8	208.4	0.0	152
5	115.0	298.0	0.0	156
6	133.8	328.2	0.0	152
7	79.6	959.6	0.0	152
8	111.0	257.6	0.0	160
9	121.6	335.6	0.0	152
10	470.0	118.0	1102.0	160
11	78.4	284.8	0.0	160
12	99.4	970.0	0.0	144
13	140.4	343.8	0.0	144

Display the numbers of crews required on each shift. You will find many fractional numbers of crews; how would you convert this solution to an optimal one in whole numbers?

(b) To be consistent, you should also require at most a change of M between the known initial number of crews (already employed in the period before the first) and the number of crews to be employed in the first planning period. Add a provision for this restriction to the model.

Re-solve with 11 initial crews. You should get the same solution.

(c) Because of the limit on the change in crews from period to period, more crews than necessary are employed in some periods. One way to deal with this is to carry inventories from one period to the next, so as to smooth out the amount of production required in each period. Add a variable for the amount of inventory of each product after each period, as in the model of Figure 4-4, and add constraints that relate inventory to production and demand. (Because inventories can be carried forward, production X_{pt} need not be $\geq$ demand d_{pt} in every period as required by the previous versions.) Also make a provision for setting initial inventories to zero. Finally, add an inventory cost per period per 1000 boxes to the objective.

Let the inventory costs be $34.56 for product 18REG, and $43.80 for 24REG and 24PRO. Solve the resulting linear program; display the crew sizes and inventory levels. How different is this solution? How much of a saving is achieved in labor cost, at how much expense in inventory cost?

(d) The demands in the given data peak at certain periods, when special discount promotions are in effect. Big inventories are built up in advance of these peaks, particularly before period 4. Baked goods are perishable, however, so that building up inventories past a certain number of periods is unrealistic.

Modify the model so that the inventory variables are indexed by product, period and age, where age runs from 1 to some specified limit. Add constraints that the inventories of age 1 after any period cannot exceed the amounts just produced, and that inventories of age $a > 1$ after period t cannot exceed the inventories of age $a - 1$ after period $t - 1$.

Verify that, with a maximum inventory age of 2 periods, you can use essentially the same solution as in (c), but that with a maximum inventory age of 1 there are some periods that require more crews.

4-5. Multiperiod linear programs can be especially difficult to develop, because they require data pertaining to the future. To hedge against the uncertainty of the future, a user of these LPs typically develops various scenarios, containing different forecasts of certain key parameters. This exercise asks you to develop what is known as a stochastic program, which finds a solution that can be considered robust over all scenarios.

(a) The revenues per ton might be particularly hard to predict, because they depend on fluctuating market conditions. Let the revenue data in Figure 4-5 be scenario 1, and also consider scenario 2:

```
param revenue:    1     2     3     4 :=
        bands    23    24    25    25
        coils    30    33    35    36 ;
```

and scenario 3:

```
param revenue:    1     2     3     4 :=
        bands    21    27    33    35
        coils    30    32    33    33 ;
```

By solving the three associated linear programs, verify that each of these scenarios leads to a different optimal production and sales strategy, even for the first week. You need one strategy, however, not three. The purpose of the stochastic programming approach is to determine a single solution that produces a good profit "on average" in a certain sense.

(b) As a first step toward formulating a stochastic program, consider how the three scenarios could be brought together into one linear program. Define a parameter S as the number of scenarios, and replicate the revenue data over the set 1..S:

```
param S > 0;
param revenue {PROD,1..T,1..S} >= 0;
```

Replicate all the variables and constraints in a similar way. (The idea is the same as earlier in this chapter, where we replicated model components over products or weeks.)

Define a new collection of parameters prob[s], to represent your estimate of the probability that a scenario s takes place:

```
param prob {1..S} >= 0, <= 1;
    check: 0.99999 < sum {s in 1..S} prob[s] < 1.00001;
```

The objective function is the expected profit, which equals the sum over all scenarios of the probability of each scenario times the optimum profit under that scenario:

```
maximize expected_profit:
    sum {s in 1..S} prob[s] *
        sum {p in PROD, t in 1..T} (revenue[p,t,s]*Sell[p,t,s] -
            prodcost[p]*Make[p,t,s] - invcost[p]*Inv[p,t,s]);
```

Complete the formulation of this multiscenario linear program, and put together the data for it. Let the probabilities of scenarios 1, 2 and 3 be 0.45, 0.35 and 0.20, respectively. Show that the solution consists of a production strategy for each scenario that is the same as the strategy in (a).

(c) The formulation in (b) is no improvement because it makes no connection between the scenarios. One way to make the model usable is to add "nonanticipativity" constraints that require each week-1 variable to be given the same value across all scenarios. Then the result will give you the best single strategy for the first week, in the sense of maximizing expected profit for all weeks. The strategies will still diverge after the first week — but a week from now you can update your data and run the stochastic program again to generate a second week's strategy.

A nonanticipativity constraint for the Make variables can be written

```
subject to Make_na {p in PROD, s in 1..S-1}:
    Make[p,1,s] = Make[p,1,s+1];
```

Add the analogous constraints for the Inv and Sell variables. Solve the stochastic program, and verify that the solution consists of a single period-1 strategy for all three scenarios.

(d) After getting your solution in (c), use the following command to look at the profits that the recommended strategy will achieve under the three scenarios:

```
display {s in 1..S}
    sum {p in PROD, t in 1..T} (revenue[p,t,s]*Sell[p,t,s] -
        prodcost[p]*Make[p,t,s] - invcost[p]*Inv[p,t,s]);
```

Which scenario will be most profitable, and which will be least profitable?

Repeat the analysis with probabilities of 0.0001, 0.0001 and 0.9998 for scenarios 1, 2 and 3. You should find that profit from strategy 3 goes up, but profits from the other two go down. Explain what these profits represent, and why the results are what you would expect.

5

Simple Sets and Indexing

The next four chapters of this book are a comprehensive presentation of AMPL's facilities for linear programming. The organization is by language features, rather than by model types as in the four preceding tutorial chapters. Since the basic features of AMPL tend to be closely interrelated, we do not attempt to explain any one feature in isolation. Rather, we assume at the outset a basic knowledge of AMPL such as Chapters 1 through 4 provide.

We begin with sets, the most fundamental components of an AMPL model. Almost all of the parameters, variables, and constraints in a typical model are indexed over sets, and many expressions contain operations (usually summations) over sets. Set indexing is the feature that permits a concise model to describe a large mathematical program.

Because sets are so fundamental, AMPL offers a broad variety of set types and operations. A set's members may be strings or numbers, ordered or unordered; they may occur singly, or as ordered pairs, triples or longer "tuples". Sets may be defined by listing or computing their members explicitly, by applying operations like union and intersection to other sets, or by specifying arbitrary arithmetic or logical conditions for membership.

Any model component or iterated operation can be indexed over any set, using a standard form of indexing expression. Even sets themselves may be declared in collections indexed over other sets.

This chapter introduces the simpler kinds of sets, as well as set operations and indexing expressions; it concludes with a discussion of ordered sets. Chapter 6 shows how these ideas are extended to compound sets, including sets of pairs and triples, and indexed collections of sets. Chapter 7 is devoted to parameters and expressions, and Chapter 8 to the variables, objectives and constraints that make up a linear program.

5.1 Unordered sets

The most elementary kind of AMPL set is an unordered collection of character strings. Usually all of the strings in a set are intended to represent instances of the same kind of entity — such as raw materials, products, factories or cities. Often the strings are chosen to have recognizable meanings (coils, FISH, New_York), but they could just as well be codes known only to the modeler (23RPFG, 486/33C). A literal string that appears in an AMPL model must be delimited by quotes, either single ('A&P') or double ("Bell+Howell"). In all contexts, upper case and lower case letters are distinct, so that for example "fish", "Fish", and "FISH" represent different set members.

The declaration of a set need only contain the keyword set and a name. For example, a model may declare

```
set PROD;
```

to indicate that a certain set will be referred to by the name PROD in the rest of the model. A name may be any sequence of letters, numerals, and underscore (_) characters that is not a legal number. A few names have special meanings in AMPL, and may only be used for specific purposes, while a larger number of names have predefined meanings that can be changed if they are used in some other way. For example, sum is reserved for the iterated addition operator; but prod is merely pre-defined as the iterated multiplication operator, so you can redefine prod as a set of products:

```
set prod;
```

A list of reserved words is given in Section A.17.

A declared set's membership is normally specified as part of the data for the model, in the manner to be described in Chapter 9; this separation of model and data is recommended for most mathematical programming applications. Occasionally, however, it is desirable to refer to a particular set of strings within a model. A literal set of this kind is specified by listing its members within braces:

```
{"bands", "coils", "plate"}
```

This expression may be used anywhere that a set is valid, for example in an assignment phrase to give the set PROD a fixed membership:

```
set PROD := {"bands", "coils", "plate"};
```

This form of declaration is best limited to cases where a set's membership is small, is a fundamental aspect of the model, or is not expected to change often.

Notice that AMPL makes a distinction between a string such as "bands" and a set like {"bands"} that has a membership of one string. The set that has no members (the empty set) is denoted { }.

5.2 Sets of numbers

Set members may also be numbers. In fact a set's members may be a mixture of numbers and strings, though this is seldom the case. In an AMPL model, a literal number is written in the customary way as a sequence of digits, optionally preceded by a sign, containing an optional decimal point, and optionally followed by an exponent; the exponent consists of a d, D, e, or E, optionally a sign, and a sequence of digits. A number (1) and the corresponding string ("1") are distinct; by contrast, different representations of the same number, such as 100 and 1E+2, stand for the same set member.

A set of numbers is often a sequence that corresponds to some progression in the situation being modeled, like a series of weeks or years. Just as for strings, the numbers in a set can be specified as part of the data, or can be specified within a model as a list between braces, such as {1,2,3,4,5,6}. This sort of set can be described more concisely by the notation 1..6. An additional by clause can be used to specify an interval other than 1 between the numbers; for instance,

```
1990 .. 2020 by 5
```

represents the set

```
{1990, 1995, 2000, 2005, 2010, 2015, 2020}
```

This kind of expression can be used anywhere that a set is appropriate, and in particular within the assignment phrase of a set declaration:

```
set YEARS := 1990 .. 2020 by 5;
```

By giving the set a short and meaningful name, this declaration may help to make the rest of the model more readable.

It is not good practice to specify all the numbers within a .. expression by literals like 2020 and 5, unless the values of these numbers are fundamental to the model or will rarely change. A better arrangement is seen in the multiperiod production example of Figures 4-4 and 4-5, where a parameter T is declared to represent the number of periods, and the expressions 1..T and 0..T are used to represent sets of periods over which parameters, variables, constraints and sums are indexed. The value of T is specified in the data, and is thus easily changed from one run to the next. As a more elaborate example, we could write

```
param start integer;
param end > start integer;
param interval > 0 integer;

set YEARS := start .. end by interval;
```

If subsequently we were to give the data as

```
param start := 1990;
param end := 2020;
param interval := 5;
```

then YEARS would be the same set as in the previous example (as it would also be if end were 2023.) You may use any arithmetic expression to represent any of the values in a .. expression.

The members of a set of numbers have the same properties as any other numbers, and hence can be used in arithmetic expressions. A simple example is seen in Figure 4-4, where the material balance constraint is declared as

```
subject to balance {p in PROD, t in 1..T}:
   Make[p,t] + Inv[p,t-1] = Sell[p,t] + Inv[p,t];
```

Because t runs over the set 1..T, we can write Inv[p,t-1] to represent the inventory at the end of the previous week. If t instead ran over a set of strings, the expression t-1 would be rejected as an error.

Set members need not be integers. AMPL attempts to store each numerical set member as the nearest representable floating-point number. You can see how this works out on your computer by trying an experiment such as the following:

```
ampl: display -5/3 .. 5/3 by 1/3;
set -5/3 .. 5/3 by 1/3 :=
-1.66666666666666666    -0.333333333333333315    1.00000000000000003
-1.33333333333333331     2.77555756156289135e-17  1.33333333333333334
-0.99999999999999999     0.33333333333333334
-0.66666666666666666     0.66666666666666671;
```

You might expect –1, 0, and 5/3 to be members of this set, but things do not work out that way due to rounding error in the floating-point computations. It is unwise to use fractional numbers in sets, if your model relies on set members having precise values. There should be no comparable problem with integer members of reasonable size; integers are represented exactly for magnitudes up to 2^{53} (approximately 10^{16}) for IEEE standard arithmetic, and up to 2^{47} (approximately 10^{14}) for almost any computer in current use.

The .. operator can also tempt you to define a set too large for your computer:

```
ampl: set HUGE := 1..1000000;
ampl: display card(HUGE);
Too much memory used -- 19316744 bytes; couldn't get 32776 more.
Highest address used = 0x12f0760 = 19859296
```

If you need to refer to a huge set like this, you should use AMPL's special notation for intervals, which is explained in Section A.6.3.

5.3 Set operations

AMPL has four operators that construct new sets from existing ones:

```
A union B        union: in either A or B
A inter B        intersection: in both A and B
A diff B         difference: in A but not B
A symdiff B      symmetric difference: in A or B but not both
```

The following excerpt from an AMPL session shows how these work:

```
ampl: set Y1 := 1990 .. 2020 by 5;
ampl: set Y2 := 2000 .. 2025 by 5;
ampl: display Y1 union Y2, Y1 inter Y2;
set Y1 union Y2 := 1990 1995 2000 2005 2010 2015 2020 2025;
set Y1 inter Y2 := 2000 2005 2010 2015 2020;

ampl: display Y1 diff Y2, Y1 symdiff Y2;
set Y1 diff Y2 := 1990 1995;
set Y1 symdiff Y2 := 1990 1995 2025;
```

The operands of set operators may be other set expressions, allowing more complex expressions to be built up:

```
ampl: display Y1 symdiff (Y1 symdiff Y2);
set Y1 symdiff (Y1 symdiff Y2) :=
2000    2005    2010    2015    2020    2025;

ampl: display (Y1 union {2025,2035,2045}) diff Y2;
set Y1 union   {2025, 2035, 2045} diff Y2 :=
1990    1995    2035    2045;

ampl: display 2000..2040 by 5 symdiff (Y1 union Y2);
set 2000 .. 2040 by 5 symdiff (Y1 union Y2) :=
2030    2035    2040    1990    1995;
```

The operands must always represent sets, however, so that for example you must write `Y1 union {2025}`, not `Y1 union 2025`.

Set operators group to the left unless parentheses are used to indicate otherwise. The union, `diff`, and `symdiff` operators have the same precedence, just below that of `inter`. Thus, for example,

```
A union B inter C diff D
```

is parsed as

```
(A union (B inter C)) diff D
```

A precedence hierarchy of all AMPL operators is given in Table A-1 of Section A.4.

Set operations are often used in the assignment phrase of a set declaration, to define a new set in terms of already declared sets. A simple example is provided by a variation on the diet model of Figure 2-1. Rather than specifying a lower limit and an upper limit on the amount of every nutrient, suppose that you want to specify a set of nutrients that have a lower limit, and a set of nutrients that have an upper limit. (Every nutrient is in one set or the other; some nutrients might be in both.) You could declare:

```
set MINREQ;    # nutrients with minimum requirements
set MAXREQ;    # nutrients with maximum requirements
set NUTR;      # all nutrients (DUBIOUS)
```

But then you would be relying on the user of the model to make sure that NUTR contains exactly all the members of MINREQ and MAXREQ. At best this is unnecessary work, and at worst it will be done incorrectly. Instead you can define NUTR as the union:

```
set NUTR := MINREQ union MAXREQ;
```

```
set MINREQ;    # nutrients with minimum requirements
set MAXREQ;    # nutrients with maximum requirements

set NUTR := MINREQ union MAXREQ;    # nutrients
set FOOD;                           # foods

param cost {FOOD} > 0;
param f_min {FOOD} >= 0;
param f_max {j in FOOD} >= f_min[j];

param n_min {MINREQ} >= 0;
param n_max {MAXREQ} >= 0;

param amt {NUTR,FOOD} >= 0;

var Buy {j in FOOD} >= f_min[j], <= f_max[j];

minimize total_cost:   sum {j in FOOD} cost[j] * Buy[j];

subject to diet_min {i in MINREQ}:
   sum {j in FOOD} amt[i,j] * Buy[j] >= n_min[i];

subject to diet_max {i in MAXREQ}:
   sum {j in FOOD} amt[i,j] * Buy[j] <= n_max[i];
```

Figure 5-1: Diet model using `union` operator (`dietu.mod`).

All three of these sets are needed, since the nutrient minima and maxima are indexed over MINREQ and MAXREQ,

```
param n_min {MINREQ} >= 0;
param n_max {MAXREQ} >= 0;
```

while the amounts of nutrients in the foods are indexed over NUTR:

```
param amt {NUTR,FOOD} >= 0;
```

The modification of the rest of the model is straightforward; the result is shown in Figure 5-1.

As a general principle, it is a bad idea to set up a model so that redundant information has to be provided. Instead a minimal necessary collection of sets should be chosen to be supplied in the data, while other relevant sets are defined by expressions in the model.

5.4 Set membership operations and functions

Two other AMPL operators, `in` and `within`, test the membership of sets. As an example, the expression

```
"B2" in NUTR
```

is true if and only if the string `"B2"` is a member of the set NUTR. The expression

```
MINREQ within NUTR
```

is true if all members of the set MINREQ are also members of NUTR — that is, if MINREQ is a subset of (or is the same as) NUTR. The in and within operators are the AMPL counterparts of $\in$ and $\subseteq$ in traditional algebraic notation. The distinction between members and sets is especially important here; the left operand of in must be an expression that evaluates to a string or number, whereas the left operand of within must be an expression that evaluates to a set.

AMPL also provides operators not in and not within, which reverse the truth value of their result.

You may apply within directly to a set you are declaring, to say that it must be a subset of some other set. Returning to the diet example, if all nutrients have a minimum requirement, but only some subset of nutrients have a maximum requirement, it would make sense to declare the sets as:

```
set NUTR;
set MAXREQ within NUTR;
```

AMPL will reject the data for this model if any member specified for MAXREQ is not also a member of NUTR.

The built-in function card computes the number of members in (or cardinality of) a set; for example, card(NUTR) is the number of members in NUTR. The argument of the card function may be any expression that evaluates to a set.

5.5 Indexing expressions

In algebraic notation, the use of sets is indicated informally by phrases such as "for all $i \in P$" or "for $t = 1, \ldots, T$" or "for all $j \in R$ such that $c_j > 0$." The AMPL counterpart is the *indexing expression* that appears within braces { ... } in nearly all of our examples. An indexing expression is used whenever we specify the set over which a model component is indexed, or the set over which a summation runs. Since an indexing expression defines a set, it can be used in any place where a set is appropriate.

The simplest form of indexing expression is just a set name or expression within braces. We have seen this in parameter declarations such as these from the multiperiod production model of Figure 4-4:

```
param rate {PROD} > 0;
param avail {1..T} >= 0;
```

Later in the model, references to these parameters are subscripted with a single set member, in expressions such as avail[t] and rate[p]. Variables can be declared and used in exactly the same way, except that the keyword var takes the place of param.

The names such as t and i that appear in subscripts and other expressions in our models are examples of *dummy indices* that have been defined by indexing expressions. In fact, any indexing expression may optionally define a dummy index that runs over the specified set. Dummy indices are convenient in specifying bounds on parameters:

```
param f_min {FOOD} >= 0;
param f_max {j in FOOD} >= f_min[j];
```

and on variables:

```
var Buy {j in FOOD} >= f_min[j], <= f_max[j];
```

They are also essential in specifying the sets over which constraints are defined, and the sets over which summations are done. We have often seen these uses together, in declarations such as

```
subject to time {t in 1..T}:
    sum {p in PROD} (1/rate[p]) * Make[p,t] <= avail[t];
```

and

```
subject to diet_min {i in MINREQ}:
    sum {j in FOOD} amt[i,j] * Buy[j] >= n_min[i];
```

An indexing expression consists of an index name, the keyword in, and a set expression as before. We have been using single letters for our index names, but this is not a requirement; an index name can be any sequence of letters, digits, and underscores that is not a valid number, just like the name for a model component.

Although a name defined by a model component's declaration is known throughout all subsequent statements in the model, the definition of a dummy index name is effective only within the *scope* of the defining indexing expression. Normally the scope is evident from the context. For instance, in the diet_min declaration above, the scope of {i in MINREQ} runs to the end of the statement, so that i can be used anywhere in the description of the constraint. On the other hand, the scope of {j in FOOD} covers only the summand amt[i,j] * Buy[j]. The scope of indexing expressions for sums and other iterated operators is discussed further in Chapter 7.

Once an indexing expression's scope has ended, its dummy index becomes undefined. Thus the same index name can be defined again and again in a model, and in fact it is good practice to use relatively few different index names. A common convention is to associate certain index names with certain sets, so that for example i always runs over NUTR and j always runs over FOOD. This is merely a convention, however, not a restriction imposed by AMPL. Indeed, when we modified the diet model so that there was a subset MINREQ of NUTR, we used i to run over MINREQ as well as NUTR. The opposite situation occurs, for example, if we want to specify a constraint that the amount of each food j in the diet is at least some fraction min_frac[j] of the total food in the diet:

```
subject to food_ratio {j in FOOD}:
    Buy[j] >= min_frac[j] * sum {jj in FOOD} Buy[jj];
```

Since the scope of j in FOOD extends to the end of the declaration, a different index jj is defined to run over the set FOOD in the summation within the constraint.

As a final option, the set in an indexing expression may be followed by a colon (:) and a logical condition. The indexing expression then represents only the subset of members that satisfy the condition. For example,

```
{j in FOOD: f_max[j] - f_min[j] < 1}
```

describes the set of all foods whose minimum and maximum amounts are nearly the same, and

```
{i in NUTR: i in MAXREQ or n_min[i] > 0}
```

describes the set of nutrients that are either in `MAXREQ` or for which `n_min` is positive. The use of operators such as `or` and `<` to form logical conditions will be fully explained in Chapter 7.

By specifying a condition, an indexing expression defines a new set. You can use the indexing expression to represent this set not only in indexed declarations and summations, but anywhere else that a set expression may appear. For example, you could say either of

```
set NUTREQ := {i in NUTR: i in MAXREQ or n_min[i] > 0};
set NUTREQ := MAXREQ union {i in MINREQ: n_min[i] > 0};
```

to define `NUTREQ` as above, and you could use either of

```
set BOTHREQ := {i in MINREQ: i in MAXREQ};
set BOTHREQ := MINREQ inter MAXREQ;
```

to define `BOTHREQ` as the set of all nutrients that have both minimum and maximum requirements. It's not unusual to find that there are several ways of describing some complicated set, depending on how you combine set operations and indexing expression conditions. Of course, some possibilities are easier to read than others, so it's worth taking some trouble to find the most readable. In Chapter 6 we also discuss efficiency considerations that sometimes make one alternative preferable to another in specifying compound sets.

In addition to being valuable within the model, indexing expressions are useful in `display` statements to summarize characteristics of the data or solution. The following example is based on the model of Figure 5-1 and the data of Figure 5-2:

```
ampl: model dietu.mod; data dietu.dat;

ampl: display MAXREQ union {i in MINREQ: n_min[i] > 0};
set MAXREQ union {i in MINREQ: n_min[i] > 0}  := A NA CAL C;

ampl: solve;
CPLEX: optimal solution; objective 74.27382022
6 iterations (2 in phase I)

ampl: display {j in FOOD: Buy[j] > f_min[j]};
set {j in FOOD: Buy[j] > f_min[j]}  := CHK MTL SPG;

ampl: display {i in MINREQ: diet_min[i].slack = 0};
set {i in MINREQ: (diet_min[i].slack) == 0}  := C CAL;
```

AMPL interactive commands are allowed to refer to variables and constraints in the condition phrase of an indexing expression, as illustrated by the last two `display` statements above. Within a model, however, only sets, parameters and dummy indices may be mentioned in any indexing expression.

```
set MINREQ := A B1 B2 C CAL ;
set MAXREQ := A NA CAL ;
set FOOD := BEEF CHK FISH HAM MCH MTL SPG TUR ;

param:    cost   f_min   f_max :=
   BEEF   3.19     2      10
   CHK    2.59     2      10
   FISH   2.29     2      10
   HAM    2.89     2      10
   MCH    1.89     2      10
   MTL    1.99     2      10
   SPG    1.99     2      10
   TUR    2.49     2      10   ;

param:    n_min   n_max :=
   A       700    20000
   C       700       .
   B1        0       .
   B2        0       .
   NA        .     50000
   CAL    16000    24000 ;

param amt (tr):    A     C    B1    B2     NA    CAL :=
            BEEF   60    20    10    15    938    295
            CHK     8     0    20    20   2180    770
            FISH    8    10    15    10    945    440
            HAM    40    40    35    10    278    430
            MCH    15    35    15    15   1182    315
            MTL    70    30    15    15    896    400
            SPG    25    50    25    15   1329    370
            TUR    60    20    15    10   1397    450 ;
```

Figure 5-2: Data for diet model (`dietu.dat`).

5.6 Ordered sets

Any set of numbers has a natural ordering, so numbers are often used to represent entities, like time periods, whose ordering is essential to the specification of a model. To describe the difference between this week's inventory and the previous week's inventory, for example, we need the weeks to be ordered so that the "previous" week is always well defined.

An AMPL model can also define its own ordering for any set of numbers or strings, by adding the keyword `ordered` or `circular` to the set's declaration. The order in which you give the set's members, in either the model or the data, is then the order in which AMPL works with them. In a set declared `circular`, the first member is considered to follow the last one, and the last to precede the first; in an `ordered` set, the first member has no predecessor and the last member has no successor.

Ordered sets of strings often provide better documentation for a model's data than sets of numbers. Returning to the multiperiod production model of Figure 4-4, we observe that there is no way to tell from the data which weeks the numbers 1 through T refer to, or even that they are weeks instead of days or months. Suppose that instead we let the weeks be represented by an ordered set that contains, say, 27sep93, 04oct93, 11oct93 and 18oct93. The declaration of T is replaced by

```
set WEEKS ordered;
```

and all subsequent occurrences of 1..T are replaced by WEEKS. In the balance constraint, the expression t−1 is replaced by prev(t), which selects the member before t in the set's ordering:

```
subject to balance {p in PROD, t in WEEKS}:
    Make[p,t] + Inv[p,prev(t)] = Sell[p,t] + Inv[p,t]; # WRONG
```

This is not quite right, however, because when t is the first week in WEEKS, the member prev(t) is not defined. When you try to solve the problem, you will get an error message like this:

```
error processing constraint balance[bands,27sep93]:
        can't compute prev(27sep93, WEEKS) --
            27sep93 is the first element
```

One way to fix this is to give a separate balance constraint for the first period, in which Inv[p,prev(t)] is replaced by the initial inventory, inv0[p]:

```
subject to balance0 {p in PROD}:
    Make[p,first(WEEKS)] + inv0[p]
        = Sell[p,first(WEEKS)] + Inv[p,first(WEEKS)];
```

The regular balance constraint is limited to the remaining weeks:

```
subject to balance {p in PROD, t in WEEKS: ord(t) > 1}:
    Make[p,t] + Inv[p,prev(t)] = Sell[p,t] + Inv[p,t];
```

The complete model and data are shown in Figures 5-3 and 5-4. As a tradeoff for more meaningful week names, we have to write a slightly more complicated model.

As our example demonstrates, AMPL provides a variety of functions that apply specifically to ordered sets. These functions are of three basic types.

First, there are functions that return a member from some absolute position in a set. You can write first(WEEKS) and last(WEEKS) for the first and last members of the ordered set WEEKS. To pick out other members, you can use member(5,WEEKS), say, for the 5th member of WEEKS. The first argument of these functions must evaluate to an ordered set; any expression that evaluates to a positive integer can appear as the second argument of member.

A second kind of function returns a member from a position relative to another member. Thus you can write prev(t,WEEKS) for the member immediately before t in WEEKS, and next(t,WEEKS) for the member immediately after. More generally, expressions such as prev(t,WEEKS,5) and next(t,WEEKS,3) refer to the 5th member before and the 3rd member after t in WEEKS. There are also "wraparound" versions prevw and nextw that work the same except that they treat the end of the set

```
set PROD;              # products
set WEEKS ordered;    # number of weeks

param rate {PROD} > 0;               # tons per hour produced
param inv0 {PROD} >= 0;              # initial inventory
param avail {WEEKS} >= 0;            # hours available in week
param market {PROD,WEEKS} >= 0;      # limit on tons sold in week

param prodcost {PROD} >= 0;          # cost per ton produced
param invcost {PROD} >= 0;           # carrying cost/ton of inventory
param revenue {PROD,WEEKS} >= 0;     # revenue/ton sold

var Make {PROD,WEEKS} >= 0;          # tons produced
var Inv {PROD,WEEKS} >= 0;           # tons inventoried
var Sell {p in PROD, t in WEEKS} >= 0, <= market[p,t]; # tons sold

maximize total_profit:
   sum {p in PROD, t in WEEKS} (revenue[p,t]*Sell[p,t] -
      prodcost[p]*Make[p,t] - invcost[p]*Inv[p,t]);

         # Objective: total revenue less costs in all weeks

subject to time {t in WEEKS}:
   sum {p in PROD} (1/rate[p]) * Make[p,t] <= avail[t];

         # Total of hours used by all products
         # may not exceed hours available, in each week

subject to balance0 {p in PROD}:
   Make[p,first(WEEKS)] + inv0[p]
      = Sell[p,first(WEEKS)] + Inv[p,first(WEEKS)];

subject to balance {p in PROD, t in WEEKS: ord(t) > 1}:
   Make[p,t] + Inv[p,prev(t)] = Sell[p,t] + Inv[p,t];

         # Tons produced and taken from inventory
         # must equal tons sold and put into inventory
```

Figure 5-3: Production model with ordered sets (steelT2.mod).

as wrapping around to the beginning; in effect, they treat all ordered sets as if their declarations were circular. In all of these functions, the first argument must evaluate to a number or string, the second argument to an ordered set, and the third to an integer. Normally the integer is positive, but zero and negative values are interpreted in a consistent way; for instance, next(t,WEEKS,0) is the same as t, and next(t,WEEKS,-5) is the same as prev(t,WEEKS,5).

Finally, there are functions that return the position of a member within a set. The expression ord(t,WEEKS) returns the numerical position of t within the set WEEKS, or gives you an error message if t is not a member of WEEKS. The alternative ord0(t,WEEKS) is the same except that it returns 0 if t is not a member of WEEKS. For these functions the first argument must evaluate to a positive integer, and the second to an ordered set.

```
set PROD := bands coils ;
set WEEKS := 27sep93 04oct93 11oct93 18oct93 ;

param avail :=  27sep93 40   04oct93 40   11oct93 32   18oct93 40 ;

param rate := bands 200  coils 140 ;
param inv0 := bands 10 coils   0 ;

param prodcost := bands 10    coils 11 ;
param invcost  := bands  2.5  coils  3 ;

param revenue: 27sep93 04oct93 11oct93 18oct93 :=
        bands       25      26      27      27
        coils       30      35      37      39 ;

param market: 27sep93 04oct93 11oct93 18oct93 :=
        bands     6000    6000    4000    6500
        coils     4000    2500    3500    4200 ;
```

Figure 5-4: Data for production model (`steelT2.dat`).

If the first argument of `next`, `nextw`, `prev`, `prevw`, or `ord` is a dummy index that runs over an ordered set, its associated indexing set is assumed if a set is not given as the second argument. Thus in the constraint

```
subject to balance {p in PROD, t in WEEKS: ord(t) > 1}:
    Make[p,t] + Inv[p,prev(t)] = Sell[p,t] + Inv[p,t];
```

the functions `ord(t)` and `prev(t)` are interpreted as if they had been written `ord(t,WEEKS)` and `prev(t,WEEKS)`.

Ordered sets can also be used with any of the AMPL operators and functions that apply to sets generally. The result of a `diff` operation preserves the ordering of the left operand, so the material balance constraint in our example could be written:

```
subject to balance {p in PROD, t in WEEKS diff {first(WEEKS)}}:
    Make[p,t] + Inv[p,prev(t)] = Sell[p,t] + Inv[p,t];
```

For `union`, `inter` and `symdiff`, however, the ordering of the result is not well defined; AMPL treats the result as an unordered set.

For a set that is contained in an ordered set, AMPL provides a way to say that the ordering should be inherited. Suppose for example that you want to try running the multiperiod production model with horizons of different lengths. In the following declarations, the ordered set `ALL_WEEKS` and the parameter `T` are given in the data, while the subset `WEEKS` is defined by an indexing expression to include only the first T weeks:

```
set ALL_WEEKS ordered;
param T > 0 integer;

set WEEKS := {t in ALL_WEEKS: ord(t) <= T} ordered by ALL_WEEKS;
```

We specify `ordered by ALL_WEEKS` so that `WEEKS` becomes an ordered set, with its members having the same ordering as they do in `ALL_WEEKS`. The `ordered by` and `circular by` phrases have the same effect as the `within` phrase of Section 5.4 together with `ordered` or `circular`, except that they also cause the declared set to inherit the ordering from the containing set. There are also `ordered by reversed` and `circular by reversed` phrases, which cause the declared set's ordering to be the opposite of the containing set's ordering. All of these phrases may be used either with a subset supplied in the data, or with a subset defined by an expression as in the example above.

5.7 Syntax summary

A set expression can have any of the following forms:

> *set-expr* :
>> *set-name*
>> { *object-list* }
>> *arith-expr* .. *arith-expr*
>> *arith-expr* .. *arith-expr* by *arith-expr*
>>
>> *set-expr* union *set-expr*
>> *set-expr* inter *set-expr*
>> *set-expr* diff *set-expr*
>> *set-expr* symdiff *set-expr*
>>
>> (*set-expr*)

where *set-name* refers to any previously declared set, and *arith-expr* is an arithmetic expression. The *object-list* may be empty, an expression, or a comma-separated list of expressions, where each expression evaluates to a number or a character string.

The declaration of a simple set has the form

> *set-declaration* :
>> set *set-name* alias_{opt} attributes_{opt} ;

where *set-name* is an unbound AMPL name and *alias* is an optional string. The *attributes* may include the following phrases, optionally separated by commas:

> *attributes* :
>> := *set-expr*
>>
>> within *set-expr*
>>
>> ordered
>> ordered by *set-expr*
>> ordered by reversed *set-expr*
>>
>> circular
>> circular by *set-expr*
>> circular by reversed *set-expr*

A declaration may have at most one of the `ordered` or `circular` attributes.

An indexing expression for a simple set is written in one of the following ways:

index-expr :

>   { *set-expr* }
>   { *dummy-name* in *set-expr* }
>   { *dummy-name* in *set-expr* : *logic-expr* }

The *dummy-name* can be any AMPL name not currently defined, and the optional *logic-expr* is a logical expression, which is considered in detail in Chapter 7.

Exercises

5-1. (a) Display the sets

```
-5/3 .. 5/3 by 1/3
0 .. 1 by .1
```

Explain any evidence of rounding error in your computer's arithmetic.

(b) Try the following commands from Section 5.2 on your computer:

```
ampl: set HUGE := 1..1000000;
ampl: display card(HUGE);
```

When AMPL runs out of memory, how many bytes does it say were available? (If your computer really does have enough memory, try `1..10000000`.) Experiment to see how big a set `HUGE` your computer can hold without running out of memory.

5-2. Revise the model of Exercise 1-6 so that it makes use of two different attribute sets: a set of attributes that have lower limits, and a set of attributes that have upper limits. Use the same approach as in Figure 5-1.

5-3. Use the `display` command, together with indexing expressions as demonstrated in Section 5.5, to determine the following sets relating to the diet model of Figures 5-1 and 5-2:

– Foods that have a unit cost greater than $2.00.

– Foods that have a sodium (`NA`) content of more than 1000.

– Foods that contribute more than $10 to the total cost in the optimal solution.

– Foods that are purchased at more than the minimum level but less than the maximum level in the optimal solution.

– Nutrients that the optimal diet supplies in exactly the minimum allowable amount.

– Nutrients that the optimal diet supplies in exactly the maximum allowable amount.

– Nutrients that the optimal diet supplies in more than the minimum allowable amount but less than the maximum allowable amount.

5-4. This exercise refers to the multiperiod production model of Figure 4-4.

(a) Suppose that we define two additional scalar parameters,

```
param Tbegin integer >= 1;
param Tend integer > Tbegin, <= T;
```

We want to solve the linear program that covers only the weeks from `Tbegin` through `Tend`. We still want the parameters to use the indexing `1..T`, however, so that we don't need to change the data tables every time we try a different value for `Tbegin` or `Tend`.

To start with, we can change every occurrence of `1..T` in the variable, objective and constraint declarations to `Tbegin..Tend`. By making these and other necessary changes, create a model that correctly covers only the desired weeks.

(b) Now suppose that we define a different scalar parameter,

```
param Tagg integer >= 1;
```

We want to "aggregate" the model, so that one "period" in our LP becomes `Tagg` weeks long, rather than one week. This would be appropriate if we have, say, a year of weekly data, which would yield an LP too large to be convenient for analysis.

To aggregate properly, we must define the availability of hours in each period to be the sum of the availabilities in all weeks of the period:

```
param avail_agg {t in 1..T by Tagg}
    := sum {u in t..t+Tagg-1} avail[u];
```

The parameters `market` and `revenue` must be similarly summed. Make all the necessary changes to the model of Figure 4-4 so that the resulting LP is properly aggregated.

(c) Re-do the models of (a) and (b) to use an ordered set of strings for the periods, as in Figure 5-3.

5-5. Extend the transportation model of Figure 3-1a to a multiperiod version, in which the periods are months represented by an ordered set of character strings such as `"Jan95"`, `"Feb95"` and so forth. Use inventories at the origins to link the periods.

5-6. Modify the model of Figure 5-3 to merge the `balance0` and `balance` constraints, as in Figure 4-4. Hint: `0..T` and `1..T` are analogous to

```
set WEEKS0 ordered;
set WEEKS := {i in WEEKS0: ord(i) > 1} ordered by WEEKS0;
```

6

Compound Sets and Indexing

Most linear programming models involve indexing over combinations of members from several different sets. Indeed, the interaction of indexing sets is often the most complicated aspect of a model; once you have the arrangement of sets worked out, the rest of the model can be written clearly and concisely.

All but the simplest models employ compound sets whose members are pairs, triples, quadruples, or even longer "tuples" of objects. This chapter begins with the declaration and use of sets of ordered pairs. We concentrate first on the set of all pairs from two sets, then move on to subsets of all pairs and to "slices" through sets of pairs. Subsequent sections explore sets of longer tuples, and extensions of AMPL's set operators and indexing expressions to sets of tuples.

The final section of this chapter introduces sets that are declared in collections indexed over other sets. An indexed collection of sets often plays much the same role as a set of tuples, but it represents a somewhat different way of thinking about the formulation. Each kind of set is appropriate in certain situations, and we offer some guidelines for choosing between them.

6.1 Sets of ordered pairs

An ordered pair of objects, whether numbers or strings, is written with the objects separated by a comma and surrounded by parentheses:

```
("PITT","STL")
("bands",5)
(3,101)
```

As the term "ordered" suggests, it makes a difference which object comes first; ("STL","PITT") is not the same as ("PITT","STL"). The same object may appear both first and second, as in ("PITT","PITT").

Pairs can be collected into sets, just like single objects. A comma-separated list of pairs may be enclosed in braces to denote a literal set of ordered pairs:

```
{("PITT","STL"),("PITT","FRE"),("PITT","DET"),("CLEV","FRE")}
{(1,1),(1,2),(1,3),(2,1),(2,2),(2,3),(3,1),(3,2),(3,3)}
```

Because sets of ordered pairs are often large and subject to change, however, they seldom appear explicitly in AMPL models. Instead they are described symbolically in a variety of ways.

The set of all ordered pairs from two given sets appears frequently in our examples. In the transportation model of Figure 3-1a, for instance, the set of all origin-destination pairs is written as either of

```
{ORIG, DEST}
{i in ORIG, j in DEST}
```

depending on whether the context requires dummy indices. The multiperiod production model of Figure 4-4 uses a set of all pairs from a set of strings (representing products) and a set of numbers (representing weeks):

```
{PROD, 1..T}
{p in PROD, t in 1..T}
```

Various collections of model components, such as the parameter revenue and the variable Sell, are indexed over this set. When individual components are referenced in the model, they must have two subscripts, as in revenue[p,t] or Sell[p,t]. The order of the subscripts is always the same as the order of the objects in the pairs; in this case the first subscript must refer to a string in PROD, and the second to a number in 1..T.

An indexing expression like {p in PROD, t in 1..T} is the AMPL transcription of a phrase like "for all p in P, $t = 1, \ldots, T$" from algebraic notation. There is no compelling reason to think in terms of ordered pairs in this case, and indeed we did not mention ordered pairs when introducing the multiperiod production model in Chapter 4. On the other hand, we can modify the transportation model of Figure 3-1a to emphasize the role of origin-destination pairs as "links" between cities, by defining this set of pairs explicitly:

```
set LINKS := {ORIG,DEST};
```

The shipment costs and amounts can then be indexed over links:

```
param cost {LINKS} >= 0;
var Trans {LINKS} >= 0;
```

In the objective, the sum of costs over all shipments can be written like this:

```
minimize total_cost:
    sum {(i,j) in LINKS} cost[i,j] * Trans[i,j];
```

Notice that when dummy indices run over a set of pairs like LINKS, they must be defined in a pair like (i,j). It would be an error to sum over {k in LINKS}. The complete model is shown in Figure 6-1, and should be compared with Figure 3-1a. The specification of the data could be the same as in Figure 3-1b.

```
set ORIG;    # origins
set DEST;    # destinations

set LINKS := {ORIG,DEST};

param supply {ORIG} >= 0;    # amounts available at origins
param demand {DEST} >= 0;    # amounts required at destinations

   check: sum {i in ORIG} supply[i] = sum {j in DEST} demand[j];

param cost {LINKS} >= 0;     # shipment costs per unit
var Trans {LINKS} >= 0;      # units to be shipped

minimize total_cost:
   sum {(i,j) in LINKS} cost[i,j] * Trans[i,j];

subject to Supply {i in ORIG}:
   sum {j in DEST} Trans[i,j] = supply[i];

subject to Demand {j in DEST}:
   sum {i in ORIG} Trans[i,j] = demand[j];
```

Figure 6-1: Transportation model with all pairs (`transp2.mod`).

6.2 Subsets and slices of ordered pairs

In many applications, we are concerned only with a subset of all ordered pairs from two sets. For example, in the transportation model, shipments may not be possible from every origin to every destination. The shipping costs per unit may be provided only for the usable origin-destination pairs, so that it is desirable to index the costs and the variables only over these pairs. In AMPL terms, we want the set LINKS defined above to contain just a subset of pairs that are given in the data, rather than all pairs from ORIG and DEST.

It is not sufficient to declare set LINKS, because that declares only a set of single members. At a minimum, we need to say

```
set LINKS dimen 2;
```

to indicate that the data must consist of members of "dimension" two — that is, pairs. Better yet, we can say that LINKS is a subset of the set of all pairs from ORIG and DEST:

```
set LINKS within {ORIG,DEST};
```

This has the advantage of making the model's intent clearer, and of helping to catch errors in the data. The subsequent declarations of parameter cost, variable Trans, and the objective function remain the same as they are in Figure 6-1. But the components cost[i,j] and Trans[i,j] will now be defined only for those pairs given in the data as members of LINKS, and the expression

```
sum {(i,j) in LINKS} cost[i,j] * Trans[i,j]
```

will represent a sum over the specified pairs only.

How are the constraints written? In the original transportation model, the supply limit constraint was:

```
subject to Supply {i in ORIG}:
    sum {j in DEST} Trans[i,j] = supply[i];
```

This does not work when LINKS is a subset of pairs, because for each i in ORIG it tries to sum Trans[i,j] over every j in DEST, while Trans[i,j] is defined only for pairs (i,j) in LINKS. If we try it, we get an error message like this:

```
error processing constraint Supply[GARY]:
        invalid subscript Trans[GARY,FRA]
```

What we want to say is that for each origin i, the sum should be over all destinations j such that (i,j) is an allowed link. This statement can be transcribed directly to AMPL, by adding a condition to the indexing expression after sum:

```
subject to Supply {i in ORIG}:
    sum {j in DEST: (i,j) in LINKS} Trans[i,j] = supply[i];
```

Rather than requiring this somewhat awkward form, however, AMPL lets us drop the j in DEST from the indexing expression to produce the following more concise constraint:

```
subject to Supply {i in ORIG}:
    sum {(i,j) in LINKS} Trans[i,j] = supply[i];
```

Because {(i,j) in LINKS} appears in a context where i has already been defined, AMPL interprets this indexing expression as the set of all j such that (i,j) is in LINKS. The demand constraint is handled similarly, and the entire revised version of the model is shown in Figure 6-2a. A small representative collection of data for this model is shown in Figure 6-2b; AMPL offers a variety of convenient ways to specify the membership of compound sets and the data indexed over them, as explained in Chapter 9.

You can see from Figure 6-2a that the indexing expression

```
{(i,j) in LINKS}
```

means something different in each of the three places where it appears. Its membership can be understood in terms of a table like this:

	FRA	DET	LAN	WIN	STL	FRE	LAF
GARY		x	x		x		x
CLEV	x	x	x	x	x		x
PITT	x			x	x	x	

The rows represent origins and the columns destinations, while each pair in the set is marked by an x. A table for {ORIG,DEST} would be completely filled in with x's, while the table shown depicts {LINKS} for the "sparse" subset of pairs defined by the data in Figure 6-2b.

At a place where i and j are not currently defined, such as in the objective

```
minimize total_cost:
    sum {(i,j) in LINKS} cost[i,j] * Trans[i,j];
```

```
set ORIG;    # origins
set DEST;    # destinations

set LINKS within {ORIG,DEST};

param supply {ORIG} >= 0;    # amounts available at origins
param demand {DEST} >= 0;    # amounts required at destinations

    check: sum {i in ORIG} supply[i] = sum {j in DEST} demand[j];

param cost {LINKS} >= 0;     # shipment costs per unit
var Trans {LINKS} >= 0;      # units to be shipped

minimize total_cost:
    sum {(i,j) in LINKS} cost[i,j] * Trans[i,j];

subject to Supply {i in ORIG}:
    sum {(i,j) in LINKS} Trans[i,j] = supply[i];

subject to Demand {j in DEST}:
    sum {(i,j) in LINKS} Trans[i,j] = demand[j];
```

Figure 6-2a: Transportation model with selected pairs (transp3.mod).

```
param: ORIG:  supply :=
    GARY  1400     CLEV  2600     PITT  2900 ;

param: DEST:  demand :=
    FRA   900    DET  1200    LAN   600    WIN  400
    STL  1700    FRE  1100    LAF  1000 ;

param: LINKS:  cost :=
    GARY DET 14    GARY LAN 11    GARY STL 16    GARY LAF  8
    CLEV FRA 27    CLEV DET  9    CLEV LAN 12    CLEV WIN  9
    CLEV STL 26    CLEV LAF 17
    PITT FRA 24    PITT WIN 13    PITT STL 28    PITT FRE 99 ;
```

Figure 6-2b: Data for transportation model (transp3.dat).

the indexing expression { (i,j) in LINKS} represents all the pairs in this table. But at a place where i has already been defined, such as in the Supply constraint

```
    subject to Supply {i in ORIG}:
        sum {(i,j) in LINKS} Trans[i,j] = supply[i];
```

the expression { (i,j) in LINKS} is associated with just the row of the table corresponding to i. You can think of it as taking a one-dimensional "slice" through the table in the row corresponding to the already-defined first component. Although in this case the first component is a previously defined dummy index, the same convention applies when the first component is any expression that can be evaluated to a valid set object; we could write

```
    {("GARY",j) in LINKS}
```

for example, to represent the pairs in the first row of the table.

Similarly, where j has already been defined, such as in the Demand constraint

```
subject to Demand {j in DEST}:
    sum {(i,j) in LINKS} Trans[i,j] = demand[j];
```

the expression `{(i,j) in LINKS}` selects pairs from the column of the table corresponding to j. Pairs in the third column of the table could be specified by `{(i,"LAN") in LINKS}`.

6.3 Sets of longer tuples

AMPL's notation for ordered pairs extends in a natural way to triples, quadruples, or ordered lists of any length. All tuples in a set must have the same dimension. A set can't contain both pairs and triples, for example, nor can the determination as to whether a set contains pairs or triples be made according to some value in the data.

The multicommodity transportation model of Figure 4-1 offers some examples of how we can use ordered triples, and by extension longer tuples. In the original version of the model, the costs and amounts shipped are indexed over origin-destination-product triples:

```
param cost {ORIG,DEST,PROD} >= 0;
var Trans {ORIG,DEST,PROD} >= 0;
```

In the objective, cost and Trans are written with three subscripts, and the total cost is determined by summing over all triples:

```
minimize total_cost:
    sum {i in ORIG, j in DEST, p in PROD}
        cost[i,j,p] * Trans[i,j,p];
```

The indexing expressions are the same as before, except that they list three sets instead of two. An indexing expression that listed k sets would similarly denote a set of k-tuples.

If instead we define LINKS as we did in Figure 6-2a, the multicommodity declarations come out like this:

```
set LINKS within {ORIG,DEST};

param cost {LINKS,PROD} >= 0;
var Trans {LINKS,PROD} >= 0;

minimize total_cost:
    sum {(i,j) in LINKS, p in PROD} cost[i,j,p] * Trans[i,j,p];
```

Here we see how a set of triples can be specified as combinations from a set of pairs (LINKS) and a set of single members (PROD). Since cost and Trans are indexed over {LINKS,PROD}, their first two subscripts must come from a pair in LINKS, and their third subscript from a member of PROD. Sets of longer tuples can be built up in an analogous way.

As a final possibility, it may be that only certain combinations of origins, destinations, and products are workable. Then it makes sense to define a set that contains only the triples of allowed combinations:

```
set ROUTES within {ORIG,DEST,PROD};
```

The costs and amounts shipped are indexed over this set:

```
param cost {ROUTES} >= 0;
var Trans {ROUTES} >= 0;
```

and in the objective, the total cost is a sum over all triples in this set:

```
minimize total_cost:
    sum {(i,j,p) in ROUTES} cost[i,j,p] * Trans[i,j,p];
```

Individual triples are written, by analogy with pairs, as a parenthesized and comma-separated list (i,j,p). Longer lists specify longer tuples.

In the three constraints of this model, the summations must be taken over three different slices through the set ROUTES:

```
subject to Supply {i in ORIG, p in PROD}:
    sum {(i,j,p) in ROUTES} Trans[i,j,p] = supply[i,p];

subject to Demand {j in DEST, p in PROD}:
    sum {(i,j,p) in ROUTES} Trans[i,j,p] = demand[j,p];

subject to Multi {i in ORIG, j in DEST}:
    sum {(i,j,p) in ROUTES} Trans[i,j,p] <= limit[i,j];
```

In the Supply constraint, for instance, indices i and p are defined before the sum, so {(i,j,p) in ROUTES} refers to all j such that (i,j,p) is a triple in ROUTES. AMPL allows comparable slices through any set of tuples, in any number of dimensions and any combination of coordinates.

When you declare a high-dimensional set such as ROUTES, a phrase like within {ORIG,DEST,PROD} may specify a set with a huge number of members. With 10 origins, 100 destinations and 100 products, for instance, this set potentially has 100,000 members. Fortunately, AMPL does not create this set when it processes the declaration, but merely checks that each tuple in the data for ROUTES has its first component in ORIG, its second in DEST, and its third in PROD. The set ROUTES can thus be handled efficiently so long as it does not itself contain a huge number of triples.

When using high-dimensional sets in other contexts, you may have to be more careful that you do not inadvertently force AMPL to generate a large set of tuples. As an example, consider how you could constrain the volume of all products shipped out of each origin to be less than some amount. You might write either

```
subject to Supply_All {i in ORIG}:
    sum {j in DEST, p in PROD: (i,j,p) in ROUTES}
        Trans[i,j,p] <= supply_all[i];
```

or, using the more compact slice notation,

```
subject to Supply_All {i in ORIG}:
    sum {(i,j,p) in ROUTES} Trans[i,j,p] <= supply_all[i];
```

In the first case, AMPL explicitly generates the set {j in DEST, p in PROD} and checks for membership of (i,j,p) in ROUTES, while in the second case it is able to use a more efficient approach to finding all (i,j,p) from ROUTES that have a given i. In our small examples this may not seem critical, but for problems of realistic size the slice version may be the only one that can be processed in a reasonable amount of time and space.

6.4 Operations on sets of tuples

Operations on compound sets are, as much as possible, the same as the operations introduced for simple sets in Chapter 5. Sets of pairs, triples, or longer tuples can be combined with union, inter, diff, and symdiff; can be tested by in and within; and can be counted with card. Dimensions of operands must match appropriately, so for example you may not form the union of a set of pairs with a set of triples. Also, compound sets in AMPL cannot be declared as ordered or circular, and hence also cannot be arguments to functions like first and next that take only ordered sets.

Another set operator, cross, gives the set of all pairs of its arguments — the cross or Cartesian product. Thus the set expression

```
ORIG cross DEST
```

represents the same set as the indexing expression {ORIG,DEST}, and

```
ORIG cross DEST cross PROD
```

is the same as {ORIG,DEST,PROD}.

Our examples so far have been constructed so that every compound set has a domain within a cross product of previously specified simple sets; LINKS lies within ORIG cross DEST, for example, and ROUTES within ORIG cross DEST cross PROD. This practice helps to produce clear and correct models. Nevertheless, if you find it inconvenient to specify the domains as part of the data, you may define them instead within the model. AMPL provides an iterated setof operator for this purpose, as in the following example:

```
set ROUTES dimen 3;

set PROD := setof {(i,j,p) in ROUTES} p;
set LINKS := setof {(i,j,p) in ROUTES} (i,j);
```

Like an iterated sum operator, setof is followed by an indexing expression and an argument, which can be any expression that evaluates to a legal set member. The argument is evaluated for each member of the indexing set, and the results are combined into a new set that is returned by the operator. Duplicate members are ignored. Thus these expressions for PROD and LINKS give the sets of all objects p and pairs (i,j) such that there is some member (i,j,p) in ROUTES.

As with simple sets, membership in a compound set may be restricted by a logical condition at the end of an indexing expression. For example, the multicommodity transportation model could define

```
set DEMAND := {j in DEST, p in PROD: demand[j,p] > 0};
```

so that DEMAND contains only those pairs (j,p) with positive demand for product p at destination j. As another example, suppose that we also wanted to model transfers of the products from one origin to another. We could simply define

```
set TRANSF := {ORIG,ORIG};
```

to specify the set of all pairs of members from ORIG. But this set would include pairs like ("PITT","PITT"); to specify the set of all pairs of *different* members from ORIG, a condition must be added:

```
set TRANSF := {i1 in ORIG, i2 in ORIG: i1 <> i2};
```

This is another case where two different dummy indices, i1 and i2, need to be defined to run over the same set; the condition selects those pairs where i1 is not equal to i2.

If a set is ordered, the condition within an indexing expression can also refer to the ordering. We could declare

```
set ORIG ordered;
set TRANSF := {i1 in ORIG, i2 in ORIG: ord(i1) < ord(i2)};
```

to define a "triangular" set of pairs from ORIG that does not contain any pair and its reverse. For example, TRANSF would contain either of the pairs ("PITT","CLEV") or ("CLEV","PITT"), depending on which came first in ORIG, but it would not contain both.

Sets of numbers can be treated in a similar way, since they are naturally ordered. Suppose that we want to accommodate inventories of different ages in the multiperiod production model of Figure 4-4, by declaring:

```
set PROD;      # products
param T > 0;   # number of weeks
param A > 0;   # maximum age of inventory

var Inv {PROD,0..T,0..A} >= 0;          # tons inventoried
```

Depending on how initial inventories are handled, we might have to include a constraint that no inventory in period t can be more than t weeks old:

```
subject to Too_Old
    {p in PROD, t in 1..T, a in 1..A: a > t}: Inv[p,t,a] = 0;
```

In this case, there is a simpler way to write the indexing expression:

```
subject to Too_Old
    {p in PROD, t in 1..T, a in t+1..A}: Inv[p,t,a] = 0;
```

Here the dummy index defined by `t in 1..T` is immediately used in the phrase `a in t+1..A`. In this and other cases where an indexing expression specifies two or more sets, the comma-separated phrases are evaluated from left to right. Any dummy index defined in one phrase is available for use in all subsequent phrases.

6.5 Indexed collections of sets

Although declarations of individual sets are most common in AMPL models, sets may also be declared in collections indexed over other sets. The principles are much the same as for indexed collections of parameters, variables or constraints.

As an example of how indexed collections of sets can be useful, let us extend the multiperiod production model of Figure 4-4 to recognize different market areas for each product. We begin by declaring:

```
set PROD;
set AREA {PROD};
```

This says that for each member p of PROD, there is to be a set AREA[p]; its members will denote the market areas in which product p is sold.

The market demands, expected sales revenues and amounts to be sold should be indexed over areas as well as products and weeks:

```
param market {p in PROD, AREA[p], 1..T} >= 0;
param revenue {p in PROD, AREA[p], 1..T} >= 0;
var Sell {p in PROD, a in AREA[p], t in 1..T}
                        >= 0, <= market[p,a,t];
```

In the declarations for `market` and `revenue`, we define only the dummy index p that is needed to specify the set AREA[p], but for the Sell variables we need to define all the dummy indices, so that they can be used to specify the upper bound `market[p,a,t]`. This is another example in which an index defined by one phrase of an indexing expression is used by a subsequent phrase; for each p from the set PROD, a runs over a different set AREA[p].

In the objective, the expression `revenue[p,t] * Sell[p,t]` from Figure 4-4 must be replaced by a sum of revenues over all areas for product p:

```
maximize total_profit:
    sum {p in PROD, t in 1..T}
       (sum {a in AREA[p]} revenue[p,a,t]*Sell[p,a,t] -
           prodcost[p]*Make[p,t] - invcost[p]*Inv[p,t]);
```

The only other change is in the `balance` constraints, where `Sell[p,t]` is similarly replaced by a summation:

```
subject to balance {p in PROD, t in 1..T}:
   Make[p,t] + Inv[p,t-1]
       = sum {a in AREA[p]} Sell[p,a,t] + Inv[p,t];
```

```
set PROD;          # products
set AREA {PROD};   # market areas for each product
param T > 0;       # number of weeks

param rate {PROD} > 0;              # tons per hour produced
param inv0 {PROD} >= 0;             # initial inventory
param avail {1..T} >= 0;            # hours available in week
param market {p in PROD, AREA[p], 1..T} >= 0;
                                   # limit on tons sold in week

param prodcost {PROD} >= 0;         # cost per ton produced
param invcost {PROD} >= 0;          # carrying cost/ton of inventory
param revenue {p in PROD, AREA[p], 1..T} >= 0;
                                   # revenue per ton sold

var Make {PROD,1..T} >= 0;          # tons produced
var Inv {PROD,0..T} >= 0;           # tons inventoried
var Sell {p in PROD, a in AREA[p], t in 1..T}   # tons sold
                >= 0, <= market[p,a,t];

maximize total_profit:
   sum {p in PROD, t in 1..T}
      (sum {a in AREA[p]} revenue[p,a,t]*Sell[p,a,t] -
         prodcost[p]*Make[p,t] - invcost[p]*Inv[p,t]);

            # Total revenue less costs for all products in all weeks
subject to time {t in 1..T}:
   sum {p in PROD} (1/rate[p]) * Make[p,t] <= avail[t];

            # Total of hours used by all products
            # may not exceed hours available, in each week

subject to initial {p in PROD}:  Inv[p,0] = inv0[p];

            # Initial inventory must equal given value

subject to balance {p in PROD, t in 1..T}:
   Make[p,t] + Inv[p,t-1]
      = sum {a in AREA[p]} Sell[p,a,t] + Inv[p,t];

            # Tons produced and taken from inventory
            # must equal tons sold and put into inventory
```

Figure 6-3: Multiperiod production with indexed sets (steelT3.mod).

The complete model is shown in Figure 6-3.

In the data for this model, each set within the indexed collection AREA is specified like an ordinary set:

```
set PROD := bands coils;
set AREA[bands] := east north ;
set AREA[coils] := east west export ;
```

The parameters revenue and market are now indexed over three sets, so their data values are specified in a series of tables. Since the indexing is over a different set

```
param T := 4;

set PROD := bands coils;
set AREA[bands] := east north ;
set AREA[coils] := east west export ;

param avail :=  1 40  2 40  3 32  4 40 ;

param rate :=  bands  200  coils  140 ;
param inv0 :=  bands   10  coils    0 ;

param prodcost :=  bands 10    coils 11 ;
param invcost  :=  bands  2.5  coils  3 ;

param revenue :=
   [bands,*,*]:    1       2       3       4 :=
        east      25.0    26.0    27.0    27.0
        north     26.5    27.5    28.0    28.5

   [coils,*,*]:    1       2       3       4 :=
        east      30      35      37      39
        west      29      32      33      35
        export    25      25      25      28 ;

param market :=
   [bands,*,*]:    1       2       3       4 :=
        east      2000    2000    1500    2000
        north     4000    4000    2500    4500

   [coils,*,*]:    1       2       3       4 :=
        east      1000     800    1000    1100
        west      2000    1200    2000    2300
        export    1000     500     500     800 ;
```

Figure 6-4: Data for multiperiod production with indexed sets (steelT3.dat).

AREA[p] for each product p, the values are most conveniently arranged as one table for each product, as shown in Figure 6-4.

We could instead have written this model with a set PRODAREA of pairs, such that product p will be sold in area a if and only if (p,a) is a member of PRODAREA. Our formulation in terms of PROD and AREA[p] seems preferable, however, because it emphasizes the hierarchical relationship between products and areas. Although the model must refer in many places to the set of all areas selling one product, it never refers to the set of all products sold in one area.

As a contrasting example, we can consider how the multicommodity transportation model might use indexed collections of sets. As shown in Figure 6-5, for each product we define a set of origins where that product is supplied, a set of destinations where the product is demanded, and a set of links that represent possible shipments of the product:

```
set orig {PROD} within ORIG;
set dest {PROD} within DEST;
set links {p in PROD} := orig[p] cross dest[p];
```

```
set ORIG;    # origins
set DEST;    # destinations
set PROD;    # products

set orig {PROD} within ORIG;
set dest {PROD} within DEST;
set links {p in PROD} := orig[p] cross dest[p];

param supply {p in PROD, orig[p]} >= 0; # available at origins
param demand {p in PROD, dest[p]} >= 0; # required at destinations
   check {p in PROD}: sum {i in orig[p]} supply[p,i]
                         = sum {j in dest[p]} demand[p,j];

param limit {ORIG,DEST} >= 0;

param cost {p in PROD, links[p]} >= 0;   # shipment costs per unit
var Trans {p in PROD, links[p]} >= 0;    # units to be shipped

minimize total_cost:
   sum {p in PROD, (i,j) in links[p]} cost[p,i,j] * Trans[p,i,j];

subject to Supply {p in PROD, i in orig[p]}:
   sum {j in dest[p]} Trans[p,i,j] = supply[p,i];

subject to Demand {p in PROD, j in dest[p]}:
   sum {i in orig[p]} Trans[p,i,j] = demand[p,j];

subject to Multi {i in ORIG, j in DEST}:
   sum {p in PROD: (i,j) in links[p]} Trans[p,i,j] <= limit[i,j];
```

Figure 6-5: Multicommodity transportation with indexed sets (`multic.mod`).

The declaration of `links` demonstrates that it is possible to have an indexed collection of compound sets, and that an indexed collection may be defined through set operations from other indexed collections. In addition to the operations previously mentioned, there are iterated union and intersection operators that apply to sets in the same way that an iterated sum applies to numbers. For example, the expressions

```
union {p in PROD} orig[p]
inter {p in PROD} orig[p]
```

represent the subset of origins that supply at least one product, and the subset of origins that supply all products.

The hierarchical relationship based on products that was observed in Figure 6-3 is seen in most of Figure 6-5 as well. The model repeatedly deals with the sets of all origins, destinations, and links associated with a particular product. The only exception comes in the last constraint, where the summation must be over all products shipped via a particular link:

```
subject to Multi {i in ORIG, j in DEST}:
   sum {p in PROD: (i,j) in links[p]} Trans[p,i,j] <= limit[i,j];
```

Here it is necessary, following `sum`, to use a somewhat awkward indexing expression to describe a set that does not match the hierarchical organization.

In general, almost any model that can be written with indexed collections of sets can also be written with sets of tuples. As our examples suggest, indexed collections are most suitable for entities such as products and areas that have a hierarchical relationship. Sets of tuples are preferable, on the other hand, in dealing with entities like origins and destinations that are related symmetrically.

6.6 Syntax summary

A set expression may be, in addition to any of the forms described at the end of Chapter 5, any of

set-expr:

$$
\begin{array}{l}
\textit{set-expr } \texttt{cross} \textit{ set-expr} \\
\texttt{union } \textit{index-expr set-expr} \\
\texttt{inter } \textit{index-expr set-expr} \\
\texttt{setof } \textit{index-expr expr} \\
\texttt{setof } \textit{index-expr } (\textit{expr}, \textit{ expr}, \dots)
\end{array}
$$

where *index-expr* is an indexing expression, which normally defines one or more dummy indices used in the *set-expr*.

The declaration of a compound set or indexed collection of sets has the form

set declaration:
   set *set-name* alias$_{opt}$ *index-expr*$_{opt}$ *attributes*$_{opt}$;

where as before *set-name* is any AMPL name, *alias* is an optional string, and *index-expr* is an optional indexing expression. The attributes may include any of the following phrases, optionally separated by commas:

attributes:
   within *set-expr*
   := *set-expr*
   default *set-expr*
   dimen *n*

where *n* is a literal positive integer. The dimension of the declared set is either *n* or the dimension of the set represented by *set-expr*. Only one of := and default may appear; the latter allows values specified by the *set-expr* to be overridden in the data, as explained (for the analogous case of parameters) in Section 7.5.

An indexing expression for a compound set has one of the forms:

index-expr:
   { *index-phrase* }
   { *index-phrase*$_1$, *index-phrase*$_2$, ... }
   { *index-phrase*$_1$, *index-phrase*$_2$, ... : *logic-expr* }

where *index-phrase* is any of:

> *index-phrase*:
>> *set-expr*
>> *dummy-name* in *set-expr*
>> (*dummy-name*$_1$, *dummy-name*$_2$, ...) in *set-expr*

The optional *logic-expr* is a logical expression (considered in detail in Chapter 7). Each *dummy-name*$_k$ can be any AMPL name not currently defined, or, for indexing over slices, an arbitrary expression; a *dummy-name* that appears in *index-phrase*$_k$ is defined and available for use in all subsequent *index-phrase*s of the indexing expression. The number of *dummy-name*s specified in an *index-phrase* must equal the dimension of the set represented by the *set-expr*.

Exercises

6-1. Return to the production and transportation model of Figures 4-6 and 4-7. Using the `display` command, together with indexing expressions as demonstrated in Section 6.4, you can determine the membership of a variety of compound sets; for example, you can use

```
ampl: display {j in DEST, p in PROD: demand[j,p] > 500};
set {j in DEST, p in PROD: demand[j,p] > 500}   :=
(DET,coils)    (STL,bands)    (STL,coils)    (FRE,coils);
```

to show the set of all combinations of products and destinations where the demand is greater than 500.

(a) Use `display` to determine the membership of the following sets, which depend only on the data:

– All combinations of origins and products for which the production rate is greater than 150 tons per hour.

– All combinations of origins, destinations and products for which there is a shipping cost of $\leq \$10$ per ton.

– All combinations of origins and destinations for which the shipping cost of coils is $\leq \$10$ per ton.

– All combinations of origins and products for which the production cost per hour is less than $30,000.

– All combinations of origins, destinations and products for which the transportation cost is more than 15% of the production cost.

– All combinations of origins, destinations and products for which the transportation cost is more than 15% but less than 25% of the production cost.

(b) Use `display` to determine the membership of the following sets, which depend on the optimal solution as well as on the data:

– All combinations of origins and products for which there is production of at least 1000 tons.

– All combinations of origins, destinations and products for which there is a nonzero amount shipped.

– All combinations of origins and products for which more than 10 hours are used in production.

– All combinations of origins and products such that the product accounts for more than 25% of the hours available at the origin.

– All combinations of origins and products such that the total amount of the product shipped from the origin is at least 1000 tons.

6-2. This exercise resembles the previous one, but asks about the ordered-pair version of the transportation model in Figure 6-2.

(a) Use `display` and indexing expressions to determine the membership of the following sets:

– Origin-destination links that have a transportation cost less than $10 per ton.

– Destinations that can be served by `GARY`.

– Origins that can serve `FRE`.

– Links that are used for transportation in the optimal solution.

– Links that are used for transportation from `CLEV` in the optimal solution.

– Destinations to which the total cost of shipping, from all origins, exceeds $20,000.

(b) Use the `display` command and the `setof` operator to determine the membership of the following sets:

– Destinations that have a shipping cost of more than 20 from any origin.

– All destination-origin pairs (`j`,`i`) such that the link from `i` to `j` is used in the optimal solution.

6-3. Use `display` and appropriate set expressions to determine the membership of the following sets from the multiperiod production model of Figures 6-3 and 6-4:

– All market areas served with any of the products.

– All combinations of products, areas and weeks such that the amount actually sold in the optimal solution equals the maximum that can be sold.

– All combinations of products and weeks such that the total sold in all areas is greater than or equal to 6000 tons.

6-4. To try the following experiment, first enter these declarations:

```
ampl: set Q := {1..10,1..10,1..10,1..10,1..10,1..10};
ampl: set S within Q;
ampl: data;
ampl: set S := 1 2 3 3 4 5  2 3 4 4 5 6  3 4 5 5 6 7  4 5 6 7 8 9 ;
```

(a) Now try the following two commands:

```
display S;
display {(a,b,c,d,e,f) in Q: (a,b,c,d,e,f) in S};
```

The two expressions in these commands represent the same set, but do you get the same response from AMPL? Explain the cause of the difference.

(b) Predict the result of the command `display Q`.

6-5. This exercise asks you to reformulate the diet model of Figure 2-1 in a variety of ways, using compound sets.

(a) Reformulate the diet model so that it uses a declaration

```
set GIVE within {NUTR,FOOD};
```

to define a subset of pairs (`i`,`j`) such that nutrient `i` can be found in food `j`.

(b) Reformulate the diet model so that it uses a declaration

```
set FN {NUTR} within FOOD;
```

to define, for each nutrient `i`, the set `FN[i]` of all foods that can supply that nutrient.

(c) Reformulate the diet model so that it uses a declaration

```
set NF {FOOD} within NUTR;
```

to define, for each food `j`, the set `NF[j]` of all nutrients supplied by that food. Explain why you find this formulation more or less natural and convenient than the one in (b).

6-6. Re-read the suggestions in Section 6.3, and complete the following reformulations of the multicommodity transportation model:

(a) Use a subset `LINKS` of origin-destination pairs.

(b) Use a subset `ROUTES` of origin-destination-product triples.

(c) Use a subset `MARKETS` of destination-product pairs, with the property that product `p` can be sold at destination `j` if and only if `(j,p)` is in the subset.

6-7. Carry through the following two suggestions from Section 6.4 for enhancements to the multicommodity transportation problem of Figure 4-1.

(a) Add a declaration

```
set DEMAND := {j in DEST, p in PROD: demand[j,p] > 0};
```

and index the variables over `{ORIG,DEMAND}`, so that variables are defined only where they might be needed to meet demand. Make all of the necessary changes in the rest of the model to use this set.

(b) Add the declarations

```
set LINKS within {ORIG,DEST};
set TRANSF := {i1 in ORIG, i2 in ORIG: i1 <> i2};
```

Define variables over `LINKS` to represent shipments to destinations, and over `TRANSF` to represent shipments between origins. The constraint at each origin now must say that total shipments out — to other origins as well as to destinations — must equal supply plus shipments in from other origins. Complete the formulation for this case.

6-8. Reformulate the model from Exercise 3-3(b) so that it uses a set `LINK1` of allowable plant-mill shipment pairs, and a set `LINK2` of allowable mill-factory shipment pairs.

6-9. As chairman of the program committee for a prestigious scientific conference, you must assign submitted papers to volunteer referees. To do so in the most effective way, you can formulate an LP model along the lines of the assignment model discussed in Chapter 3, but with a few extra twists.

After looking through the papers and the list of referees, you can compile the following data:

```
set Papers;
set Referees;
set Categories;

set PaperKind within {Papers,Categories};
set Willing within {Referees,Categories};
```

The contents of the first two sets are self-evident, while the third set contains subject categories into which papers may be classified. The set `PaperKind` contains a pair `(p,c)` if paper p falls into category c; in general, a paper can fit into several categories. The set `Willing` contains a pair `(r,c)` if referee r is willing to handle papers in category c.

(a) What is the dimension of the set

```
{(r,c) in Willing, (p,c) in PaperKind}
```

and what is the significance of the tuples contained in this set?

(b) Based on your answer to (a), explain why the declaration

```
set CanHandle := setof {(r,c) in Willing, (p,c) in PaperKind} (r,p);
```

gives the set of pairs `(r,p)` such that referee `r` can be assigned paper `p`.

Your model could use parameters `ppref` and variables `Review` indexed over `CanHandle`; `ppref[r,p]` would be the preference of referee `r` for paper `p`, and `Review[r,p]` would be 1 if referee `r` were assigned paper `p`, or 0 otherwise. Assuming higher preferences are better, write out the declarations for these components and for an objective function to maximize the sum of preferences of all assignments.

(c) Unfortunately, you don't have the referees' preferences for individual papers, since they haven't seen any papers yet. What you have are their preferences for different categories:

```
param cpref {Willing} integer >= 0, <= 5;
```

Explain why it would make sense to replace `ppref[r,p]` in your objective by

```
max {(r,c) in Willing: (p,c) in PaperKind} cpref[r,c]
```

(d) Finally, you must define the following parameters that indicate how much work is to be done:

```
param nreferees integer > 0;       # referees needed per paper
param minwork integer > 0;         # min papers to each referee
param maxwork integer > minwork;   # max papers to each referee
```

Formulate the appropriate assignment constraints. Complete the model, by formulating constraints that each paper must have the required number of referees, and that each referee must be assigned an acceptable number of papers.

7

Parameters and Expressions

A large optimization model invariably uses many numerical values. As we have explained before, only a concise symbolic description of these values need appear in an AMPL model, while the explicit data values are given in separate data statements, to be described in Chapter 9.

In AMPL a single named numerical value is called a *parameter*. Although some parameters are defined as individual scalar values, most occur in vectors or matrices or other collections of numerical values indexed over sets. We will thus loosely refer to an indexed collection of parameters as ''a parameter'' when the meaning is clear. To begin this chapter, Section 7.1 describes the rules for declaring parameters and for referring to them in an AMPL model.

Parameters and other numerical values are the building blocks of the expressions that make up a model's objective and constraints. Sections 7.2 and 7.3 describe arithmetic expressions, which have a numerical value, and logical expressions, which evaluate to true or false. Along with the standard unary and binary operators of conventional algebraic notation, AMPL provides iterated operators like sum and prod, and a conditional (if-then-else) operator that chooses between two expressions, depending on the truth of a third expression.

The expressions in objectives and constraints necessarily involve variables, whose declaration and use will be discussed in Chapter 8. There are several common uses for expressions that involve only sets and parameters, however. Section 7.4 describes how logical expressions are used to test the validity of data, either directly in a parameter declaration, or separately in a check statement. Section 7.5 introduces features for defining new parameters through arithmetic expressions in previously declared parameters and sets.

Although the key purpose of parameters is to represent numerical values, they can also represent logical values or arbitrary strings. These possibilities are covered in Section 7.6.

7.1 Parameter declarations

A parameter declaration describes certain data required by a model, and indicates how the model will refer to data values in subsequent expressions.

The simplest parameter declaration consists of the keyword `param` and a name:

```
param T;
```

At any point after this declaration, `T` can be used to refer to a numerical value.

More often, the name in a parameter declaration is followed by an indexing expression:

```
param avail {1..T};
param demand {DEST,PROD};
param revenue {p in PROD, AREA[p], 1..T};
```

One parameter is defined for each member of the set specified by the indexing expression. Thus a parameter is uniquely determined by its name and its associated set member; throughout the rest of the model, you would refer to this parameter by writing the name and bracketed "subscripts":

```
avail[i]
demand[j,p]
revenue[p,a,t]
```

If the indexing is over a simple set of objects as described in Chapter 5, there is one subscript. If the indexing is over a set of pairs, triples, or longer tuples as described in Chapter 6, there must be a corresponding pair, triple, or longer list of subscripts separated by commas. The subscripts can be any expressions, so long as they evaluate to members of the underlying index set.

An unindexed parameter is a scalar value, but a parameter indexed over a simple set has the characteristics of a vector or an array; when the indexing is over a sequence of integers, like

```
param avail {1..T};
```

the individual subscripted parameters are `avail[1]`, `avail[2]`, ..., `avail[T]`, and there is an obvious analogy to the vectors of linear algebra or the arrays of a programming language like Fortran or Pascal. AMPL's concept of a vector is more general, however, since parameters may also be indexed over sets of strings, which need not even be ordered. Indexing over sets of strings is best suited for parameters that correspond to places, products and other entities for which no numbering is especially natural. Indexing over sequences of numbers is more appropriate for parameters that correspond to weeks, stages, and the like, which by their nature tend to be ordered and numbered; even for these, you may prefer to use ordered sets of strings as described in Section 5.6.

A parameter indexed over a set of pairs is like a two-dimensional array or matrix. If the indexing is over all pairs from two sets, as in

```
set ORIG;
set DEST;
param cost {ORIG,DEST};
```

then there is a parameter cost[i,j] for every combination of i from ORIG and j from DEST, and the analogy to a matrix is strongest — although again the subscripts are more likely to be strings than numbers. If the indexing is over a subset of pairs, however:

```
set ORIG;
set DEST;
set LINKS within {ORIG,DEST};
param cost {LINKS};
```

then cost[i,j] exists only for those i from ORIG and j from DEST such that (i,j) is a member of LINKS. In this case, you can think of cost as being a ''sparse'' matrix.

Similar comments apply to parameters indexed over triples and longer tuples, which resemble arrays of higher dimension in programming languages.

7.2 Arithmetic expressions

Arithmetic expressions in AMPL are much the same as in other computer languages. Literal numbers consist of an optional sign preceding a sequence of digits, which may or may not include a decimal point (for example, −17 or 2.71828 or +.3). At the end of a literal there may also be an exponent, consisting of the letter d, D, e, or E and an optional sign followed by digits (1e30 or 7.66439D−07).

Literals, parameters, and variables are combined into expressions by the standard operations of addition (+), subtraction (−), multiplication (*), division (/), and exponentiation (**). The familiar conventions of arithmetic apply. Exponentiation has higher precedence than multiplication and division, which have higher precedence than addition and subtraction; successive operations of the same precedence group to the left, except for exponentiation, which groups to the right. Parentheses may be used to change the order of evaluation.

Arithmetic expressions may also use the div operator, which returns the truncated quotient when its left operand is divided by its right operand; the mod operator, which computes the remainder; and the less operator, which returns its left operand minus its right operand if the result is positive, or zero otherwise. For purposes of precedence and grouping, AMPL treats div and mod like division, and less like subtraction.

A list of arithmetic operators (and logical operators, to be described shortly) is given in Table 7-1. As much as possible, AMPL follows common programming languages in its choice of operator symbols, such as * for multiplication and / for division. There is sometimes more than one standard, however, as with exponentiation, where some languages use ^ while others use **. In this and other cases, AMPL provides alternate forms. Table 7-1 shows the more common forms to the left, and the alternatives (if any)

Usual style	alternative style	type of operands	type of result
if-then-else		logical, arithmetic	arithmetic
or	||	logical	logical
exists forall		logical	logical
and	&&	logical	logical
not (unary)	!	logical	logical
< <= = <> > >=	< <= == != > >=	arithmetic	logical
in not in		object, set	logical
+ - less		arithmetic	arithmetic
sum prod min max		arithmetic	arithmetic
* / div mod		arithmetic	arithmetic
+ - (unary)		arithmetic	arithmetic
^	**	arithmetic	arithmetic

Exponentiation and if-then-else are right-associative; the other operators are left-associative. The logical operand of if-then-else appears after if, and the arithmetic operands after then and (optionally) else.

Table 7-1: Arithmetic and logical operators, in increasing precedence.

to the right; you can mix them as you like, but your models will be easier to read and understand if you are consistent in your choices.

Another way to build arithmetic expressions is by applying functions to other expressions. A function reference consists of a name followed by a parenthesized argument or comma-separated list of arguments; in the case of arithmetic functions, each argument can be any arithmetic expression. Here are a few examples, which compute the minimum, absolute value, and square root of their arguments, respectively:

```
min(T,20)
abs(sum {i in ORIG} supply[i] - sum {j in DEST} demand[j])
sqrt((floor[j]-floor[k])^2)
```

Table 7-2 lists the built-in arithmetic functions. Except for min and max, the names of any of these functions may be redefined, but their original meanings will become inaccessible. For example, a model may declare a parameter named floor as in the last example above, but then it cannot also refer to the function floor.

The set functions card and ord, which were described in Chapter 5, also produce an arithmetic result. In addition, AMPL provides several "rounding" functions (Section 10.8) and a variety of random-number functions (Appendix A.4.1). A mechanism for "importing" functions defined by your own programs is described in Appendix A.14.

Finally, the indexed operators such as Σ and Π from algebraic notation are generalized in AMPL by expressions for iterating operations over sets. In particular, most large-scale linear programming models contain iterated summations:

abs (x)	absolute value, $\lvert x \rvert$
acos (x)	inverse cosine, $\cos^{-1}(x)$
acosh (x)	inverse hyperbolic cosine, $\cosh^{-1}(x)$
asin (x)	inverse sine, $\sin^{-1}(x)$
asinh (x)	inverse hyperbolic sine, $\sinh^{-1}(x)$
atan (x)	inverse tangent, $\tan^{-1}(x)$
atan2 (y, x)	inverse tangent, $\tan^{-1}(y/x)$
atanh (x)	inverse hyperbolic tangent, $\tanh^{-1}(x)$
ceil (x)	ceiling of x (next higher integer)
cos (x)	cosine
exp (x)	exponential, e^{x}
floor (x)	floor of x (next lower integer)
log (x)	natural logarithm, $\log_{e}(x)$
log10 (x)	common logarithm, $\log_{10}(x)$
max $(x, y, ...)$	maximum (2 or more arguments)
min $(x, y, ...)$	minimum (2 or more arguments)
sin (x)	sine
sinh (x)	hyperbolic sine
sqrt (x)	square root
tan (x)	tangent
tanh (x)	hyperbolic tangent

Table 7-2: Built-in arithmetic functions.

```
sum {i in ORIG} supply[i,p]
```

The keyword sum may be followed by any indexing expression. The subsequent arithmetic expression is evaluated once for each member of the index set, and all the resulting values are added. Thus the sum above, from the multicommodity transportation model of Figure 4-1, represents the total supply, at all origins, of product p. The sum operator has lower precedence than *, so the objective of the same model can be written

```
sum {i in ORIG, j in DEST, p in PROD}
   cost[i,j,p] * Trans[i,j,p]
```

to represent the total of cost[i,j,p] * Trans[i,j,p] over all combinations of origins, destinations and products. The precedence of sum is higher than that of + or −, however, so for the objective of the multiperiod production model in Figure 6-3 we must write

```
sum {p in PROD, t in 1..T}
   (sum {a in AREA[p]} revenue[p,a,t]*Sell[p,a,t] −
      prodcost[p]*Make[p,t] − invcost[p]*Inv[p,t]);
```

The outer sum applies to the entire parenthesized expression following it, while the inner sum applies only to the term revenue[p,a,t] * Sell[p,a,t].

Other iterated arithmetic operators are `prod` for multiplication, `min` for minimum, and `max` for maximum. As an example, we could use

```
max {i in ORIG} supply[i,p]
```

to describe the greatest supply of product p at any origin.

Bear in mind that, while an AMPL arithmetic function or operator may be applied to parameters or to numeric members of sets in any context, most operations on variables are not linear. AMPL's requirements for arithmetic expressions in a linear program are described in Section 8.2.

7.3 Logical and conditional expressions

The values of arithmetic expressions can be tested against each other by comparison operators:

=	equal to
<>	not equal to
<	less than
<=	less than or equal to
>	greater than
>=	greater than or equal to

The result of a comparison is either "true" or "false". Thus T > 1 is true if the parameter T has a value greater than 1, and is false otherwise; and

```
sum {i in ORIG} supply[i] = sum {j in DEST} demand[j]
```

is true if and only if total supply equals total demand.

Comparisons are one example of AMPL's logical expressions, which evaluate to true or false. Set membership tests using `in` and `within`, described in Section 5.4, are another example. More complex logical expressions can be built up with logical operators. The `and` operator returns true if and only if both its operands are true, while `or` returns true if and only if at least one of its operands is true; the unary operator `not` returns false for true and true for false. Thus the expression

```
T >= 0 and T <= 10
```

is only true if T lics in the interval [0, 10], while the following from Section 5.5,

```
i in MAXREQ or n_min[i] > 0
```

is true if i is a member of MAXREQ, or n_min[i] is positive, or both. Where several operators are used together, any comparison, membership or arithmetic operator has higher precedence than the logical operators; `and` has higher precedence than `or`, while `not` has higher precedence than either. Thus the expression

```
not i in MAXREQ or n_min[i] > 0 and n_min[i] <= 10
```

is interpreted as

```
(not (i in MAXREQ)) or ((n_min[i] > 0) and (n_min[i] <= 10))
```

The precedences are summarized in Table 7-1, which also gives alternative forms.

Like + and *, the operators or and and have iterated versions. The iterated or is denoted by exists, and the iterated and by forall. Thus, as an example, the expression

```
exists {i in ORIG} demand[i] > 10
```

is true if and only if at least one origin has a demand greater than 10, while

```
forall {i in ORIG} demand[i] > 10
```

is true if and only if every origin has demand greater than 10.

Another use for a logical expression is as an operand to the conditional or if-then-else operator, which returns one of two different arithmetic values depending on whether the logical expression is true or false. Consider the two collections of inventory balance constraints in the multiperiod production model of Figure 5-3:

```
subject to balance0 {p in PROD}:
   Make[p,first(WEEKS)] + inv0[p]
      = Sell[p,first(WEEKS)] + Inv[p,first(WEEKS)];

subject to balance {p in PROD, t in WEEKS: ord(t) > 1}:
   Make[p,t] + Inv[p,prev(t)] = Sell[p,t] + Inv[p,t];
```

The balance0 constraints are basically the balance constraints with t set to first(WEEKS). The only difference is in the second term, which represents the previous week's inventory; it is given as inv0[p] for the first week (in the balance0 constraints) but is represented by the variable Inv[p,prev(t)] for subsequent weeks (in the balance constraints). We would like to combine these constraints into one declaration, by having a term that takes the value inv0[p] when t is the first week, and takes the value Inv[p,prev(t)] otherwise. Such a term is written in AMPL as:

```
if t = first(WEEKS) then inv0[p] else Inv[p,prev(t)]
```

Placing this expression into the constraint declaration, we can write

```
subject to balance {p in PROD, t in WEEKS}:
   Make[p,t] + (if t = first(WEEKS)
                  then inv0[p] else Inv[p,prev(t)])
      = Sell[p,t] + Inv[p,t];
```

This form communicates the inventory balance constraints more concisely and directly than two separate declarations.

The general form of a conditional expression is

```
if a then b else c
```

where a is a logical expression. If a evaluates to true, the conditional expression takes the value of b; if a is false, the expression takes the value of c. If c is zero, the else c part can be dropped. Most often b and c are arithmetic expressions, but they can also be strings or set expressions. Because then and else have lower precedence than any other operators, a conditional expression needs to be parenthesized (as in the example above) unless it occurs at the end of a statement.

Many programming languages have an if-then-else that conditionally executes one or another block of statements. By contrast, AMPL's if-then-else is merely an expression, whose value is conditionally determined. It therefore belongs inside of a declaration, in a place where an expression would normally be evaluated.

7.4 Restrictions on parameters

If T is intended to represent the number of weeks in a multiperiod model, it should be an integer and greater than 1. By including these conditions in T's declaration,

```
param T > 1 integer;
```

you instruct AMPL to reject your data if you inadvertently set T to 1:

```
error processing param T:
        failed check: param T is not > 1;
```

or to 2.5:

```
error processing param T:
        failed check: param T is not an integer;
```

In the declaration of an indexed collection of parameters, a simple restriction such as integer or >= 0 applies to every parameter defined. Our examples often use this option to specify that vectors and arrays are nonnegative:

```
param demand {DEST,PROD} >= 0;
```

If you include dummy indices in the indexing expression, however, you can use them to specify a different restriction for each parameter:

```
param f_min {FOOD} >= 0;
param f_max {j in FOOD} >= f_min[j];
```

The effect of these declarations is to define a pair of parameters f_max[j] >= f_min[j] for every j in the set FOOD.

A restriction phrase for a parameter declaration may be the word integer or binary or a comparison operator followed by an arithmetic expression. While integer restricts a parameter to integral values, binary restricts it to zero or one. The arithmetic expression may refer to sets and parameters previously defined in the model, and to dummy indices defined by the current declaration. There may be several restriction phrases in the same declaration, in which case they may optionally be separated by commas.

In special circumstances, a restriction phrase may even refer to the parameter in whose declaration it appears. Some multiperiod production models, for example, are defined in terms of a parameter cum_market[p,t] that represents the cumulative demand for product p in weeks 1 through t. Since cumulative demand does not decrease, you might try to write a restriction phrase like this:

```
param cum_market {p in PROD, t in 1..T}
   >= cum_market[p,t-1];   # ERROR
```

For the parameters cum_market[p,1], however, the restriction phrase will refer to cum_market[p,0], which is undefined; AMPL will reject the declaration with an error message. What you need here again is a conditional expression that handles the first period specially:

```
param cum_market {p in PROD, t in 1..T}
   >= if t = 1 then 0 else cum_market[p,t-1];
```

The same thing could be written a little more compactly as

```
param cum_market {p in PROD, t in 1..T}
   >= if t > 1 then cum_market[p,t-1];
```

since "else 0" is assumed. Almost always, some form of if-then-else expression is needed to make this kind of self-reference possible.

As you might suspect from this last example, sometimes it is desirable to place a more complex restriction on the model's data than can be expressed by a restriction phrase within a declaration. This is the purpose of the check statement. For example, in the transportation model of Figure 3-1a, total supply must equal total demand:

```
check: sum {i in ORIG} supply[i] = sum {j in DEST} demand[j];
```

The multicommodity version, in Figure 4-1, uses an indexed check to say that total supply must equal total demand for each product:

```
check {p in PROD}:
   sum {i in ORIG} supply[i,p] = sum {j in DEST} demand[j,p];
```

Here the restriction is tested once for each member p of PROD. If the check fails for any member, AMPL prints an error message and rejects all of the data.

You can think of the check statement as specifying a kind of constraint, but only on the data. The restriction clause is a logical expression, which may use any previously defined sets and parameters as well as dummy indices defined in the indexing expression. After the data values have been read, the logical expression must evaluate to true; if an indexing expression has been specified, the logical expression is evaluated separately for each assignment of set members to the dummy indices, and must be true for each.

We strongly recommend the use of restriction phrases and check statements to validate a model's data. These features will help you to catch data errors at an early stage, when they are easy to fix. Data errors not caught will, at best, cause errors in the generation of the variables and constraints, so that you will get some kind of error message from AMPL. In other cases, data errors lead to the generation of an incorrect linear program. If you are fortunate, the incorrect linear program will have a meaningless optimal solution, so that — possibly after a good deal of effort — you will be able to work backward to find the error in the data. At worst, the incorrect linear program will have a plausible solution, and the error will go undetected.

7.5 Computed parameters

It is seldom possible to arrange that the data values available to a model are precisely the coefficient values required by the objective and constraints. Even in the simple production model of Figure 1-4, for example, we wrote the constraint as

```
sum {p in PROD} (1/rate[p]) * Make[p] <= avail;
```

because production rates were given in tons per hour, while the coefficient of `Make[p]` had to be in hours per ton. Any parameter expression may be used in the constraints and objective, but the expressions are best kept simple. When more complex expressions are needed, the model is usually easier to understand if new computed parameters are defined in terms of the data parameters.

The declaration of a computed parameter has an assignment phrase, which resembles the restriction phrase described in the previous section except for the use of a `:=` operator to represent assignment. As a first example, suppose that the data provided to the multi-commodity transportation model of Figure 4-1 consist of the total demand for each product, together with each destination's share of demand. The destinations' shares are percentages between zero and 100, but their sum over all destinations might not exactly equal 100%, because of rounding and approximation. Thus we declare data parameters to represent the shares, and a computed parameter equal to their sum:

```
param share {DEST} >= 0, <= 100;
param tot_sh := sum {j in DEST} share[j];
```

We can then declare a data parameter to represent total demands, and a computed parameter that equals demand at each destination:

```
param tot_dem {PROD} >= 0;
param demand {j in DEST, p in PROD}
       := share[j] * tot_dem[p] / tot_sh;
```

The division by `tot_sh` acts as a correction factor for a sum not equal to 100%. Once `demand` has been defined in this way, the model can use it as in Figure 4-1:

```
subject to Demand {j in DEST, p in PROD}:
   sum {i in ORIG} Trans[i,j,p] = demand[j,p];
```

We could avoid computed parameters by substituting the formulas for `tot_sh` and `demand[j,p]` directly into this constraint:

```
subject to Demand {j in DEST, p in PROD}:
   sum {i in ORIG} Trans[i,j,p]
     = share[j] * tot_dem[p] / sum {k in DEST} share[k];
```

This alternative makes the model a little shorter, but the computation of the demand and the structure of the constraint are both harder to follow.

As another example, consider a scenario for the multiperiod production model (Figure 4-4) in which minimum inventories are computed. Specifically, suppose that the inventory of product p for week t must be at least a certain fraction of `market[p,t+1]`, the maximum that can be sold in the following week. We thus use the following declarations for the data to be supplied:

```
param frac > 0;
param market {PROD,1..T+1} >= 0;
```

and then declare

```
param mininv {p in PROD, t in 0..T} := frac * market[p,t+1];
var Inv {p in PROD, t in 0..T} >= mininv[p,t];
```

to define and use parameters `mininv[p,t]` that represent the minimum inventory of product p for week t.

If you define a computed parameter as in the examples above, then you cannot also specify a data value for it. An attempt to do so will result in an error message:

```
mininv was defined in the model
context:  param  >>> mininv <<<  :=  bands 2   3000
```

However, if you define a computed parameter using the operator `default` in place of `:=`, any data values that you specify explicitly will override the ones that would otherwise be computed. Thus, by declaring

```
param mininv {p in PROD, t in 0..T}
   default frac * market[p,t+1];
```

you can allow a few exceptional minimum inventories to be specified as part of the data for the model, either in a list:

```
param mininv :=  bands 2   3000
                 coils 2   2000
                 coils 3   2000 ;
```

or in a table:

```
param market:    1     2     3     4 :=
        bands    .   3000    .     .
        coils    .   2000  2000    . ;
```

AMPL uses ``.'' in a data statement to indicate an omitted entry, as explained in Chapter 9.

An assignment phrase in a `param` declaration normally consists of `:=` or `default` followed by an arithmetic expression in previously defined sets and parameters and currently defined dummy indices. When you declare an indexed collection of parameters, however, some may be computed in terms of others. As an example, you can smooth out some of the variation in the minimum inventories by a running average like this:

```
param mininv {p in PROD, t in 0..T} :=
   if t = 0 then inv0[p]
                else 0.5 * (mininv[p,t-1] + frac * market[p,t+1]);
```

The values of `mininv` for week 0 are set explicitly to the initial inventories, while the values for each subsequent week `t` are defined in terms of the previous week's values. AMPL permits any "recursive" definition of this kind, but will signal an error if it detects a circular reference that causes a parameter's value to depend directly or indirectly on itself.

7.6 Logical and symbolic parameters

Although parameters normally represent numeric values, they can optionally be used to stand for true-false values or for character strings.

The current version of AMPL does not support a full-fledged "logical" type of parameter that would stand for only the values true and false, but a parameter of type `binary` may be used to the same effect. As an illustration, we describe an application of the preceding inventory example to consumer goods. Certain products in each week may be specially promoted, in which case they require a higher inventory fraction. Using parameters of type `binary`, we can represent this situation by the following declarations:

```
param fr_reg > 0;        # regular inventory fraction
param fr_pro > fr_reg;   # fraction for promoted items

param promote {PROD,1..T+1} binary;
param market {PROD,1..T+1} >= 0;

param mininv {p in PROD, t in 0..T} :=
   (if promote[p,t] = 1 then fr_pro else fr_reg)
      * market[p,t+1];
```

The binary parameters `promote[p,t]` are 0 when there is no promotion, and 1 when there is a promotion. Rather than explicitly comparing their values to 1 or 0 in the `if-then-else` expression, however, we may write:

```
param mininv {p in PROD, t in 0..T} :=
   (if promote[p,t] then fr_pro else fr_reg) * market[p,t+1];
```

If an arithmetic expression appears where a logical expression is required, AMPL interprets any nonzero value as true, and zero as false. You do need to exercise a little caution to avoid being tripped up by this implicit conversion. For example, in Section 7.4 we used the expression

```
if t = 1 then 0 else cum_market[p,t-1]
```

If you accidentally write

```
if t then 0 else cum_market[p,t-1]     # DIFFERENT
```

it's perfectly legal, but it doesn't mean what you intended.

You may permit a parameter to represent character string values, by including the keyword symbolic in its declaration. A symbolic parameter's values may be strings or numbers, just like a set's members, but the string values may not participate in arithmetic.

A major use of symbolic parameters is to designate individual set members that are to be treated specially. For example, in a model of traffic flow, there is a set of intersections, two of whose members are designated as the entrance and exit. Symbolic parameters can be used to represent these two members:

```
set INTER;

param entr symbolic in INTER;
param exit symbolic in INTER, <> entr;
```

In the data statements, an appropriate string is assigned to each symbolic parameter:

```
set INTER := a b c d e f g ;

param entr := a ;
param exit := g ;
```

These parameters are subsequently used in defining the objective and constraints; the complete model is developed in Section 11.2.

Another use of symbolic parameters is to associate descriptive strings with set members. Consider for example the set of "origins" in the transportation model of Figure 3-1a. When we introduced this set at the beginning of Chapter 3, we described each originating city by means of a 4-character string and a longer descriptive string. The short strings became the members of the AMPL set ORIG, while the longer strings played no further role. To make both available, we could declare

```
set ORIG;
param orig_name {ORIG} symbolic;
param supply {ORIG} >= 0;
```

Then in the data we could specify

```
param: ORIG:    orig_name                    supply :=
        GARY    "Gary, Indiana"              1400
        CLEV    "Cleveland, Ohio"            2600
        PITT    "Pittsburgh, Pennsylvania"   2900 ;
```

Since the long strings do not have the form of AMPL names, they do need to be quoted. They still play no role in the model or the resulting linear program, but they can be retrieved for documentary purposes by the display and printf commands described in Chapter 10.

7.7 Syntax summary

A member of a simple set is either a number or a string, and an expression that yields such a value is denoted by *expr*.

A simple *parameter* is a name that stands for a possible set member. An indexed parameter is a collection of such values, one for each member of the indexing set specified in the parameter's declaration. A reference to a parameter's value is written as:

> *param-ref*:
>   *param-name*
>   *param-name* [*expr*]
>   *param-name* [*expr*$_1$, *expr*$_2$, ...]

where the number of *exprs* must equal the dimension of the indexing set.

An arithmetic expression yields a number and can have any of these forms:

> *arith-expr*:
>   *number*
>   *param-ref*
>   *dummy-name*
>   *func* (*arg-list*)
>
>   *unary-arith-op arith-expr*   (+ -)
>   *arith-expr binary-arith-op arith-expr* (+ - less * / div mod ** ^)
>   *iterated-arith-op index-expr arith-expr* (sum prod max min)
>
>   if *logic-expr* then *arith-expr*
>   if *logic-expr* then *arith-expr* else *arith-expr*
>
>   (*arith-expr*)

In function references, *func* may be any of the built-in functions listed in Appendix A.4.1, or a previously declared imported function, as explained in Appendix A.14; the *arg-list* may be empty, or a single expression, or a comma-separated list of expressions, as appropriate for the particular function.

A logical expression has one of the forms:

> *logic-expr*:
>   *arith-expr*
>
>   *expr equality-op expr*   (= == <> !=)
>   *arith-expr inequality-op arith-expr* (< <= > >=)
>
>   *expr* in *set-expr*
>   (*expr*$_1$, *expr*$_2$, ...) in *set-expr*
>   *set-expr* within *set-expr*
>
>   *unary-logic-op logic-expr*   (not !)
>   *logic-expr binary-logic-op logic-expr* (and && or ||)
>   *iterated-logic-op index-expr logic-expr* (forall exists)
>
>   (*logic-expr*)

An *arith-expr* is interpreted as a *logic-expr* if and only if it appears in one of the contexts above where a *logic-expr* is required. A value of zero is interpreted as false, while a nonzero value is interpreted as true. The in and within operators may be preceded by not to invert the truth value of the test.

A parameter declaration is written

param-declaration:
> param *param-name* alias_{opt} *index-expr*_{opt} *attributes*_{opt} ;

where *param-name* is any AMPL name, *alias* is an optional string, and *index-expr* is an optional indexing expression as defined in Chapters 5 and 6. The attributes may include

attributes:
> *type-phrase*
> *restriction-phrase*
> *assignment-phrase*

type-phrase:
> `binary`
> `integer`
> `symbolic`

restriction-phrase:
> *equality-op* *expr*
> *inequality-op* *arith-expr*
> `in` *set-expr*

assignment-phrase:
> := *expr*
> `default` *expr*

Only one *type-phrase* and one *assignment-phrase* may be specified. The parameter's value(s) must be numeric unless the *type-phrase* is `symbolic`. Expressions may involve references only to sets and parameters declared previously, or to the parameter currently being declared, or to dummy indices in the current declaration's *index-expr*.

A `check` statement has the form

check-statement:
> `check` *index-expr*_{opt} : *logic-expr* ;

The *logic-expr* may only refer to previously declared sets and parameters, and to dummy indices defined in optional *index-expr*.

Exercises

7-1. Show how the multicommodity transportation model of Figure 4-1 could be modified so that it applies the following restrictions to the data. Use either a restriction phrase in a `set` or `param` declaration, or a `check` statement, whichever is appropriate.

– No city is a member of both `ORIG` and `DEST`.

– The number of cities in `DEST` must be greater than the number in `ORIG`.

– Demand does not exceed 1000 at any one city in `DEST`.

– Total supply for each product at all origins must equal total demand for that product at all destinations.

– Total supply for all products at all origins must equal total demand for all products at all destinations.

– Total supply of all products at an origin must not exceed total capacity for all shipments from that origin.

– Total demand for all products at a destination must not exceed total capacity for all shipments to that destination.

7-2. Show how the multiperiod production model of Figure 4-4 could be modified so that it applies the following restrictions to the data.

– The number of weeks is a positive integer greater than 1.

– The initial inventory of a product does not exceed the total market demand for that product over all weeks.

– The inventory cost for a product is never more than 10% of the expected revenue for that product in any one week.

– The number of hours in a week is between 24 and 40, and does not change by more than 8 hours from one week to the next.

– For each product, the expected revenue never decreases from one week to the next.

7-3. The solutions to the following exercises involve the use of an `if-then-else` operator to formulate a constraint.

(a) In the example of the constraint `balance` in Section 7.3, we used an expression beginning

```
if t = first(WEEKS) then ...
```

Find an equivalent expression that uses the function `ord(t)`.

(b) Combine the `diet_min` and `diet_max` constraints of Figure 5-1's diet model into one constraint declaration.

(c) In the multicommodity transportation model of Figure 4-1, imagine that there is more demand at the destinations than we can meet from the supply produced at the origins. To make up the difference, a limited number of additional tons can be purchased (rather than manufactured) for shipment at certain origins.

To model this situation, suppose that we declare a subset of origins,

```
set BUY_ORIG within ORIG;
```

where the additional tons can be bought. The relevant data values and decision variables could be indexed over this subset:

```
param buy_supply {BUY_ORIG,PROD} >= 0;   # available for purchase
param buy_cost {BUY_ORIG,PROD} > 0;      # purchase cost per ton

var Buy {i in BUY_ORIG, p in PROD} >= 0, <= buy_supply[i,p];
                                         # amount to buy
```

Revise the objective function to include the purchase costs. Revise the `Supply` constraints to say that, for each origin and each product, total tons shipped out must equal tons of supply from production plus (if applicable) tons purchased.

(d) Formulate the same model as in (c), but with `BUY_ORIG` being the set of pairs `(i,p)` such that product `p` can be bought at origin `i`.

7-4. This exercise is concerned with the following sets and parameters from Figure 4-1:

```
set ORIG;    # origins
set DEST;    # destinations
set PROD;    # products

param supply {ORIG,PROD} >= 0;
param demand {DEST,PROD} >= 0;
```

(a) Write `param` declarations, using the `:=` operator, to compute parameters having the following definitions:

- `prod_supply[p]` is the total supply of product p at all origins.

- `dest_demand[j]` is the total demand for all products at destination j.

- `true_limit[i,j,p]` is the largest quantity of product p that can be shipped from i to j — that is, the largest value that does not exceed `limit[i,j]`, or the supply of p at i, or the demand for p at j.

- `max_supply[p]` is the largest supply of product p available at any origin.

- `max_diff[p]` is the largest difference, over all combinations of origins and destinations, between the supply and demand for product p.

(b) Write `set` declarations, using the `:=` operator, to compute these sets:

- Products p whose demand is at least 500 at some destination j.

- Products p whose demand is at least 250 at all destinations j.

- Products p whose demand is equal to 500 at some destination j.

7-5. AMPL parameters can be defined to contain many kinds of series, especially by using recursive definitions. For example, we can make `s[j]` equal the sum of the first j integers, for j from 1 to some given limit N, by writing

```
param N;
param s {j in 1..N} := sum {jj in 1..j} jj;
```

or, using a formula for the sum,

```
param s {j in 1..N} := j * (j+1) / 2;
```

or, using a recursive definition,

```
param s {j in 1..N} := if j = 1 then 1 else s[j-1] + j;
```

This exercise asks you to play with some other possibilities.

(a) Define `fact[n]` to be n factorial, the product of the first n integers. Give both a recursive and a nonrecursive definition as above.

(b) The Fibonacci numbers are defined mathematically by $f_0 = f_1 = 1$ and $f_n = f_{n-1} + f_{n-2}$. Using a recursive declaration, define `fib[n]` in AMPL to equal the n-th Fibonacci number.

Use another AMPL declaration to verify that the nth Fibonacci number equals the closest integer to $(\frac{1}{2} + \frac{1}{2}\sqrt{5})^n / \sqrt{5}$.

(c) Here's another recursive definition, called Ackermann's function, for positive integers i and j:

$$A(i, 0) = i + 1$$
$$A(0, j + 1) = A(1, j)$$
$$A(i + 1, j + 1) = A(A(i, j + 1), j)$$

Using a recursive declaration, define ack[i,j] in AMPL so that it will equal $A(i, j)$. Use display to print ack[0,0], ack[1,1], ack[2,2] and so forth. What difficulty do you encounter?

(d) What are the values odd[i] defined by the following odd declaration?

```
param odd {i in 1..N} :=
    if i = 1 then 3 else
        min {j in odd[i-1]+2..odd[i-1]*2 by 2:
            not exists {k in 1..i-1} j mod odd[k] = 0} j;
```

Once you've figured it out, create a simpler and more efficient declaration that gives a set rather than an array of these numbers.

(e) A "tree" consists of a collection of nodes, one of which we designate as the "root". Each node except the root has a unique predecessor node in the tree, such that if you work backwards from a node to its predecessor, then to its predecessor's predecessor, and so forth, you always eventually reach the root. A tree can be drawn like this, with the root at the left and an arrow from each node to its successors:

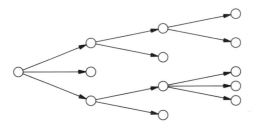

We can store the structure of a tree in AMPL sets and parameters as follows:

```
set NODES;
param Root symbolic in NODES;
param pred {i in NODES diff {Root}} symbolic in NODES diff {i};
```

Every node i, except Root, has a predecessor pred[i].

The depth of a node is the number of predecessors that you encounter on tracing back to the root; the depth of the root is 0. Give an AMPL definition for depth[i] that correctly computes the depth of each node i. To check your answer, apply your definition to AMPL data for the tree depicted above; after reading in the data, use display to view the parameter depth.

An error in the data could give a tree plus a disconnected cycle, like this:

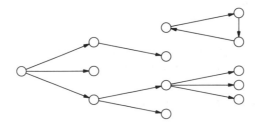

If you enter such data, what will happen when you try to display depth?

8

Linear Programs: Variables, Objectives and Constraints

The best-known kind of optimization model, which has served for all of our examples so far, is the linear program. The variables of a linear program take values from some continuous range; the objective and constraints must use only linear functions of the variables. Previous chapters have described these requirements informally or implicitly; here we will be more specific.

Linear programs are particularly important because they accurately represent many practical applications of optimization. The simplicity of linear functions makes linear models easy to formulate, interpret, and analyze. They are also easy to solve; if you can express your problem as a linear program, even in thousands of constraints and variables, then you can be confident of finding an optimal solution accurately and quickly.

This chapter describes how variables are declared, defines the expressions that AMPL recognizes as being linear in the variables, and gives the rules for declaring linear objectives and constraints. Much of the material on variables, objectives and constraints is basic to other AMPL models as well, and will be used in later chapters.

Because AMPL is fundamentally an algebraic modeling language, this chapter concentrates on features for expressing linear programs in terms of algebraic objectives and constraints. For linear programs that have certain special structures, such as networks, AMPL offers alternative notations that may make models easier to write, read and solve. Such special structures are the topics of Chapters 11, 12 and 14.

8.1 Variables

The variables of a linear program have much in common with its numerical parameters. Both are symbols that stand for numbers, and that may be used in arithmetic expressions. Parameter values are supplied by the modeler or computed from other values, while the values of variables are determined by an optimizing algorithm (as implemented in one of the packages that we refer to as solvers).

Syntactically, variable declarations are the same as the parameter declarations defined in Chapter 7, except that they begin with the keyword `var` rather than `param`. The meaning of qualifying phrases within the declaration may be different, however, when these phrases are applied to variables rather than to parameters.

Phrases beginning with >= or <= are by far the most common in declarations of variables for linear programs. They have appeared in all of our examples, beginning with the production model of Figure 1-4:

```
var Make {p in PROD} >= 0, <= market[p];
```

This declaration creates an indexed collection of variables `Make[p]`, one for each member p of the set `PROD`; the rules in this respect are exactly the same as for parameters. The effect of the two qualifying phrases is to impose a restriction, or constraint, on the permissible values of the variables. Specifically, >= 0 implies that all of the variables `Make[p]` must be assigned nonnegative values by the optimizing algorithm, while the phrase <= `market[p]` says that, for each product p, the value given to `Make[p]` may not exceed the value of the parameter `market[p]`.

In general, either >= or <= may be followed by any arithmetic expression in previously defined sets and parameters and currently defined dummy indices. Most linear programs are formulated in such a way that every variable must be nonnegative; an AMPL variable declaration can specify nonnegativity either directly by >= 0, or indirectly as in the diet model of Figure 5-1, which is repeated in Figure 8-1:

```
param f_min {FOOD} >= 0;
param f_max {j in FOOD} >= f_min[j];

var Buy {j in FOOD} >= f_min[j], <= f_max[j];
```

The values following >= and <= are lower and upper *bounds* on the variables. Because these bounds represent a kind of constraint, they could just as well be imposed by the constraint declarations described later in this chapter. By placing bounds in the `var` declaration instead, you may be able to make the model shorter or clearer, although you will not make the optimal solution any different or easier to find. Some solvers do treat bounds specially in order to speed up their algorithms, but with AMPL all bounds are identified automatically, no matter how they are expressed in your model.

Variable declarations may not use the comparison operators <, > or <> in restriction phrases. For linear programming it makes no sense to constrain a variable to be, say, < 3, since it could always be chosen as 2.99999... or as close to 3 as you like.

The = operator is allowed in a variable declaration. If you were to write, for example,

```
var Buy {j in FOOD} = f_min[j];
```

then each variable `Buy[j]` would have to equal the value of `f_min[j]` in any optimal solution; in linear programming jargon, these would be "fixed" variables. AMPL does not treat an = phrase as specifying a constraint or a bound, but rather uses it to substitute the variables out of the linear program. In our example, `f_min[j]` would be substituted

```
set MINREQ;    # nutrients with minimum requirements
set MAXREQ;    # nutrients with maximum requirements

set NUTR := MINREQ union MAXREQ;    # nutrients
set FOOD;                           # foods

param cost {FOOD} > 0;
param f_min {FOOD} >= 0;
param f_max {j in FOOD} >= f_min[j];

param n_min {MINREQ} >= 0;
param n_max {MAXREQ} >= 0;

param amt {NUTR,FOOD} >= 0;

var Buy {j in FOOD} >= f_min[j], <= f_max[j];

minimize total_cost:  sum {j in FOOD} cost[j] * Buy[j];

subject to diet_min {i in MINREQ}:
   sum {j in FOOD} amt[i,j] * Buy[j] >= n_min[i];

subject to diet_max {i in MAXREQ}:
   sum {j in FOOD} amt[i,j] * Buy[j] <= n_max[i];
```

Figure 8-1: Diet model from Chapter 5 (dietu.mod).

for every occurrence of Buy[j] in subsequent declarations. Since only a substitution is involved, AMPL can allow references to previously defined variables in an = phrase. This feature is more useful for nonlinear than linear programming, however, so we will defer discussion of it to Chapter 13.

For developing and analyzing a model, = phrases are a convenient way to fix certain variables temporarily. If you want some variables to be fixed in all runs, however, then they are not really variables; you should instead treat them as parameters whose values are assigned by a := phrase, as explained in Section 7.5. On the other hand, you would hardly want to fix all of the Buy variables, since they are the only variables in the diet model! Instead you would do best to go back to the original declaration,

```
var Buy {j in FOOD} >= f_min[j], <= f_max[j];
```

and let f_min[j] equal f_max[j] in the data for those foods whose purchase amounts you want to fix. AMPL automatically recognizes that a variable is fixed when its upper and lower bounds are the same.

All of the phrases described above are concerned with restrictions on the values of the variables in any optimal solution. They do not assign values to the variables — that is the job of the optimizing algorithm. If you want to assign *initial* values to the variables, you can do so in the same ways as for parameters: by adding a := phrase as explained in Section 7.5, or by specifying values with the data as described in Chapter 9. Initial values of variables are normally changed — ideally to optimal values — when a solver is invoked. Still, you may want to set the initial values to some obvious or known solution, in order to evaluate the objective function or other quantities before asking the solver to try to do better.

As an example of assigning values to variables, suppose we declare the variables of Figure 8-1 as

```
var Buy {j in FOOD} >= f_min[j], <= f_max[j], := f_min[j];
```

This initializes all variables to their minimum values; we can use display to show the total cost and nutrient supplies at these levels, *before* we solve the problem:

```
ampl: model dietu1.mod;
ampl: data dietu.dat;

ampl: display total_cost;
total_cost = 38.64

ampl: display Buy;

Buy [*] :=
BEEF   2
 CHK   2
FISH   2
 HAM   2
 MCH   2
 MTL   2
 SPG   2
 TUR   2
;

ampl: display diet_min.body, diet_min.lb;
:    diet_min.body diet_min.lb    :=
A            572          700
B1           300            0
B2           220            0
C            410          700
CAL         6940        16000
;
```

Here we can see that if we buy only the minimum amount of each food, which costs $38.64, the diet is short of the requirements for vitamin A, vitamin C and calcium. (The expression diet_min.body refers to the total supplies of nutrients as expressed in the constraints diet_min, while diet_min.lb gives the nutrient lower bounds specified by the constraints; the use of these and other qualified names is explained in more detail in Section 10.7.)

After viewing this information, we can proceed to solve the linear program:

```
ampl: solve;
OSL 1.2: optimal solution; objective 74.27382022
3 simplex iterations
```

```
ampl: display Buy;
Buy [*] :=
BEEF   2
 CHK   10
FISH   2
 HAM   2
 MCH   2
 MTL   6.23596
 SPG   5.25843
 TUR   2
;

ampl: display diet_min.body, diet_min.lb;
:    diet_min.body diet_min.lb    :=
A       1013.98         700
B1        605             0
B2        492.416         0
C         700           700
CAL     16000         16000
;
```

At this point the variables have been changed to their optimal values, and all constraints are satisfied.

Initial values can also serve as a "starting guess" for some optimizers, which is particularly important in nonlinear programming; this is discussed further in Chapter 13.

Finally, variables may be declared as `integer` so that they must take whole number values in any optimal solution, or as `binary` so that they may only take the values 0 and 1. Models that contain any such variables are integer programs, which are the topic of Chapter 15.

8.2 Linear expressions

An arithmetic expression is *linear* in a given variable if, for every unit increase or decrease in the variable, the value of the expression increases or decreases by some fixed amount. An expression that is linear in all its variables is called a linear expression.

AMPL recognizes as a linear expression any sum of terms of the form

> *constant-expr*
> *var-ref*
> (*constant-expr*) * *variable-ref*

provided that each *constant-expr* is an arithmetic expression that contains no variables, while *var-ref* is a reference (possibly subscripted) to a variable. The parentheses around the *constant-expr* may be omitted if the result is the same according to the rules of operator precedence (Table 7-1). The following examples, from the constraints in the multi-period production model of Figure 6-3, are all linear expressions under this definition:

```
avail[t]
Make[p,t] + Inv[p,t-1]
sum {p in PROD} (1/rate[p]) * Make[p,t]
sum {a in AREA[p]} Sell[p,a,t] + Inv[p,t]
```

The model's objective,

```
sum {p in PROD, t in 1..T}
   (sum {a in AREA[p]} revenue[p,a,t] * Sell[p,a,t] -
      prodcost[p] * Make[p,t] - invcost[p] * Inv[p,t])
```

is also linear because subtraction of a term is the addition of its negative, and a sum of sums is itself a sum.

Various kinds of expressions are equivalent to a sum of terms of the forms above, and are also recognized as linear by AMPL. Division by an arithmetic expression is equivalent to multiplication by its inverse, so

```
(1/rate[p]) * Make[p,t]
```

may be written in a linear program as

```
Make[p,t] / rate[p]
```

The order of multiplications is irrelevant, so the *variable-ref* need not come at the end of a term; for instance,

```
revenue[p,a,t] * Sell[p,a,t]
```

is equivalent to

```
Sell[p,a,t] * revenue[p,a,t]
```

As an example combining these principles, imagine that revenue[p,a,t] is in dollars per metric ton, while Sell remains in tons. If we define conversion factors

```
param mt_t := 0.90718474;    # metric tons per ton
param t_mt := 1 / mt_t;      # tons per metric ton
```

then both

```
sum {a in AREA[p]} mt_t * revenue[p,a,t] * Sell[p,a,t]
```

and

```
sum {a in AREA[p]} revenue[p,a,t] * Sell[p,a,t] / t_mt
```

are linear expressions for total revenue.

To continue our example, if costs are also in dollars per metric ton, the objective could be written as

```
mt_t * sum {p in PROD, t in 1..T}
   (sum {a in AREA[p]} revenue[p,a,t] * Sell[p,a,t] -
      prodcost[p] * Make[p,t] - invcost[p] * Inv[p,t])
```

or as

```
sum {p in PROD, t in 1..T}
    (sum {a in AREA[p]} revenue[p,a,t] * Sell[p,a,t] -
        prodcost[p] * Make[p,t] - invcost[p] * Inv[p,t]) / t_mt
```

Multiplication and division distribute over any summation to yield an equivalent linear sum of terms. Notice that in the first form, `mt_t` multiplies the entire `sum {p in PROD, t in 1..T}`, while in the second `t_mt` divides only the summand that follows `sum {p in PROD, t in 1..T}`, because the `/` operator has higher precedence than the sum operator. In these examples the effect is the same, however.

Finally, an `if-then-else` operator produces a linear result if the expressions following `then` and `else` are both linear. The following example appeared in Section 7.3:

```
Make[j,t] + (if t = first(WEEKS)
                then inv0[j] else Inv[j,prev(t)])
```

The variables in a linear expression may not appear as the operands to any other operators, or in the arguments to any functions. This rule applies to iterated operators like `max`, `min`, `abs`, `forall`, and `exists`, as well as `**` and standard numerical functions such as `sqrt`, `log`, and `cos`.

To summarize, a linear expression may be any sum of terms in the forms

> *constant-expr*
> *var-ref*
> (*constant-expr*) `*` (*linear-expr*)
> (*linear-expr*) `*` (*constant-expr*)
> (*linear-expr*) `/` (*constant-expr*)
> `if` *logical-expr* `then` *linear-expr* `else` *linear-expr*

where *constant-expr* is any arithmetic expression that contains no references to variables, and *linear-expr* is any other (simpler) linear expression. Parentheses may be omitted if the result is the same by the rules of operator precedence in Table 7-1. AMPL automatically performs the transformations that convert any such expression to a simple sum of linear terms.

8.3 Objectives

The declaration of an objective function consists of one of the keywords `minimize` or `maximize`, a name, a colon, and a linear expression in previously defined sets, parameters and variables. We have seen examples such as

```
minimize total_cost:  sum {j in FOOD} cost[j] * Buy[j];
```

and

```
maximize total_profit:
    sum {p in PROD, t in 1..T}
        (sum {a in AREA[p]} revenue[p,a,t] * Sell[p,a,t] -
            prodcost[p] * Make[p,t] - invcost[p] * Inv[p,t]);
```

```
set MINREQ;     # nutrients with minimum requirements
set MAXREQ;     # nutrients with maximum requirements

set NUTR := MINREQ union MAXREQ;    # nutrients
set FOOD;                           # foods
set STORE;                          # stores

param cost {STORE,FOOD} > 0;
param f_min {FOOD} >= 0;
param f_max {j in FOOD} >= f_min[j];

param n_min {MINREQ} >= 0;
param n_max {MAXREQ} >= 0;

param amt {NUTR,FOOD} >= 0;

var Buy {j in FOOD} >= f_min[j], <= f_max[j];

minimize total_cost {s in STORE}:
   sum {j in FOOD} cost[s,j] * Buy[j];

minimize total_number:  sum {j in FOOD} Buy[j];

subject to diet_min {i in MINREQ}:
   sum {j in FOOD} amt[i,j] * Buy[j] >= n_min[i];

subject to diet_max {i in MAXREQ}:
   sum {j in FOOD} amt[i,j] * Buy[j] <= n_max[i];
```

Figure 8-2: Diet model with several objectives (diet.obj.mod).

The name of the objective plays no further role in the model, with the exception of certain "columnwise" declarations to be introduced in Chapters 11 and 12. Within AMPL's command environment, the objective's name refers to its value; thus in one of the examples in Section 8.1 above, we were able to issue the command display total_cost to show the cost of all foods.

Although a particular linear program must have one objective function, a model may contain more than one declaration of an objective. Moreover, any minimize or maximize declaration may define an indexed collection of objective functions, by including an indexing expression after the objective name. In these cases, you may issue an objective command, before typing solve, to indicate which objective is to be optimized.

As an example, we can expand the diet model of Figure 8-1 so that costs are indexed over stores as well as foods, as shown in Figure 8-2. Then the model can declare a variety of objectives:

```
minimize total_cost {s in STORE}:
   sum {j in FOOD} cost[s,j] * Buy[j];

minimize total_number:
   sum {j in FOOD} Buy[j];
```

For each member of the set STORE, the first declaration creates a separate objective that is the cost of the food if you do your shopping in that store. The second declaration gives a different objective that minimizes the total number of packages purchased, regardless of price.

To show how this model might be used, we need only augment the data to give the membership of STORE and the costs of foods at different stores:

```
set STORE := "A&P" JEWEL VONS ;

param cost:  BEEF   CHK  FISH   HAM   MCH   MTL   SPG   TUR :=
     "A&P"   3.19  2.59  2.29  2.89  1.89  1.99  1.99  2.49
     JEWEL   3.09  2.79  2.29  2.59  1.59  1.99  2.09  2.30
      VONS   2.59  2.99  2.49  2.69  1.99  2.29  2.00  2.69 ;
```

To minimize total cost at the A&P, we use the objective command to select the objective function named total_cost["A&P"]:

```
ampl: model dietobj.mod; data dietobj.dat;
ampl: objective total_cost["A&P"];
ampl: solve;
OSL 1.2: optimal solution; objective 74.27382022
3 simplex iterations

ampl: display total_cost,total_number;
total_cost [*] :=
'A&P'  74.2738
JEWEL  75.0197
 VONS  79.5972
;

total_number = 31.4944
```

The display command shows the values of all objectives at the resulting solution, even though only one has been minimized. ('A&P' and "A&P" are the same string.) By issuing a different objective command, we can minimize total_number instead:

```
ampl: objective total_number;
ampl: solve;
OSL 1.2: optimal solution; objective 30.92537313
4 simplex iterations

ampl: display total_cost,total_number;
total_cost [*] :=
'A&P'  78.2057
JEWEL  77.0675
 VONS  81.1133
;

total_number = 30.9254
```

As you would expect, while the number of packages goes down, the cost at each of the stores is higher than in the previous run.

As another example, here's how we could experiment with different optimal solutions for the office assignment problem of Figure 3-2. First we solve the original problem:

```
ampl: model transp.mod; data assign.dat; solve;
CPLEX 2.0: optimal solution; objective 28
8 iterations (2 in phase I)

ampl: option display_1col 1000, omit_zero_rows 1;
ampl: option display_eps .000001;

ampl: display total_cost,
ampl?    {i in ORIG, j in DEST} cost[i,j] * Trans[i,j];
total_cost = 28

cost[i,j]*Trans[i,j] :=
Coullard 1021    6
Hazen    2053    4
Hopp     1055    1
Hurter   2009    1
Jones    1049    2
Mehrotra 1083    3
Rieders  2019    2
Spearman 1053    4
Sun      2083    1
Tamhane  1087    3
Zazanis  2087    1
;
```

To keep the objective value at this optimal level while we experiment, we add a constraint that fixes the expression for the objective equal to the current value, 28:

```
ampl: subject to stay_optimal:
ampl?    sum {i in ORIG, j in DEST} cost[i,j] * Trans[i,j] = 28;
```

Next, recall that cost[i,j] is the ranking that person i has given to office j, while Trans[i,j] is set to 1 if it's optimal to put person i in office j, or 0 otherwise. Thus

```
sum {j in DEST} cost[i,j] * Trans[i,j]
```

always equals the ranking of person i for the office to which i is assigned. We declare this expression to be the new objective function pref_of[i]:

```
ampl: minimize pref_of {i in ORIG}:
ampl?    sum {j in DEST} cost[i,j] * Trans[i,j];
```

Now we can select any one person and optimize his or her ranking in the assignment:

```
ampl: objective pref_of["Coullard"];
ampl: solve;
CPLEX 2.0: optimal solution; objective 3
29 iterations (23 in phase I)
```

Looking at the new assignment, we see that the original objective is unchanged, and that the selected individual's situation is in fact improved, although of course at the expense of others:

```
ampl: display total_cost,
ampl?    {i in ORIG, j in DEST} cost[i,j] * Trans[i,j];
total_cost = 28

cost[i,j]*Trans[i,j] :=
Coullard 2053    3
Hazen    2009    1
Hopp     1055    1
Hurter   1087    5
Jones    1049    2
Mehrotra 1083    3
Rieders  2019    2
Spearman 1053    4
Sun      2083    1
Tamhane  1021    5
Zazanis  2087    1
;
```

We were able to make this change because there are several optimal solutions to the original total-ranking objective. A solver arbitrarily returns one of these, but by use of a second objective we can force it toward others.

8.4 Constraints

The simplest kind of constraint declaration begins with the keywords subject to, a name, and a colon. Even the subject to is optional; AMPL assumes that any declaration not beginning with a keyword is a constraint. Following the colon is an algebraic description of the constraint, in terms of previously defined sets, parameters and variables. Thus in the production model introduced by Figure 1-4, we have the following constraint imposed by limited processing time:

```
subject to Time:
    sum {p in PROD} (1/rate[p]) * Make[p] <= avail;
```

The name of a constraint, like the name of an objective, is not used anywhere else in an algebraic model, though it figures in alternative "columnwise" formulations (Chapter 12) and is used in the AMPL command environment to specify the constraint's dual value and other associated quantities (Chapter 10).

Most of the constraints in large linear programming models are defined as indexed collections, by giving an indexing expression after the constraint name. The constraint Time, for example, is generalized in subsequent examples to say that the production time may not exceed the time available in each processing stage s (Figure 1-6a):

```
subject to Time {s in STAGE}:
    sum {p in PROD} (1/rate[p,s]) * Make[p] <= avail[s];
```

or in each week t (Figure 4-4):

```
subject to Time {t in 1..T}:
    sum {p in PROD} (1/rate[p]) * Make[p,t] <= avail[t];
```

Another constraint from the latter example says that production, sales and inventories must balance for each product `p` in each week `t`:

```
subject to Balance {p in PROD, t in 1..T}:
    Make[p,t] + Inv[p,t-1] = Sell[p,t] + Inv[p,t];
```

A constraint declaration can specify any valid indexing expression, which defines a set (as explained in Chapters 5 and 6); there is one constraint for each member of this set. The constraint name can be subscripted, so that `Time[1]` or `Balance[p,t+1]` refers to a particular constraint from an indexed collection.

The indexing expression in a constraint declaration should specify a number of dummy indices (like `s`, `t` and `p` in the preceding examples) that equals the dimension of the indexing set. Then when the constraint corresponding to a particular set member is processed by AMPL, the dummy indices take their values from that member. This use of dummy indices is what permits a single constraint expression to represent many constraints; the indexing expression is AMPL's translation of a phrase such as "for all products p and weeks $t = 1$ to T" that might be seen in an algebraic statement of the model.

By using more complex indexing expressions, you can specify more precisely the constraints to be included in a model. Consider, for example, the following variation on the production time constraint:

```
subject to Time {t in 1..T: avail[t] > 0}:
    sum {p in PROD} (1/rate[p]) * Make[p,t] <= avail[t];
```

This says that if `avail[t]` is specified as zero in the data for any week `t`, it is to be interpreted as meaning "no constraint on time available in week `t`" rather than "limit of zero on time available in week `t`". In the simpler case where there is just one `Time` constraint not indexed over weeks, you can specify an analogous conditional definition as follows:

```
subject to Time {if avail > 0}:
    sum {p in PROD} (1/rate[p]) * Make[p] <= avail;
```

The pseudo-indexing expression `{if avail > 0}` causes one constraint, named `Time`, to be generated if the condition `avail > 0` is true, and no constraint at all to be generated if the condition is false. The same notation can be used to conditionally define an unindexed variable, objective, set, parameter, or `check`.

AMPL's algebraic description of a constraint may consist of any two linear expressions separated by an equality or inequality operator:

> *linear-expr* `<=` *linear-expr*
> *linear-expr* `=` *linear-expr*
> *linear-expr* `>=` *linear-expr*

While it is customary in mathematical descriptions of linear programming to place all terms containing variables to the left of the operator and all other terms to the right (as in constraint `Time`), AMPL imposes no such requirement (as seen in constraint `Balance`).

Convenience and readability should determine what terms you place on each side of the operator.

AMPL also allows double-inequality constraints such as the following from the diet model of Figure 2-1:

```
subject to diet {i in NUTR}:
    n_min[i] <= sum {j in FOOD} amt[i,j] * Buy[j] <= n_max[i];
```

This says that the middle expression, the amount of nutrient i supplied by all foods, must be greater than or equal to `n_min[i]` and also less than or equal to `n_max[i]`. The permissible forms for a constraint of this kind are

const-expr <= *linear-expr* <= *const-expr*
const-expr >= *linear-expr* >= *const-expr*

where *const-expr* may contain no variables. The effect is to give upper and lower bounds on the value of the *linear-expr*. If your model requires variables in the left-hand or right-hand *const-expr*, you must define two different constraints in separate declarations.

For most applications of linear programming, you need not worry about the form of the constraints. If you simply write the constraints in the most convenient way, they will be recognized as proper linear constraints according to the rules in this chapter. There do exist situations, however, in which your choice of formulation will determine whether AMPL recognizes your model as linear. Imagine that we want to further constrain the production model so that no product p may represent more than a certain fraction of total production. We define a parameter `max_frac` to represent the limiting fraction; the constraint then says that production of p divided by total production must be less than or equal to `max_frac`:

```
subject to Limit {p in PROD}:
    Make[p] / sum {q in PROD} Make[q] <= max_frac;
```

This is not a linear constraint to AMPL, because its left-hand expression contains a division by a sum of variables. But if we rewrite it as

```
subject to Limit {p in PROD}:
    Make[p] <= max_frac * sum {q in PROD} Make[q];
```

then AMPL does recognize it as linear.

AMPL simplifies and canonicalizes constraints as it prepares the model and data for handing to a solver. For example, it may combine linear terms involving the same variable, move variables from one side of a constraint to the other, or eliminate variables fixed at a value. Entire constraints may be dropped because they can be combined with the simple bounds on the variables, or because a mathematical test shows that they are implied by other constraints. This work is carried out by a "presolve" phase that is discussed further in Section 10.2. To see a summary of the actions of presolve and of the size of the resulting linear program, give the command `option show_stats 1`. If you want AMPL to generate the constraints exactly as you wrote them, without simplification, you can turn off the presolve phase with the command `option presolve 0`.

8.5 Syntax summary

A reference to a variable is written

> *var-ref*:
>> *var-name*
>> *var-name* [*expr*]
>> *var-name* [*expr*$_1$, *expr*$_2$, ...]

where the *var-name* has been defined by a previous variable declaration, and *expr* or (*expr*$_1$, *expr*$_2$, ...) evaluates to a member of the set over which the variable is indexed.

A variable declaration has the form

> *var-declaration*:
>> var *var-name* *alias*$_{opt}$ *index-expr*$_{opt}$ *var-attrib*$_{opt}$;

where *var-name* is an AMPL name, *alias* is an optional string, and *index-expr* is an optional indexing expression as defined in Chapters 5 and 6. The optional *var-attrib* may include:

> *var-attrib*:
>> binary
>> integer
>> >= *const-expr*
>> <= *const-expr*
>> = *arith-expr*
>> := *const-expr*
>> default *const-expr*

Each attribute may be specified at most once in a declaration, with successive attributes optionally separated by commas. The following attributes are mutually exclusive: binary and integer, >= or <= and =, and := and default. An *arith-expr* is an arithmetic expression as defined in Chapter 7, while a *const-expr* is the same but contains no *var-ref*s.

The declaration of an objective has one of the forms

> *objective-declaration*:
>> minimize *objective-name* *alias*$_{opt}$ *index-expr*$_{opt}$: *arith-expr* ;
>> maximize *objective-name* *alias*$_{opt}$ *index-expr*$_{opt}$: *arith-expr* ;

For a linear program, the *arith-expr* must obey the rules for linear expressions given earlier in this chapter.

A constraint declaration has the form

> *constraint-declaration*:
>> subject to *constraint-name* *alias*$_{opt}$
>>> *index-expr*$_{opt}$ *constraint-attrib*$_{opt}$: *constraint-expr* ;

The optional *constraint-attrib* has the form

> *constraint-attrib*:
>> := *const-expr*
>> default *const-expr*

to specify initial dual values associated with the constraints.

The *constraint-expr* has one of the following forms:

constraint-expr:
        arith-expr <= *arith-expr*
        arith-expr = *arith-expr*
        arith-expr >= *arith-expr*

        const-expr <= *arith-expr* <= *const-expr*
        const-expr >= *arith-expr* >= *const-expr*

For a linear program, the *arith-exprs* must obey the rules for linear expressions.

Exercises

8-1. In the diet model of Figure 5-1, add a := phrase to the var declaration (as explained in Section 8.1) to initialize each variable to a value midway between its lower and upper bounds.

Read this model into AMPL along with the data from Figure 5-2. Using display commands, determine which constraints (if any) the initial solution fails to satisfy, and what total cost this solution gives. Is the total cost more or less than the optimal total cost?

8-2. This exercise asks you to reformulate various kinds of constraints to make them linear.

(a) The following constraint says that the inventory Inv[p,t] for product p in any period t must not exceed the smallest one-period production Make[p,t] of product p:

```
subject to Inv_Limit {p in PROD, t in 1..T}:
    Inv[p,t] <= min {tt in 1..T} Make[p,tt];
```

This constraint is not recognized as linear by AMPL, because it applies the min operator to variables. Formulate a linear constraint that has the same effect.

(b) The following constraint says that the change in total inventories from one period to the next may not exceed a certain parameter max_change:

```
subject to Max_Change {t in 1..T}:
    abs(sum {p in PROD} Inv[p,t-1] - sum {p in PROD} Inv[p,t])
        <= max_change;
```

This constraint is not linear because it applies the abs function to an expression involving variables. Formulate a linear constraint that has the same effect.

(c) The following constraint says that the ratio of total production to total inventory in a period may not exceed max_inv_ratio:

```
subject to Max_Inv_Ratio {t in 1..T}:
    (sum {p in PROD} Inv[p,t]) / (sum {p in PROD} Make[p,t])
        <= max_inv_ratio;
```

This constraint is not linear because it divides one sum of variables by another. Formulate a linear constraint that has the same effect.

(d) What can you say about formulation of an alternative linear constraint for the following cases?

– In (a), min is replaced by max.

– In (b), <= max_change is replaced by >= min_change.

– In (c), the parameter max_inv_ratio is replaced by a new variable, Ratio[t].

8-3. This exercise deals with possibilities for using more than one objective function in a model.

(a) Continuing the diet example of Section 8.3, use AMPL to find the minimum-cost diets for the remaining two stores. Which optimal solution uses the fewest packages?

(b) Continuing the assignment example of Section 8.3, what is the best-ranking office that you can assign to each individual, given that the total of the rankings must stay at the optimal value of 28? How many different optimal assignments do there seem to be, and which individuals get different offices in different assignments?

(c) Modify the assignment example so that it will find the best-ranking office that you can assign to each individual, given that the total of the rankings may increase from 28, but may not exceed 30.

(d) After making the modification suggested in (c), the person in charge of assigning offices has tried again to minimize the objective pref_of["Coullard"]. This time, the reported solution is as follows:

```
ampl: display total_cost,
ampl?    {i in ORIG,  j in DEST} cost[i,j]*Trans[i,j];
total_cost = 30

cost[i,j]*Trans[i,j]  :=
Coullard 2087    1
Hazen    2053    4
Hopp     1055    1
Hurter   1087    2.5
Hurter   2009    0.5
Jones    1049    2
Mehrotra 1083    3
Rieders  2019    2
Spearman 1053    4
Sun      2083    1
Tamhane  1021    5
Zazanis  1087    2.5
Zazanis  2009    1.5
;
```

Coullard is now assigned her first choice, but what is the difficulty with the overall solution? Why doesn't it give a useful resolution to the assignment problem as we have stated it?

8-4. Return to the assignment version of the transportation model, in Figures 3-1a and 3-2.

(a) Add parameters worst[i] for each i in ORIG, and constraints saying that Trans[i,j] must equal 0 for every combination of i in ORIG and j in DEST such that cost[i,j] is greater than worst[i]. (See the constraint Time in Section 8.4 for a similar example.) In the assignment interpretation of this model, what do the new constraints mean?

(b) Use the model from (a) to show that there is an optimal solution, with the objective equal to 28, in which no one gets an office worse than their fifth choice.

(c) Use the model from (a) to show that at least one person must get an office worse than fourth choice.

(d) Use the model from (a) to show that if you give Spearman his first choice, without any restrictions on the other individuals' choices, the objective cannot be made smaller than 31. Determine similarly how small the objective can be made if each other individual is given first choice.

9

Specifying Data

As we emphasize throughout this book, there is a distinction between an AMPL model for an optimization problem, and the data values that define a particular instance of the problem. Chapters 5 through 8 have focused on the declarations of sets, parameters, variables, objectives and constraints that are necessary to describe models. In this chapter we take a closer look at the statements that specify the data.

AMPL's data statements offer several formats for set and parameter values, making use of one-dimensional lists and two-dimensional tables. Some formats are most naturally created and maintained in a text editing or word processing environment; other formats are particularly easy to generate from data-handling programs such as database management systems and spreadsheets. AMPL's display command (Chapter 10) also produces output in these formats. Wherever possible, similar syntax and concepts are used for both sets and parameters.

Examples of AMPL data statements appear in almost every chapter of this book. This chapter explains the various formats systematically, first for set member data, and then for parameter values. A short final section shows how the parameter data format can also be used to assign initial values to variables. A summary of the data statement rules appears in Section A.12 of the Appendix.

AMPL reads data statements in a "data mode" that is initiated by the data command; for example, the command

    ampl: *data diet.dat;*

reads data statements from a file named diet.dat. The data command is discussed in Section 10.2.

While reading data statements, AMPL treats any sequence of space, tab and "newline" characters as a single space. Judicious use of these characters can help to arrange data into easy-to-read tables; our examples use a combination of spaces and newlines. If your data statements are produced as output from other data management software and are sent directly to AMPL, however, then you may ignore the niceties of arrangement and use whatever format is convenient.

The end of any statement in data mode is indicated by a semicolon.

9.1 Set data

We begin with sets of simple (one-dimensional) objects, for which the only option is to list the members. Then we proceed to sets of (two-dimensional) pairs, whose membership may be specified in lists, slices or tables. These options are finally extended to higher-dimensional cases, with a set of triples serving as an example.

Rather than give the membership of a set explicitly in any format, you may instruct AMPL to infer the members from a table of numerical data. This alternative is discussed in Section 9.2.

One-dimensional sets

A simple set is specified by listing its members. In data for the steel production model of Figure 4-7, for example, the sets of origins, destinations, and products are given by

```
set ORIG := GARY CLEV PITT ;
set DEST := FRA DET LAN WIN STL FRE LAF ;
set PROD := bands coils plate ;
```

If a set has been declared with the attribute `ordered` or `circular` (Chapter 5), you must list its members in order; thus the set `WEEKS` in Figure 5-4 is given as

```
set WEEKS := 27sep93 04oct93 11oct93 18oct93 ;
```

If a string in the list includes any character other than letters, digits, underscores, period, + and −, it must be enclosed in quotes, as in this example from Section 8.3:

```
set STORE := "A&P" JEWEL VONS ;
```

Also a string that looks like a number (for example `"+1"` or `"3e4"`) must be quoted, to distinguish it from a member that is actually a number. You may use a pair of either single quotes (`'A&P'`) or double quotes (`"A&P"`), unless the string contains a quote, in which case the other kind of quote must surround it (`"DOMINICK'S"`) or the surrounding quote must be doubled inside (`"DOMINICK""S"`).

Members of a set must all be different; AMPL will warn you of duplicates:

```
ampl: set DEST := FRA DET LAN WIN STL FRE LAF DET;

duplicate member DET for set DEST
context:  set DEST := FRA DET LAN WIN STL FRE LAF DET >>> ; <<<
```

The context bracketed by >>> and <<< isn't the exact point of the error, but the message makes the situation clear. Numbers that have the same representation in your computer are considered to be the same, so that for example 2, 2.00, 2.e0 and 0.02E+2 all denote the same member.

For an indexed collection of sets, the members of each set in the collection are specified individually. In the example of Figure 6-3, the sets named `AREA` are indexed by the set `PROD`:

```
set PROD;          # products
set AREA {PROD};   # market areas for each product
```

The membership of these sets is given in Figure 6-4 by:

```
set PROD := bands coils ;
set AREA[bands] := east north ;
set AREA[coils] := east west export ;
```

You may specify explicitly that one or more of these sets is empty, by giving an empty list; put the semicolon right after the `:=` operator. If you want AMPL to assume that each set is empty unless specified otherwise, include a default phrase to this effect in the model:

```
set AREA {PROD} default {};
```

Otherwise you will be warned about any set whose specification is missing.

Two-dimensional sets

For a set of ordered pairs, the membership may be specified in a variety of ways. As an example, consider the following sets from Figure 6-2a:

```
set ORIG;    # origins
set DEST;    # destinations

set LINKS within {ORIG,DEST};    # transportation links
```

The members of ORIG and DEST are as given earlier in this section. In the simplest format, the membership of LINKS may be specified as a list of tuples such as you would find in a model's indexing expressions,

```
set LINKS :=
    (GARY,DET)  (GARY,LAN)  (GARY,STL)  (GARY,LAF)  (CLEV,FRA)
    (CLEV,DET)  (CLEV,LAN)  (CLEV,WIN)  (CLEV,STL)  (CLEV,LAF)
    (PITT,FRA)  (PITT,WIN)  (PITT,STL)  (PITT,FRE)  ;
```

or as a list of pairs, without the parentheses and commas:

```
set LINKS :=
    GARY DET    GARY LAN    GARY STL    GARY LAF    CLEV FRA
    CLEV DET    CLEV LAN    CLEV WIN    CLEV STL    CLEV LAF
    PITT FRA    PITT WIN    PITT STL    PITT FRE ;
```

The order of members within each pair is significant — the first must be from ORIG, and the second from DEST — but the pairs themselves may appear in any order.

A set of pairs may also be specified in a table:

```
set LINKS: FRA DET LAN WIN STL FRE LAF :=
        GARY  -   +   +   -   +   -   +
        CLEV  +   +   +   +   +   -   +
        PITT  +   -   -   +   +   +   -   ;
```

A + indicates a pair that is in the set, and a − indicates a pair that is not. Normally the rows are labeled with the first component, and the columns with the second. If you prefer the opposite, you can indicate a transposed table by adding (tr) after the set name:

```
set LINKS (tr):
                GARY CLEV PITT :=
        FRA      −    +    +
        DET      +    +    −
        LAN      +    +    −
        WIN      −    +    +
        STL      +    +    +
        FRE      −    −    +
        LAF      +    +    −   ;
```

Tables tend to be more convenient for sets that are relatively dense, in the sense that a fairly large fraction of the pairs are in the set. Otherwise the list-of-pairs form works better; it is also easier to generate from a program.

   Yet another way to describe a set of pairs is by listing all second components that go with each first component:

```
set LINKS :=
    (GARY,*) DET LAN STL LAF
    (CLEV,*) FRA DET LAN WIN STL LAF
    (PITT,*) FRA WIN STL FRE ;
```

Comparing this to the tabular format above, you can see that each list is a "slice" through a member of ORIG, corresponding to a row of the table or a column of the transposed table.

   We could also specify the set by slicing through the second component, so that each list corresponds to a column of the table or a row of the transposed table:

```
set LINKS :=
    (*,FRA) CLEV PITT    (*,DET) GARY CLEV    (*,LAN) GARY CLEV
    (*,WIN) CLEV PITT    (*,LAF) GARY CLEV    (*,FRE) PITT
    (*,STL) GARY CLEV PITT ;
```

An expression such as (GARY,*) or (*,FRA), resembling a pair but with a component replaced by a *, is a *template* for a slice. Each template is followed by a list, whose entries are substituted for the * to generate pairs in the set.

   A tuple without any *'s, like (GARY,DET), is in effect a template that specifies only itself, so it is not followed by any values; conversely, in a table like

```
set LINKS :=
    GARY DET  GARY LAN  GARY STL  GARY LAF
    CLEV FRA  CLEV DET  CLEV LAN  CLEV WIN  CLEV STL  CLEV LAF
    PITT FRA  PITT WIN  PITT STL  PITT FRE ;
```

a default template (*,*) applies to all entries.

Higher-dimensional sets

The alternatives above extend in natural ways to sets of triples and longer tuples. In Section 6.3, for example, we suggested a version of the multicommodity transportation model that defines a set of triples as follows:

```
set ROUTES within {ORIG,DEST,PROD};
```

If `ORIG` and `DEST` are as above, and `PROD` has members `bands` and `coils`, the membership of `ROUTES` could be given simply by a list of triples, either

```
set ROUTES :=
    (GARY,LAN,coils) (GARY,STL,coils) (GARY,LAF,coils)
    (CLEV,FRA,bands) (CLEV,FRA,coils) (CLEV,DET,bands)
    (CLEV,DET,coils) (CLEV,LAN,bands) (CLEV,LAN,coils)
    (CLEV,WIN,coils) (CLEV,STL,bands) (CLEV,STL,coils)
    (CLEV,LAF,bands) (PITT,FRA,bands) (PITT,WIN,bands)
    (PITT,STL,bands) (PITT,FRE,bands) (PITT,FRE,coils) ;
```

or

```
set ROUTES :=
    GARY LAN coils    GARY STL coils    GARY LAF coils
    CLEV FRA bands    CLEV FRA coils    CLEV DET bands
    CLEV DET coils    CLEV LAN bands    CLEV LAN coils
    CLEV WIN coils    CLEV STL bands    CLEV STL coils
    CLEV LAF bands    PITT FRA bands    PITT WIN bands
    PITT STL bands    PITT FRE bands    PITT FRE coils ;
```

Using templates, we can break the specification into slices through one component. In the following example, we slice through the second component:

```
set ROUTES :=
    (CLEV,*,bands) FRA DET LAN STL LAF
    (PITT,*,bands) FRA WIN STL FRE

    (GARY,*,coils) LAN STL LAF
    (CLEV,*,coils) FRA DET LAN WIN STL
    (PITT,*,coils) FRE ;
```

Because the set contains no members with origin `GARY` and product `bands`, the template `(GARY,*,bands)` is omitted.

When the set's dimension is more than two, the slices can also be through more than one component. Here we slice through both the first and third:

```
set ROUTES :=
    (*,FRA,*)  CLEV bands   CLEV coils   PITT bands
    (*,DET,*)  CLEV bands   CLEV coils
    (*,LAN,*)  GARY coils   CLEV bands   CLEV coils
    (*,WIN,*)  CLEV coils   PITT bands
    (*,STL,*)  GARY coils   CLEV bands   CLEV coils   PITT bands
    (*,FRE,*)  PITT bands   PITT coils
    (*,LAF,*)  GARY coils   CLEV bands ;
```

Since the templates in this case have two *'s in them, they must be followed by pairs of components, which are substituted from left to right to generate the set members. As an alternative, when there are exactly two *'s, the template may be followed by a table:

```
set ROUTES :=

(*,*,bands): FRA DET LAN WIN STL FRE LAF :=
       GARY   -   -   -   -   -   -   -
       CLEV   +   +   +   -   +   -   +
       PITT   +   -   -   +   +   +   -

(*,*,coils): FRA DET LAN WIN STL FRE LAF :=
       GARY   -   -   +   -   +   -   +
       CLEV   +   +   +   +   +   -   -
       PITT   -   -   -   -   -   +   -   ;
```

The table is interpreted as previously explained, to give pairs of components that are substituted for the *'s in the template. The row labels replace the first * and the column labels the second, unless the reverse is specified by the notation (tr) directly after the template. Since the first row above has all − entries, it could be omitted; the same goes for any column that has all − entries.

Looking back to the two simple list formats for this set, you may recognize the first as a sequence of templates having no *'s in them, and the second as a list for the template (*,*,*) that is assumed when no template is given. Thus our examples can be viewed as each using one template arrangement, and one format (either list or table). It is possible, however, to combine different templates and formats in specifying the same set, as long as each member is specified only once. In the end, it is up to you to choose whichever format you find most convenient to generate or maintain.

9.2 Parameter data

For an unindexed (scalar) parameter, a data statement assigns one value:

```
param avail := 40;
```

Most of a typical model's parameters are indexed over sets, however, and their values are specified in a variety of lists and tables. We start with data statements for one or more (one-dimensional) parameters indexed over a single set, optionally with a simultaneous specification of the indexing set itself. Then we consider (two-dimensional) parameters that are indexed over sets of pairs, and that can be represented through a variety of tables, lists and slices. These formats are finally generalized to higher-dimensional parameters.

A special character can be used in any table to indicate entries that are undefined or that take a specified default value. This feature is discussed at the end of the section.

All parameter data statements can also be used to give initial values to variables, as explained in Section 9.3.

One-dimensional parameters

The simplest way to give data for an indexed parameter is by a list. For a parameter indexed over a simple set,

```
set PROD;
param rate {PROD} > 0;
```

each item in the data list consists of a set member and a value:

```
set PROD := bands coils plate ;

param rate :=
        bands   200
        coils   140
        plate   160 ;
```

Line breaks are ignored, and commas are optional, so the param statement could equally well be written

```
param rate := bands 200, coils 140, plate 160 ;
```

Tab and newline characters anywhere in a data statement are treated like spaces, which may be helpful if you are generating your data file from spreadsheet or database software.

Often you need to supply data for several parameters indexed over the same set, such as the parameters rate, profit and market that are all indexed over PROD in Figure 1-4a. Rather than write a separate data statement for each parameter,

```
param rate   := bands   200   coils   140   plate   160 ;
param profit := bands    25   coils    30   plate    29 ;
param market := bands 6000   coils 4000   plate 3500 ;
```

you can combine these statements into one by listing all three parameter names after the keyword param:

```
param:     rate  profit   market :=
   bands    200    25     6000
   coils    140    30     4000
   plate    160    29     3500 ;
```

The colon after param is required; it indicates that several parameters are being given values. If you follow it with the set name PROD and another colon,

```
param: PROD:    rate  profit   market :=
        bands    200    25     6000
        coils    140    30     4000
        plate    160    29     3500 ;
```

then the data for PROD are also taken from this statement, rather from a set PROD statement as previously shown; the effect is to combine the specifications of the set and of the three parameters indexed over it. (As in all param statements, you may place the line breaks anywhere you like; the neat tabular arrangement shown above uses one line for each set member and its associated data, but this is not required.)

Two-dimensional parameters

Data values for a parameter indexed over two sets, such as the shipping cost data from the transportation model of Figure 3-1a,

```
set ORIG;
set DEST;
param cost {ORIG,DEST} >= 0;
```

are most naturally specified in a table (Figure 3-1b):

```
param cost:   FRA  DET  LAN  WIN  STL  FRE  LAF :=
      GARY    39   14   11   14   16   82    8
      CLEV    27    9   12    9   26   95   17
      PITT    24   14   17   13   28   99   20 ;
```

The row labels give the first index and the column labels the second index, so that for example `cost["GARY","FRA"]` is set to 39. So that AMPL can recognize this as a table, a colon must follow the parameter name, while the `:=` operator follows the list of column labels.

For larger index sets, the columns of tables become impossible to view within the width of a single screen or page. To deal with this situation, AMPL offers several alternatives, which we will illustrate on the small table above. When only one of the index sets is uncomfortably large, the table may be transposed so that the column labels correspond to the smaller set:

```
param cost (tr):
        GARY CLEV PITT :=
   FRA   39   27   24
   DET   14    9   14
   LAN   11   12   17
   WIN   14    9   13
   STL   16   26   28
   FRE   82   95   99
   LAF    8   17   20 ;
```

The notation `(tr)` after the parameter name indicates a transposed table, in which the column labels give the first index and the row labels the second index. When both of the index sets are large, either the table or its transpose may be divided up in some way. Since line breaks are ignored, each row may be divided across several lines:

```
param cost:   FRA  DET  LAN  WIN
              STL  FRE  LAF       :=
      GARY    39   14   11   14
              16   82    8
      CLEV    27    9   12    9
              26   95   17
      PITT    24   14   17   13
              28   99   20        ;
```

Or the table may be divided columnwise into several smaller ones:

```
param cost:   FRA   DET   LAN   WIN  :=
        GARY   39    14    11    14
        CLEV   27     9    12     9
        PITT   24    14    17    13

           :   STL   FRE   LAF  :=
        GARY   16    82     8
        CLEV   26    95    17
        PITT   28    99    20 ;
```

A colon indicates the start of each new sub-table; each has the same row labels, but a different subset of the column labels.

In the alternative formulation of this model presented in Figure 6-2a, cost is not indexed over all combinations of members of ORIG and DEST, but over a subset of pairs from these sets:

```
set LINKS within {ORIG,DEST};
param cost {LINKS} >= 0;
```

We showed in the previous section how the membership of this subset could be specified by a table:

```
set LINKS:   FRA   DET   LAN   WIN   STL   FRE   LAF  :=
      GARY    -     +     +     -     +     -     +
      CLEV    +     +     +     +     +     -     +
      PITT    +     -     -     +     +     +     -  ;
```

The values of the parameter indexed over this set can be given in an analogous table:

```
param cost:   FRA   DET   LAN   WIN   STL   FRE   LAF  :=
       GARY    .    14    11     .    16     .     8
       CLEV   27     9    12     9    26     .    17
       PITT   24     .     .    13    28    99     .  ;
```

Where a + indicated a member of the set, the table for cost gives a value. Where a − indicated no member, the table contains a dot (.) as a place-holder. The dot can appear in any AMPL table to indicate "no value specified here". You can use a different symbol, say −−, by including the following statement in the data:

```
defaultsym "--";
```

See the end of this section for a discussion of how unspecified entries can be filled in with a default value.

The contents of the table above can also be given in the simpler list format previously described for one-dimensional data. The only difference is that two set members must be associated with each value:

```
param cost :=
    GARY DET 14   GARY LAN 11   GARY STL 16   GARY LAF  8
    CLEV FRA 27   CLEV DET  9   CLEV LAN 12   CLEV WIN  9
    CLEV STL 26   CLEV LAF 17   PITT FRA 24   PITT WIN 13
    PITT STL 28   PITT FRE 99 ;
```

When a parameter is indexed over a sparse subset of pairs, such a list can be more compact and readable than the tabular representation, which would be mostly dots. The list format is also easier for a program to generate. Another advantage of list format is that, as in the one-dimensional case, the data for several components may be given together:

```
param: LINKS:    cost   limit :=
       GARY DET   14    1000
       GARY LAN   11     800
       GARY STL   16    1200
       GARY LAF    8    1100
       CLEV FRA   27    1200
       CLEV DET    9     600
       CLEV LAN   12     900
       CLEV WIN    9     950
       CLEV STL   26    1000
       CLEV LAF   17     800
       PITT FRA   24    1500
       PITT WIN   13    1400
       PITT STL   28    1500
       PITT FRE   99    1200 ;
```

This table simultaneously gives the membership of LINKS and the values for cost, as well as the values for another parameter, limit, that is also indexed over LINKS.

Finally, the list data for cost may be written more concisely by organizing it into slices, much as we did for the membership of the set LINKS in the previous section. For example, corresponding to this data statement for LINKS,

```
set LINKS :=
    (GARY,*) DET LAN STL LAF
    (CLEV,*) FRA DET LAN WIN STL LAF
    (PITT,*) FRA WIN STL FRE ;
```

there is the following statement for the values of cost:

```
param cost :=
    [GARY,*] DET 14  LAN 11  STL 16  LAF  8
    [CLEV,*] FRA 27  DET  9  LAN 12  WIN  9  STL 26  LAF 17
    [PITT,*] FRA 24  WIN 13  STL 28  FRE 99 ;
```

A template such as [GARY,*] indicates that the ensuing entries will be for values of cost that have a first index of GARY. An entry such as DET 14 thus gives cost["GARY","DET"] a value of 14. Comparing this to the table format for cost, you can see that each slice consists of the numbers in one row of the table. Templates such as [*,DET] would give slices corresponding to the columns.

By analogy with sets, a tuple like [GARY,DET] acts as a template standing for itself. At the other extreme, a list of values like

```
param cost :=
    GARY DET 14  GARY LAN 11  GARY STL 16  GARY LAF  8
    ...
```

has the default template [*,*].

Higher-dimensional parameters

Parameters that have three or more indices are handled in much the same way as lower-dimensional ones, except that they cannot be represented all in one table. The simple list format may be used, or a series of slices may be represented as lists or tables. As an example consider a parameter indexed over a set of triples as suggested in Section 6.3:

```
set ORIG;
set DEST;
set PROD;

set ROUTES within {ORIG,DEST,PROD};
param cost {ROUTES} >= 0;
```

We assume that the members of ROUTES have been given as in the previous section of this chapter, and that now the data values for cost must be specified.

The simple list format works as well for three or more dimensions as for one or two:

```
param cost :=
    CLEV DET bands  9  CLEV DET coils  8  CLEV FRA bands 27
    CLEV FRA coils 23  CLEV LAF bands 17  CLEV LAN bands 12
    CLEV LAN coils 10  CLEV STL bands 26  CLEV STL coils 21
    CLEV WIN coils  9  GARY LAF coils  8  GARY LAN coils 11
    GARY STL coils 16  PITT FRA bands 24  PITT FRE bands 99
    PITT FRE coils 81  PITT STL bands 28  PITT WIN bands 13 ;
```

The number of indices before each value is equal to the dimension of the set, or three in this case. The options for combining two or more list declarations into one table are the same as previously described for lower dimensions.

As in the two-dimensional case, lists of data can be organized by use of slices, but there are more possibilities. If the templates have one *, then each slice is one-dimensional and the lists give one index before each value:

```
param cost :=
    [CLEV,*,bands] FRA 27  DET  9  LAN 12  STL 26  LAF 17
    [PITT,*,bands] FRA 24  WIN 13  STL 28  FRE 99

    [GARY,*,coils] LAN 11  STL 16  LAF  8
    [CLEV,*,coils] FRA 23  DET  8  LAN 10  WIN  9  STL 21
    [PITT,*,coils] FRE 81 ;
```

If the templates have two *'s, then each slice is two-dimensional and the lists give two indices before each value:

```
param cost :=
    [*,FRA,*]  CLEV bands 27  CLEV coils 23  PITT bands 24
    [*,DET,*]  CLEV bands  9  CLEV coils  8
    [*,LAN,*]  GARY coils 11  CLEV bands 12  CLEV coils 10
    [*,WIN,*]  CLEV coils  9  PITT bands 13
    [*,STL,*]  GARY coils 16  CLEV bands 26  CLEV coils 21
               PITT bands 28
    [*,FRE,*]  PITT bands 99  PITT coils 81
    [*,LAF,*]  GARY coils  8  CLEV bands 17 ;
```

The indices are substituted for the *'s from left to right, so that for example the first entry specifies cost ["CLEV", "FRA", "bands"] as 27. By placing the *'s in different positions within the templates, you can slice one-dimensionally in any of three different ways, or two-dimensionally in any of three different ways. (The template [*,*,*] would specify a three-dimensional list like

```
param cost :=
    CLEV DET bands  9  CLEV DET coils  8  CLEV FRA bands 27
    ...
```

as already shown above.) Different kinds of slices may be used together in the same data statement, so long as each parameter is assigned a value only once.

When specifying a two-dimensional slice, you also have the option of giving the data in a table rather than a list:

```
param cost :=

[*,*,bands]: FRA   DET   LAN   WIN   STL   FRE   LAF :=
      CLEV   27     9    12     .    26     .    17
      PITT   24     .     .    13    28    99     .

[*,*,coils]: FRA   DET   LAN   WIN   STL   FRE   LAF :=
      GARY    .     .    11     .    16     .     8
      CLEV   23     8    10     9    21     .     .
      PITT    .     .     .     .     .    81     . ;
```

There must be two *'s in the template for this format; a value's row label is substituted for the first *, and its column label for the second, unless the opposite is specified by (tr) right after the template. You can omit any rows or columns that would have no significant entries, such as the row for GARY in the [*,*,bands] table above.

Your choice of a format for multi-dimensional data has no effect upon how the data values are used in your model. Thus you can choose whatever format you find most convenient. For the cost parameter above, it is appealing to slice along the third index, so that the data values are organized into one shipping-cost table for each product. As another example, consider the revenue parameter from the production model of Figure 6-3:

```
set PROD;          # products
set AREA {PROD};   # market areas for each product
param T > 0;       # number of weeks

param revenue {p in PROD, AREA[p], 1..T} >= 0;
```

Because the index set AREA[p] is potentially different for each product p, slices through the first (PROD) index are most attractive. In our sample data from Figure 6-4, they look like this:

```
param T := 4 ;
set PROD := bands coils ;
set AREA[bands] := east north ;
set AREA[coils] := east west export ;
```

```
param revenue :=

  [bands,*,*]:    1       2       3       4    :=
       east      25.0    26.0    27.0    27.0
       north     26.5    27.5    28.0    28.5

  [coils,*,*]:    1       2       3       4    :=
       east       30      35      37      39
       west       29      32      33      35
       export     25      25      25      28 ;
```

We have a separate revenue table for each product p, with market areas from AREA[p] labeling the rows, and weeks from 1..T labeling the columns.

Default values

AMPL checks that your data statements provide values for exactly the parameters in your model. You will receive an error message if you give a value for a nonexistent parameter,

```
error processing param cost:
        invalid subscript cost[PITT,DET,coils]
```

or if you fail to give a value for a parameter that does exist:

```
error processing objective total_cost:
        no value for cost[CLEV,LAN,coils]
```

The error message appears the first time that AMPL tries to use the offending parameter, usually after you type solve.

If the same value would appear many times in a data statement, you can avoid specifying it repeatedly by including a default phrase. For example, suppose that the parameter cost above is indexed over all possible triples,

```
set ORIG;
set DEST;
set PROD;
param cost {ORIG,DEST,PROD} >= 0;
```

but that an arbitrarily high cost is assigned to routes that should not be used. This can be expressed as

```
param cost   default 9999   :=

  [*,*,bands]: FRA  DET  LAN  WIN  STL  FRE  LAF :=
        CLEV   27    9   12    .   26    .   17
        PITT   24    .    .   13   28   99    .

  [*,*,coils]: FRA  DET  LAN  WIN  STL  FRE  LAF :=
        GARY    .    .   11    .   16    .    8
        CLEV   23    8   10    9   21    .    .
        PITT    .    .    .    .    .   81    . ;
```

Missing parameters like cost ["GARY", "FRA", "bands"], as well as those explic-itly marked ''omitted'' by use of a dot (like cost ["GARY", "FRA", "coils"]), are given the value 9999. In total, 24 values of 9999 are assigned.

The default feature is especially useful when you want all parameters of an indexed collection to be assigned the same value. For instance, in Figure 3-2, we apply a transportation model to an assignment problem by setting all supplies and demands to 1. The model declares

```
param supply {ORIG} >= 0;
param demand {DEST} >= 0;
```

but in the data we give only a default value:

```
param supply default 1 ;
param demand default 1 ;
```

Since no other values are specified, the default of 1 is automatically assigned to every ele-ment of supply and demand.

As explained in Chapter 7, a parameter declaration in the model may include a default expression. This offers an alternative way to specify a single default value:

```
param cost {ORIG,DEST,PROD} >= 0, default 9999;
```

If you just want to avoid storing a lot of 9999's in your data file, however, it is better to put the default phrase in the data statement. The default phrase should go in the model when you want the default value to depend in some way on other data. For instance, a different arbitrarily large cost could be given for each product by specifying:

```
param huge_cost {PROD} > 0;
param cost {ORIG, DEST, p in PROD} >= 0, default huge_cost[p];
```

Another example is given in Section 7.5.

9.3 Variable data

You may optionally assign initial values to the variables in a model, using any of the options for assigning values to parameters. A variable's name stands for its value, and a constraint's name stands for the associated dual variable's value. (See Section 10.7 for a short explanation of dual variables.)

Any param data statement may specify initial values for variables. The variable or constraint name is simply used in place of a parameter name, in any of the formats described by the previous section of this chapter. To help clarify the intent, the keyword var may be substituted for param at the start of any data statement. For example, the following data table gives initial values for the variable Trans of Figure 3-1a:

```
var Trans:   FRA  DET  LAN  WIN  STL  FRE  LAF :=
       GARY  100  100  800  100  100  500  200
       CLEV  900  100  100  500  500  200  200
       PITT  100  900  100  500  100  900  200 ;
```

As another example, in the model of Figure 1-4, a single table can give values for the parameters `rate`, `profit` and `market`, and initial values for the variables `Make`:

```
param:      rate  profit  market  Make :=
    bands    200     25     6000   3000
    coils    140     30     4000   2500
    plate    160     29     3500   1500 ;
```

All of the previously described features for default values also apply to variables.

Initial values of variables (as well as the values of expressions involving these initial values) may be viewed before you type `solve`, using the `display`, `print` or `printf` commands described in Sections 10.3 through 10.6. Initial values are also optionally passed to the solver, as explained in Section 10.8. After a solution is returned, the variables no longer have their initial values; but even then you can refer to the initial values by placing an appropriate suffix after the variable's name, as shown in Appendix A.11.

The most common use of initial values is to give a good starting guess to a solver for nonlinear optimization. The importance of the starting guess is discussed in Chapter 13.

9.4 Syntax summary

An *object* in a data specification may be either a number, or any character string. As in a model, a character string may be specified by surrounding it with quotes (' or "). Since so many strings appear in data, however, AMPL allows data statements to drop the quotes around any string that consists only of characters that may occur in a name or number.

The general form of a set data statement is

> *set-data-statement*:
>   `set` *set-name* `:=` *set-spec set-spec* ... `;`
>
> *set-spec*:
>   *set-template*_{opt} *member-list*
>   *set-template*_{opt} *member-table member-table* ...

The *set-name* must be the name of an individually declared set, or the subscripted name of a set from an indexed collection. The optional template has the form

> *set-template*:
>   (*templ-item*, *templ-item*, ...)
>
> *templ-item*:
>   *object*
>   *

where the number of *templ-item*s must equal the dimension of the named set. If no template is given, a template of all *'s is assumed.

The list format of *set-spec* is

>*member-list*:
>>*member-item member-item* ...
>
>*member-item*:
>>*object object* ...

The number of *object*s in a *member-item* must match the number of *'s in the preceding template; the *object*s are substituted for the *'s, from left to right, to produce a member that is added to the set being specified. In the special case that the template contains no *'s, the *member-list* should be empty, while the template itself specifies one member to be added.

The table format of *set-spec* looks like this:

>*member-table*:
>>(tr)_{opt} : *col-label   col-label   col-label*   ...   :=
>>*row-label*   ±   ±   ±
>>*row-label*   ±   ±   ±
>>...
>
>*row-label, col-label*:
>>*object*

Each table entry shown as ± must be either a + or a − symbol. The preceding template must contain exactly two *'s; each + entry's *row-label* and *col-label* are substituted for the *'s to produce a set member, while − entries are ignored. The *row-label* is substituted for the first * unless the optional (tr) is present, in which case the *col-label* is substituted for the first *.

There are two forms of the statement that specifies parameter or variable data. The first form is analogous to the set data statement,

>*param-data-statement*:
>>param *param-name param-default*_{opt} := *param-spec param-spec* ... ;
>
>*param-spec*:
>>*param-template*_{opt} *value-list*
>>*param-template*_{opt} *value-table value-table* ...

with the addition of an optional *param-default* that will be described below. The *param-name* is usually the name of a parameter declared in the model, but may also be the name of a variable or constraint; the keyword var may be used instead of param to make the distinction clear.

The param statement's templates have the same content as in the set data statement, but are given in brackets (like subscripts) rather than parentheses:

param-template :
> [*templ-item*, *templ-item*, ...]

templ-item :
> *object*
>
> *

The *value-list* is like the previously defined *member-list*, except that it also specifies a parameter or variable value:

value-list :
> *value-item* *value-item* ...

value-item :
> *object object* ... *entry*

The objects are substituted for *'s in the template to define a set member, and the parameter or variable indexed by this set member is assigned the value associated with the *entry* (see below). The *value-table* is like the previously defined *member-table*, except that its *entry*s are values rather than + or −:

value-table :
> (tr) *opt* : *col-label*   *col-label*   *col-label*   ...   :=
> *row-label*   *entry*   *entry*   *entry*
> *row-label*   *entry*   *entry*   *entry*
> ...

row-label, *col-label* :
> *object*

entry :
> *number*
> *string*
> *default-symbol*

Each *entry*'s *row-label* and *col-label* are substituted for *'s in the template to define a set member, and the parameter or variable indexed by this set member is assigned the value specified by the *entry*. The *entry* may be a number for variables and for parameters that take numerical values, or a string for parameters declared with the attribute symbolic. An *entry* that is the default symbol (see below) is ignored.

The second form of parameter data statement provides for the definition of multiple parameters, and also optionally the set over which they are indexed:

param-data-alternate :
> param *param-default* *opt* :
> > *param-name param-name* ... := *value-item value-item* ... ;
>
> param *param-default* *opt* : *set-name* :
> > *param-name param-name* ... := *value-item value-item* ... ;

The named parameters must all have the same dimension. If the optional *set-name* is specified, its membership is also defined by this statement. Each *value-item* consists of a list of objects followed by a list of values:

value-item :
> > *object* ... *entry entry* ...

The objects must be equal in number to the dimension of the parameters; taken together, they define a set member. If a set is being defined, this member is added to it. The parameters indexed by this member are assigned the values associated with the subsequent *entry*s, which obey the same rules as the table *entry*s previously described. Values are assigned in the order in which the parameters' names appeared at the beginning of the statement; the number of *entry*s must equal the number of named parameters.

A `param` data statement's optional `default` phrase has the form:

param-default :
> > `default` *number*

If this phrase is present, any parameter named but not explicitly assigned a value in the statement is given the value of *number*.

The *default-symbol* is initially a dot (.). It can be changed to a different string of one or more characters by the statement

    defaultsym *string* ;

and can be turned off entirely by the statement

    nodefaultsym ;

These statements maintain a stack of default symbols; the mechanism is described in Section A.12.2.

Exercises

9-1. Section 9.1 gave a variety of data statements for a three-dimensional set, `ROUTES`. Construct some other alternatives for this set as follows:

(a) Use templates that look like `(CLEV,FRA,*)`.

(b) Use templates that look like `(*,*,bands)`, with the list format.

(c) Use templates that look like `(CLEV,*,*)`, with the table format.

(d) Specify some of the set's members using templates with one `*`, and some using templates with two `*`'s.

9-2. Rewrite the production model data of Figure 5-4 so that it consists of just three data statements arranged as follows:

The set `PROD` and parameters `rate`, `inv0`, `prodcost` and `invcost` are given in one table.

The set `WEEKS` and parameter `avail` are given in one table.

The parameters `revenue` and `market` are given in one table.

9-3. For the assignment problem whose data is depicted in Figure 3-2, suppose that the only information you receive about people's preferences for offices is as follows:

Coullard	2087 2083 2053 2009 2019
Hazen	2009 2083 2087 2053 2019 1055 1053
Hopp	1055 2009 2083 2087 1083 1087 2019
Hurter	2009 2083 2087 2053 1087 1083
Jones	2009 1049 1021 2053 2019
Mehrotra	2083 2087 1083 1087 1055 2009
Rieders	2009 2019 2087 2083 2053 1021 1087
Spearman	2009 2083 2087
Sun	2083 2087 2019 2053 1087 1083 2009
Tamhane	2087 2083 1087 1083 1021 1049 2009
Zazanis	2087 2083 2009 1055

This means that, for example, Coullard's first choice is 2087, her second choice is 2083, and so on through her fifth choice, 2019, but she hasn't given any preference for the other offices.

To use this information with the transportation model of Figure 3-1a as explained in Chapter 3, you must set cost["Coullard",2087] to 1, cost["Coullard",2083] to 2, and so forth. For an office not ranked, such as 1055, you can set cost["Coullard",1055] to 99, to indicate that it is a highly undesirable assignment.

(a) Using the list format and a default phrase, convert the information above to an appropriate AMPL data statement for the parameter cost.

(b) Do the same, but with a table format.

9-4. Section 9.2 gave a variety of data statements for a three-dimensional parameter, cost, indexed over the set ROUTES of triples. Construct some other alternatives for this parameter as follows:

(a) Use templates that look like [CLEV,FRA,*].

(b) Use templates that look like [*,*,bands], employing the list format.

(c) Use templates that look like [CLEV,*,*], employing the table format.

(d) Specify some of the parameter values using templates with one *, and some using templates with two *'s.

9-5. For the three-dimensional parameter revenue of Figure 6-4, construct alternative data statements as follows:

(a) Use templates that look like [*,east,*], employing the table format.

(b) Use templates that look like [*,*,1], employing the table format.

(c) Use templates that look like [bands,*,1].

9-6. Given the following declarations,

```
set ORIG;
set DEST;
var Trans {ORIG, DEST} >= 0;
```

how could you use a data statement to assign an initial value of 300 to all of the Trans variables?

10

Command Environment

AMPL provides a variety of commands like model, solve, and display that tell the AMPL modeling system what to do with models and data. Although these commands may use AMPL expressions, they are not part of the modeling language itself. They are intended to be used in an interactive environment, where you type a command, wait for the system to display a response, then decide what command to type next.

This chapter begins by describing the general principles of the command environment. Sections 10.2 then presents the commands that you are likely to use most for setting up and solving optimization problems.

AMPL offers a rich variety of commands and options to help you examine and report the results of optimization. Section 10.3 introduces display, the most convenient command for arranging set members and numerical values into lists and tables; Sections 10.4 and 10.5 provide a detailed account of display options that give you more control over how the lists and tables are arranged and how numbers appear in them. Section 10.6 describes print and printf, two related commands that are useful for preparing data to be sent to other programs, and for formatting simple reports. Although our examples are based on the display of sets, parameters and variables — and expressions involving them — you can use the same commands to inspect dual values, slacks, reduced costs, and other quantities associated with an optimal solution; the rules for doing so are explained in Section 10.7.

After solving a problem and looking at the results, the next step is often to make a change and solve again. Section 10.8 describes commands that let you drop constraints, fix variables, reread data, or modify specific data values, without restarting the AMPL session from the beginning.

You will probably find that your most intensive use of the command environment occurs during the initial development of a model, when the results are unfamiliar and changes are frequent. When the formulation eventually settles down, you may find yourself typing the same series of commands over and over to solve for different collections of data. To accelerate this process, you can arrange to have AMPL read often-used sequences of commands from files, as described in the last section of this chapter.

10.1 General principles

To begin an interactive AMPL session, you must start up the AMPL program from your computer's operating system. In most cases, all you have to do is type the command *ampl* in response to your system's prompt. The startup procedure necessarily varies somewhat from one operating system to another; for details, you should refer to the system-specific instructions that come with your AMPL software.

You communicate with AMPL in two ways: by typing commands, and by setting options that influence subsequent commands. The basic rules for commands and options are discussed in this section.

Commands

After starting AMPL, the first thing you should see is AMPL's prompt:

```
ampl:
```

Whenever you see this prompt, AMPL is ready to read and interpret what you type. As with most command interpreters, AMPL waits until you press the ''enter'' or ''return'' key, then processes everything you typed on the line.

An AMPL command ends with a semicolon. If you enter one or more complete commands on a line, AMPL processes them, prints any appropriate messages in response, and issues the `ampl:` prompt again. If you end a line in the middle of a command, you are prompted to continue on the next line; you can tell that AMPL is prompting you to continue a command, because its prompt ends with a question mark rather than a colon:

```
ampl: display {i in ORIG, j in DEST}
ampl? sum {p in PROD} Trans[i,j,p];
```

You can type any number of characters on a line (up to whatever limit your operating system might impose), and can continue a command on any number of lines.

Several commands use filenames for reading or writing information. A filename can be any sequence of printing characters (except for semicolon ; and quotes " or ') or any sequence of any characters enclosed in matching quotes. The rules for correct filenames are determined by your computer's operating system, however, not by AMPL. For the examples in this book we have used filenames like diet.mod that are acceptable to almost any operating system.

To conclude an AMPL session, type end or quit.

Options

The behavior of AMPL commands depends not only on what you type directly, but on a variety of options that you can specify for purposes such as choosing alternative solvers and controlling the display of results. Even the AMPL prompts can be changed by resetting an option.

Each option has a name, and a value that may be a number or a character string. For example, the options `prompt1` and `prompt2` are strings that specify the prompts. The option `display_width` has a numeric value, which says how many characters wide the output produced by the `display` command may be.

The `option` command displays and sets option values. If `option` is followed by a list of option names, AMPL replies with the current values:

```
ampl: option prompt1, display_width;
option prompt1 'ampl: ';
option display_width 79;
ampl:
```

A `*` in an option name is a "wild card" that matches any sequence of characters:

```
ampl: option prom*;
option prompt1 'ampl: ';
option prompt2 'ampl? ';
ampl:
```

The command `option *`, or just `option` alone, lists all current options and values.

When `option` is followed by a name and a value, it resets the named option to the specified value. In the following example we change the prompt and the display width, and then verify that the latter has been changed:

```
ampl: option prompt1 "A> ", display_width 60;
A> option display_width;
option display_width 60;
A>
```

You can specify any string value by surrounding the string in matching quotes '...' or "..." as above; the quotes may be omitted if the string looks like a name or number. Two consecutive quotes (' ' or " ") denote an empty string, which is a meaningful value for some options. At the other extreme, if you want to spread a long string over several lines, place the backslash character \ at the end of each intermediate line.

When AMPL starts, it sets many options to initial, or default, values. The `prompt1` option is initialized to `'ampl: '`, for instance, so prompts appear in the standard way; the `display_width` option has a default of 79. Other options, especially ones that pertain to particular solvers, are initially unset:

```
ampl: option osl_options;
option osl_options ''; #not defined
```

To return all options to their default values, use the command `reset options`.

AMPL maintains no master list of valid options, but rather accepts any new option that you define. Thus if you mis-type an option name, you will most likely define a new option by mistake, as the following example demonstrates:

```
ampl: option display_wdith 60;
ampl: option display_w*;
option display_wdith 60;
option display_width 79;
```

The `option` statement also doesn't check to see if you have assigned a meaningful value to an option. You will be informed of a value error only when an option is used by some subsequent command. In these respects, AMPL options are much like the "environment variables" of the DOS or UNIX operating systems. In fact you can use the settings of environment variables to override AMPL's option defaults; see your system-specific documentation for details.

10.2 Setting up and solving models

To apply a solver to an instance of a model, the examples in this book use `model`, `data`, and `solve` commands:

```
ampl: model diet.mod; data diet.dat; solve;
MINOS 5.4: optimal solution found.
6 iterations, objective 88.2
```

The `model` command names a file that contains model declarations (Chapters 5 through 8), and the `data` command names a file that contains data values for model components (Chapter 9). The `solve` command causes a description of the optimization problem to be sent to a solver, and the results to be retrieved for examination. This section takes a closer look at the main AMPL features for setting up and solving models. Features for subsequently changing and re-solving models are covered in Section 10.8.

Entering models and data

AMPL maintains a "current" model, which is the one that will be sent to the solver if you type `solve`. At the beginning of an interactive session, the current model is empty. A `model` command reads declarations from a file and adds them to the current model; a `data` command reads data statements from a file to supply values for components already in the current model. Thus you may use several `model` or `data` commands to build up the description of an optimization problem, reading different parts of the model and data from different files.

You can also type parts of a model and its data directly at an AMPL prompt. Model declarations such as `param`, `var` and `subject to` act as commands that add components to the current model. The data statements of Chapter 9 also act as commands, which supply data values for already defined components such as sets and parameters. Because model and data statements look much alike, however, you need to tell AMPL which you will be typing. AMPL always starts out in "model mode"; the statement `data` (without a filename) switches the interpreter to "data mode", and the statement `model` (without a filename) switches it back. Any command (like `option`, `solve` or `subject to`) that does not begin like a data statement also has the effect of switching data mode back to model mode.

In Chapter 1 we showed an example of typing a whole model in response to prompts, but this is practical only for very small test models. You might occasionally find it convenient to type in a few data statements, after the rest of a model and data have been read from files.

If a model declares more than one objective function, AMPL by default passes all of them to the solver; most solvers deal only with one objective function and usually select the first by default. The `objective` command lets you select a single objective function to pass to the solver; it consists of the keyword `objective` followed by a name from a `minimize` or `maximize` declaration:

```
objective total_number;
```

If a model has an indexed collection of objectives, you must supply a subscript to indicate which one is to be chosen:

```
objective total_cost["A&P"];
```

The uses of multiple objectives are illustrated by two examples in Section 8.3.

Solving a model

The `solve` command sets in motion a series of activities. First, it causes AMPL to generate a specific optimization problem from the model and data that you have supplied. If you have neglected to provide some needed data, an error message is printed; you will also get error messages if your data values violate any restrictions imposed by qualification phrases in `var` or `param` declarations or by `check` statements. AMPL waits to verify data restrictions until you type `solve`, because a restriction may depend in a complicated way on many different data values. Some arithmetic errors such as dividing by zero are also caught at this stage.

After the optimization problem is generated, AMPL enters a "presolve" phase that tries to make the problem easier for the solver. Presolve finds all constraints that "fix" a variable, such as these initial inventory constraints from Figure 4-4:

```
subject to initial {p in PROD}:  Inv[p,0] = inv0[p];
```

The fixed value is substituted everywhere that the variable appears, so that both the variable and the fixing constraint may be dropped from the problem. Constraints that express simple upper or lower bounds are also handled specially, since most solvers can enforce bounds implicitly in a very efficient way. In fact, presolve applies a number of tests that can tighten various bounds on a problem without changing the optimum; conventions for viewing such bounds are explained in Appendix A.13.3.

If presolve tightens bounds so much that some upper bound becomes less than the corresponding lower bound, then it has already determined that no feasible solution is possible, without any work on the part of the solver:

```
ampl: solve;
presolve: constraint Time['reheat'] cannot hold:
          body <= 20 cannot be >= 22.5; difference = -2.5
MINOS 5.4: infeasible problem.
0 iterations
```

The error message says that the constraint `Time['reheat']` was determined to have the form "body <= 20" where the body is the sum of the variables; but presolve's tests established that the same sum of variables would have to be at least 22.5 in any solution that satisfied all the other constraints and bounds. Hence this constraint can't possibly be satisfied. (The discrepancy of –2.5 is also reported in the message. If presolve reports a tiny discrepancy that you would like it to ignore, change the option `presolve_eps` from its default of 0 to a positive value that represents the smallest discrepancy you would like it to report.)

An error like this may be the result of an incorrect model formulation that is causing some nonsensical constraints to be generated. In other cases, the formulation is correct and the difficulty is with the data; in the example above, the constraint `Time['reheat']` could not be satisfied because the time available at the reheat facility was less than the minimum time needed to satisfy all lower bounds on sales. The source of an error like this is seldom immediately obvious, but you can usually get a good start toward finding it by examining any constraint or variable named in the error message.

When the model and data are correct, the work of presolve will normally be invisible. To see a summary of its actions and the resulting size of the optimization problem, change the option `show_stats` to 1 from its default of 0:

```
ampl: option show_stats 1;
ampl: model steelT.mod;
ampl: data steelT.dat;
ampl: solve;

Presolve eliminates 2 constraints and 2 variables.
Adjusted problem:
24 variables, all linear
12 constraints, all linear; 38 nonzeros
1 linear objective; 24 nonzeros.

MINOS 5.4: optimal solution found.
20 iterations, objective 515033
```

In rare cases, presolve can substantially affect the optimal values of the variables — when there is more than one optimal solution — or can interfere with other preprocessing routines that are built into your solver software. To turn off presolve, set option `presolve` to 0.

The generated optimization problem, as possibly modified by presolve, is finally sent by AMPL to the solver of your choice. Every version of AMPL is distributed with some default solver that will be used automatically if you give no other instructions; type `option solver` to see its name:

```
ampl: option solver;
option solver minos;
```

If you have more than one solver, you can switch among them by changing the `solver` option. Virtually any solver can handle a linear program:

```
ampl: model steelT.mod; data steelT.dat;
ampl: solve;
MINOS 5.4: optimal solution found.
15 iterations, objective 515033

ampl: option solver cplex;
ampl: solve;
CPLEX 2.0: optimal solution; objective 515033
17 iterations (0 in phase I)

ampl: option solver osl;
ampl: solve;
OSL 1.2:
  The primal algorithm has been chosen
OSL 1.2: optimal solution; objective 515033
19 simplex iterations
```

Other kinds of problems, such as nonlinear (Chapter 13) and integer (Chapter 15), can be handled only by solvers designed for them. A message such as "ignoring integrality" or "can't handle nonlinearities" is an indication that you have not chosen a solver appropriate for your model.

If the optimization problems generated by your application are not too difficult, you should be able to use AMPL without referring to instructions for a specific solver; set the `solver` option appropriately, type `solve`, and wait for the results. If your solver takes a very long time to return with a solution, or returns to AMPL without any "optimal solution" message, then it's time to read further. Each solver is a sophisticated collection of algorithms and algorithmic strategies, from which many combinations of choices can be made. For most problems the solver makes good choices automatically, but you can also pass along your own choices through AMPL options. The details may vary with each solver, so for more information you must look to the solver-specific instructions that accompany your AMPL software.

If your problem takes a long time to optimize, you may want some evidence of the solver's progress to appear on your screen. Options for this purpose are also described in the solver-specific instructions.

10.3 Browsing through results: the `display` command

The easiest way to examine data and result values is to type `display` and a description of what you want to look at. The `display` command automatically exhibits the values in a pleasing and familiar arrangement; as much as possible, it uses the same list and table formats as the data statements described in Chapter 9. If the arrangement is not to your liking, you can set options, described in Sections 10.4 and 10.5, that influence how `display` chooses a list or table format and how it presents numeric values. The related `print` and `printf` commands, presented in Section 10.6, retrieve the same data without formatting or with your own specified format.

To keep things simple, our examples use parameters and variables from models defined in other chapters, but the same principles apply to displaying objectives, dual prices, slacks, reduced costs and other solution information, as explained in Section 10.7.

Displaying sets

The contents of sets are shown by typing `display` and a list of set names. This example is taken from the model of Figure 6-2a:

```
ampl: display ORIG,DEST,LINKS;
set ORIG := GARY CLEV PITT;

set DEST := FRA DET LAN WIN STL FRE LAF;

set LINKS :=
(GARY,DET)   (GARY,LAF)   (CLEV,LAN)   (CLEV,LAF)   (PITT,STL)
(GARY,LAN)   (CLEV,FRA)   (CLEV,WIN)   (PITT,FRA)   (PITT,FRE)
(GARY,STL)   (CLEV,DET)   (CLEV,STL)   (PITT,WIN) ;
```

If you specify the name of an indexed collection of sets, each set in the collection is shown (from Figure 6-3):

```
ampl: display PROD,AREA;
set PROD := bands coils;

set AREA[bands] := east north;
set AREA[coils] := east west export;
```

Particular members of an indexed collection can be viewed by use of subscripting, as in `display AREA["bands"]`.

The argument of `display` need not be a declared set; it can be any of the expressions described in Chapter 5 or 6 that evaluate to sets. For example, you can show the union of all the sets `AREA[p]`:

```
ampl: display union {p in PROD} AREA[p];
set  union {p in PROD}  AREA[p] := east north west export;
```

or the set of all transportation links on which the total shipping cost is greater than 500:

```
ampl: display {(i,j) in LINKS: cost[i,j] * Trans[i,j] > 500};
set {(i,j) in LINKS: cost[i,j]*Trans[i,j] > 500}   :=
(GARY,STL)   (CLEV,DET)   (CLEV,WIN)   (PITT,FRA)   (PITT,FRE)
(GARY,LAF)   (CLEV,LAN)   (CLEV,LAF)   (PITT,STL) ;
```

Because the membership of this set depends upon the current values of the variables `Trans[i,j]`, you could not refer to it in a model, but it is legal in a `display` command, where variables are treated the same as parameters.

Displaying parameters and variables

The display command can show the value of an unindexed model component (from Figure 4-4),

```
ampl: display T;
T = 4
```

or the values of individual components from an indexed collection (Figure 1-6b):

```
ampl: display avail["reheat"], avail["roll"];
avail['reheat'] = 35
avail['roll'] = 40
```

The major use of display, however, is to show whole indexed collections of data. For "one-dimensional" data — parameters or variables indexed over a simple set — AMPL uses a columnar format:

```
ampl: display avail;
avail [*] :=
reheat  35
  roll  40
;
```

For "two-dimensional" parameters or variables — indexed over a set of pairs or two simple sets — AMPL forms a table. Normally the rows correspond to the first index, and the columns to the second (Figure 4-1):

```
ampl: display supply;
supply [*,*]
:      bands  coils plate     :=
CLEV   700    1600   300
GARY   400     800   200
PITT   800    1800   300
;
```

To promote readability, however, some tables are transposed, as indicated by (tr) in the heading (Figure 4-4):

```
ampl: display Make;
Make [*,*] (tr)
:  bands   coils     :=
1   5990    1407
2   6000    1400
3   1400    3500
4   2000    4200
;
```

Because the table has been transposed, the value 5990, for example, is associated with Make["bands",1], not with Make[1,"bands"].

A parameter or variable indexed over a set of ordered pairs is also considered to be two-dimensional. Here is the table for a parameter indexed over the set LINKS that was displayed earlier in this section (from Figure 6-2a):

```
ampl: display cost;
cost [*,*] (tr)
:    CLEV GARY PITT      :=
DET    9    14     .
FRA   27     .    24
FRE    .     .    99
LAF   17     8     .
LAN   12    11     .
STL   26    16    28
WIN    9     .    13
;
```

A dot (.) entry indicates a nonexistent combination in the index set. Thus in the GARY column of the table, there is a dot in the FRA row because the pair (GARY, FRA) is not a member of LINKS; no cost["GARY", "FRA"] is defined for this problem. On the other hand, LINKS does contain the pair (GARY, LAF), and cost["GARY", "LAF"] is shown as 8 in the table.

To display data in three or more dimensions, AMPL "slices" it into two-dimensional tables, as in the case of this variable from Figure 4-6:

```
ampl: display Trans;
Trans [CLEV,*,*]
:    bands coils plate      :=
DET    0    750     0
FRA    0      0     0
FRE    0      0     0
LAF    0    500     0
LAN    0    400     0
STL    0     50     0
WIN    0    250     0

   [GARY,*,*]
:    bands coils plate      :=
DET    0      0     0
FRA    0      0     0
FRE  225    850   100
LAF  250      0     0
LAN    0      0     0
STL  650    900   200
WIN    0      0     0

   [PITT,*,*]
:    bands coils plate      :=
DET  300      0   100
FRA  300    500   100
FRE    0      0     0
LAF    0      0   250
LAN  100      0     0
STL    0      0     0
WIN   75      0    50
;
```

At the head of the first table, the template [CLEV, *, *] indicates that the slice is through CLEV in the first component, so the entry in row LAF and column coils says that Trans["CLEV","LAF","coils"] is 500. Since the first index of Trans is always CLEV, GARY or PITT in this case, there are three slice tables in all. But AMPL does not always slice through the first component; it picks the slices so that the display will contain the fewest possible tables.

A display of two or more components of the same dimensionality is always presented in a columnar format, whether the components are one-dimensional (Figure 4-4),

```
ampl: display inv0,prodcost,invcost;
:       inv0 prodcost invcost    :=
bands    10     10       2.5
coils     0     11       3
;
```

or two-dimensional (Figure 4-6),

```
ampl: display rate,make_cost,Make;
:            rate make_cost   Make      :=
CLEV bands   190     190        0
CLEV coils   130     170      1950
CLEV plate   160     185        0
GARY bands   200     180      1125
GARY coils   140     170      1750
GARY plate   160     180       300
PITT bands   230     190       775
PITT coils   160     180       500
PITT plate   170     185       500
;
```

or any higher dimension. The indices appear in a list to the left, with the last one changing most rapidly.

As you can see from these examples, display normally arranges row and column labels in alphabetical or numerical order, regardless of the order in which they might have been given in your data file. When the labels come from an ordered set, however, the original ordering is honored (Figure 5-3):

```
ampl: display avail;
avail [*] :=
27sep93   40
04oct93   40
11oct93   32
18oct93   40
;
```

For this reason, it is a good idea to declare sets of time periods as ordered, even if their ordering plays no explicit role in your model.

Displaying indexed expressions

The display command can show the value of any arithmetic expression that is valid in an AMPL model. Single-valued expressions pose no difficulty, as in the case of these three profit components from Figure 4-4:

```
ampl: display sum {p in PROD, t in 1..T} revenue[p,t]*Sell[p,t],
ampl?             sum {p in PROD, t in 1..T} prodcost[p]*Make[p,t],
ampl?             sum {p in PROD, t in 1..T} invcost[p]*Inv[p,t];

sum{p in PROD, t in 1 .. T} revenue[p,t]*Sell[p,t] = 787810
sum{p in PROD, t in 1 .. T} prodcost[p]*Make[p,t] = 269477
sum{p in PROD, t in 1 .. T} invcost[p]*Inv[p,t] = 3300
```

Suppose, however, that you want to see all the individual values of revenue[p,t] * Sell[p,t]. Since you can type display revenue, Sell to display the separate values of revenue[p,t] and Sell[p,t], you might want to ask for the products of these values by typing:

```
ampl: display revenue * Sell;
syntax error
context:  display revenue  >>> *  <<< Sell;
```

AMPL does not recognize this kind of array arithmetic. To display an indexed collection of expressions, you must explicitly specify the indexing:

```
ampl: display {p in PROD, t in 1..T} revenue[p,t] * Sell[p,t];
revenue[p,t]*Sell[p,t] [*,*] (tr)
:    bands     coils     :=
1    150000      9210
2    156000     87500
3     37800    129500
4     54000    163800
;
```

To apply the same indexing to two or more expressions, enclose a list of them in parentheses after the indexing expression:

```
ampl: display {p in PROD, t in 1..T}
ampl?     (revenue[p,t]*Sell[p,t], prodcost[p]*Make[p,t]);
:        revenue[p,t]*Sell[p,t] prodcost[p]*Make[p,t]    :=
bands 1          150000               59900
bands 2          156000               60000
bands 3           37800               14000
bands 4           54000               20000
coils 1            9210               15477
coils 2           87500               15400
coils 3          129500               38500
coils 4          163800               46200
;
```

An indexing expression followed by an expression or parenthesized list of expressions is treated as a single display item, which specifies some indexed collection of values. A display command may contain one of these items as above, or a comma-separated list of them.

The presentation of the values for indexed expressions follows the same rules as described above for individual parameters and variables. In fact, you can regard a command like display revenue, Sell as shorthand for

```
ampl: display {p in PROD, t in 1..T} (revenue[p,t], Sell[p,t]);
:          revenue[p,t] Sell[p,t]      :=
bands 1        25          6000
bands 2        26          6000
bands 3        27          1400
bands 4        27          2000
coils 1        30           307
coils 2        35          2500
coils 3        37          3500
coils 4        39          4200
;
```

If you rearrange the indexing expression so that t in 1..T comes first, however, the rows of the list are instead sorted first on the members of 1..T:

```
ampl: display {t in 1..T, p in PROD} (revenue[p,t], Sell[p,t]);
:          revenue[p,t] Sell[p,t]      :=
1 bands        25          6000
1 coils        30           307
2 bands        26          6000
2 coils        35          2500
3 bands        27          1400
3 coils        37          3500
4 bands        27          2000
4 coils        39          4200
;
```

This change in the default presentation can only be achieved by placing an explicit indexing expression after display.

In addition to indexing individual display items, you can specify a set over which the whole display command is indexed — that is, you can ask that the command be executed once for each member of an indexing set. This feature is particularly useful for rearranging slices of multi-dimensional tables. When, earlier in this section, we displayed the 3-dimensional variable Trans indexed over {ORIG, DEST, PROD}, AMPL chose to slice the values through members of ORIG to produce a series of two-dimensional tables.

What if you want to display slices through PROD? Rearranging the indexing expression, as in our previous example, will not reliably have the desired effect, since AMPL always slices through one of the smallest indexing sets. Instead, you can say explicitly that you want a separate display for each p in PROD:

```
ampl: display {p in PROD}: {i in ORIG, j in DEST} Trans[i,j,p];
Trans[i,j,'bands'] [*,*] (tr)
:    CLEV  GARY  PITT     :=
DET   0      0    300
FRA   0      0    300
FRE   0    225      0
LAF   0    250      0
LAN   0      0    100
STL   0    650      0
WIN   0      0     75
;

Trans[i,j,'coils'] [*,*] (tr)
:    CLEV  GARY  PITT     :=
DET  750     0      0
FRA    0     0    500
FRE    0   850      0
LAF  500     0      0
LAN  400     0      0
STL   50   900      0
WIN  250     0      0
;

Trans[i,j,'plate'] [*,*] (tr)
:    CLEV  GARY  PITT     :=
DET   0      0    100
FRA   0      0    100
FRE   0    100      0
LAF   0      0    250
LAN   0      0      0
STL   0    200      0
WIN   0      0     50
;
```

As this example shows, if a display command specifies an indexing expression right at
the beginning, followed by a colon, the indexing set applies to the whole command. For
each member of the set, the expressions following the colon are evaluated and displayed
separately.

Output to a file

To direct the output from display to a file rather than the screen, add a > and the
name of the file:

```
ampl: display ORIG, DEST, PROD >multi.out;
ampl: display supply >multi.out;
ampl: display demand >multi.out;
ampl: display cost, Trans >multi.out;
```

The first display that specifies >multi.out creates a new file by that name (or over-
writes any existing file of the same name). Subsequent commands add to the end of the
file, until the end of the session or a matching close command:

```
ampl: close multi.out;
```

To open a file and append output to whatever is already there (rather than overwriting), use >> instead of >. Once a file is open, subsequent uses of > and >> have the same effect.

10.4 Formatting options for `display`

The `display` command uses a few simple rules for choosing a good arrangement of data. By changing several options, you can control overall arrangement, handling of zero values, and line width. These options are summarized in Table 10-1, with default values shown in parentheses.

Arrangement of lists and tables

The display of a one-dimensional parameter or variable can produce a very long list, as in this example from the scheduling model of Figure 12-4:

```
ampl: display required;
required [*] :=
Mon1   100
Tue1   100
Wed1   100
(12 lines omitted)
Thu3    52
Fri3    52
;
```

The option `display_1col` can be used to request a more compact format:

```
ampl: option display_1col 0;
ampl: display required;
required [*] :=
Mon1 100    Thu1 100    Mon2  78    Thu2  78    Mon3  52    Thu3  52
Tue1 100    Fri1 100    Tue2  78    Fri2  78    Tue3  52    Fri3  52
Wed1 100    Sat1 100    Wed2  78    Sat2  78    Wed3  52
;
```

The one-column list format is used when the number of values to be displayed is less than or equal to `display_1col`, and the compact format is used otherwise. The default for `display_1col` is 20; set it to zero to force the compact format, or to a very large number to force the list format.

Multi-dimensional displays are affected by `display_1col` in an analogous way (from Figure 4-1):

display_1col	maximum elements for a table to be displayed in list format (20)
display_transpose	transpose tables if rows − columns < display_transpose (0)
display_width	maximum line width (79)
gutter_width	separation between table columns (3)
omit_zero_cols	if not 0, omit all-zero columns from displays (0)
omit_zero_rows	if not 0, omit all-zero rows from displays (0)

Table 10-1: Formatting options for display (with default values).

```
ampl: option display_1col 10;
ampl: display supply;

supply :=
CLEV bands     700
CLEV coils    1600
CLEV plate     300
GARY bands     400
GARY coils     800
GARY plate     200
PITT bands     800
PITT coils    1800
PITT plate     300
;

ampl: display demand;

demand [*,*]
:    bands coils plate     :=
DET   300   750   100
FRA   300   500   100
FRE   225   850   100
LAF   250   500   250
LAN   100   400     0
STL   650   950   200
WIN    75   250    50
;
```

Again the one-column list format is used when the number of values is less than or equal to display_1col, while the appropriate compact format — in this case, a table — is used otherwise.

In choosing an orientation for tables, the display command by default favors rows over columns; that is, if the number of columns would exceed the number of rows, the table is transposed. Thus the table for demand above has rows labeled by the first coordinate and columns by the second.

By contrast, the following table for limit is transposed to have rows labeled by the second coordinate and columns by the first:

```
ampl: display limit;
limit [*,*] (tr)
:     CLEV  GARY  PITT      :=
DET   625   625   625
FRA   625   625   625
FRE   625   625   625
LAF   625   625   625
LAN   625   625   625
STL   625   625   625
WIN   625   625   625
;
```

These choices can be reversed by changing the display_transpose option:

```
ampl: option display_transpose 5;
ampl: display demand;
demand [*,*] (tr)
:       DET   FRA   FRE   LAF   LAN   STL   WIN     :=
bands   300   300   225   250   100   650    75
coils   750   500   850   500   400   950   250
plate   100   100   100   250     0   200    50
;

ampl: option display_transpose -5;
ampl: display limit;
limit [*,*]
:       DET   FRA   FRE   LAF   LAN   STL   WIN     :=
CLEV    625   625   625   625   625   625   625
GARY    625   625   625   625   625   625   625
PITT    625   625   625   625   625   625   625
;
```

The rule is as follows: a table is transposed only when the number of rows minus the number of columns would be less than display_transpose. At its default value of zero, display_transpose gives the previously described default behavior. Large positive or negative values will force or suppress transposition, respectively.

Control of line width

The option display_width gives the maximum number of characters on a line generated by display (Figure 12-4):

```
ampl: option display_width 40;
ampl: display required;

required [*] :=
Mon1 100    Mon2  78    Mon3  52
Tue1 100    Tue2  78    Tue3  52
Wed1 100    Wed2  78    Wed3  52
Thu1 100    Thu2  78    Thu3  52
Fri1 100    Fri2  78    Fri3  52
Sat1 100    Sat2  78
;
```

When a table would be wider than `display_width`, it is cut vertically into two or more tables. The rows in each table are the same, but the columns are different:

```
ampl: option display_width 50; display cost;
cost [*,*]
:          1021 1049 1053 1055 1083 1087 2009 2019 :=
Coullard      6    9    8    7   11   10    4    5
Hazen        11    8    7    6    9   10    1    5
Hopp          9   10   11    1    5    6    2    7
Hurter       11    9    8   10    6    5    1    7
Jones         3    2    8    9   10   11    1    5
Mehrotra     11    9   10    5    3    4    6    7
Rieders       6   11   10    9    8    7    1    2
Spearman     11    5    4    6    7    8    1    9
Sun          11    9   10    8    6    5    7    3
Tamhane       5    6    9    8    4    3    7   10
Zazanis      11    9    8    4    6    5    3   10

:          2053 2083 2087      :=
Coullard      3    2    1
Hazen         4    2    3
Hopp          8    3    4
Hurter        4    2    3
Jones         4    6    7
Mehrotra      8    1    2
Rieders       5    4    3
Spearman     10    2    3
Sun           4    1    2
Tamhane      11    2    1
Zazanis       7    2    1
;
```

If a table's column headings are much wider than the values, `display` introduces abbreviations to keep all columns together (Figure 4-4):

```
ampl: display {p in PROD, t in 1..T} (revenue[p,t]*Sell[p,t],
ampl?     prodcost[p]*Make[p,t], invcost[p]*Inv[p,t]);
# $1 = revenue[p,t]*Sell[p,t]
# $2 = prodcost[p]*Make[p,t]
# $3 = invcost[p]*Inv[p,t]

:           $1      $2     $3     :=
bands 1   150000   59900     0
bands 2   156000   60000     0
bands 3    37800   14000     0
bands 4    54000   20000     0
coils 1     9210   15477  3300
coils 2    87500   15400     0
coils 3   129500   38500     0
coils 4   163800   46200     0
;
```

On the other hand, where the headings are narrower than the values, you may be able to squeeze more on a line by reducing the option gutter_width — the number of spaces between columns — from its default value of 3 to 2 or 1.

Suppression of zeros

In some kinds of linear programs that have many more variables than constraints, most of the variables have an optimal value of zero. In the model of Figure 12-4:

```
ampl: option solver cplex;
ampl: option show_stats 1;
ampl: model sched.mod; data sched.dat; solve;

126 variables, all integer
17 constraints, all linear; 630 nonzeros
1 linear objective; 126 nonzeros.

CPLEX 2.0: optimal integer solution; objective 266
27 integer iterations

ampl: option display_width 65; display Work;
Work [*] :=
 1 28      19 0     37 0     55 0     73 6     91 0     109 0
 2  0      20 0     38 0     56 0     74 0     92 0     110 0
 3  0      21 0     39 0     57 0     75 0     93 0     111 30
 4  0      22 0     40 0     58 0     76 0     94 0     112 22
 5  0      23 0     41 30    59 0     77 0     95 0     113 0
 6  0      24 0     42 0     60 0     78 0     96 0     114 0
 7 30      25 0     43 0     61 0     79 0     97 0     115 0
 8  0      26 0     44 0     62 0     80 0     98 0     116 0
 9  0      27 0     45 0     63 0     81 22    99 0     117 0
10  0      28 0     46 0     64 0     82 0     100 0    118 0
11  0      29 0     47 0     65 0     83 0     101 36   119 0
12  0      30 0     48 0     66 0     84 0     102 0    120 0
13  6      31 0     49 0     67 0     85 0     103 0    121 0
14  0      32 0     50 0     68 0     86 0     104 0    122 2
15  0      33 0     51 0     69 0     87 0     105 0    123 0
16  0      34 0     52 0     70 0     88 0     106 0    124 0
17  0      35 0     53 6     71 8     89 6     107 0    125 34
18  0      36 0     54 0     72 0     90 0     108 0    126 0
;
```

To see only the nonzeros, change option omit_zero_rows from its default of 0 to 1:

```
ampl: option omit_zero_rows 1; display Work;
Work [*] :=
 1 28      13 6     53 6     73 6     89 6     111 30   122 2
 7 30      41 30    71 8     81 22    101 36   112 22   125 34
;
```

This option is equally useful for multi-dimensional data. In our example of an assignment problem (Figure 3-2), the optimal values of all the variables form this table:

```
ampl: display Trans;
Trans [*,*]
:         1021 1049 1053 1055 1083 1087 2009 2019 2053 2083 2087    :=
Coullard    1    0    0    0    0    0    0    0    0    0    0
Hazen       0    0    0    0    0    0    0    0    1    0    0
Hopp        0    0    0    1    0    0    0    0    0    0    0
Hurter      0    0    0    0    0    0    1    0    0    0    0
Jones       0    1    0    0    0    0    0    0    0    0    0
Mehrotra    0    0    0    0    1    0    0    0    0    0    0
Rieders     0    0    0    0    0    0    0    1    0    0    0
Spearman    0    0    1    0    0    0    0    0    0    0    0
Sun         0    0    0    0    0    0    0    0    0    1    0
Tamhane     0    0    0    0    0    1    0    0    0    0    0
Zazanis     0    0    0    0    0    0    0    0    0    0    1
;
```

When we set omit_zero_rows to 1, the list comes down to the entries of interest:

```
ampl: option omit_zero_rows 1;
ampl: display Trans;
Trans :=
Coullard 1021    1
Hazen    2053    1
Hopp     1055    1
Hurter   2009    1
Jones    1049    1
Mehrotra 1083    1
Rieders  2019    1
Spearman 1053    1
Sun      2083    1
Tamhane  1087    1
Zazanis  2087    1
;
```

The format has changed to a list, because the number of nonzero entries is less than the value of display_1col.

If the number of nonzeros were greater than display_1col, the table format would be retained, and the omit_zero_rows option would only suppress table rows that contain all zero entries. For example, the display of the three-dimensional variable Trans from earlier in this chapter would be condensed to the following:

```
ampl: display Trans;
Trans [CLEV,*,*]
:    bands coils plate    :=
DET     0   750    0
LAF     0   500    0
LAN     0   400    0
STL     0    50    0
WIN     0   250    0
```

display_eps	Smallest magnitude displayed differently from zero (0)
display_precision	Digits of precision to which displayed numbers are rounded; full precision if 0 (6)
display_round	Digits left or (if negative) right of decimal place to which displayed numbers are rounded, overriding display_precision ("")
solution_precision	Digits of precision to which solution values are rounded; full precision if 0 (0)
solution_round	Digits left or (if negative) right of decimal place to which solution values are rounded, overriding solution_precision ("")

Table 10-2: Numeric options for display (with default values).

```
   [GARY,*,*]
:    bands coils plate     :=
FRE    225    850    100
LAF    250      0      0
STL    650    900    200

   [PITT,*,*]
:    bands coils plate     :=
DET    300      0    100
FRA    300    500    100
LAF      0      0    250
LAN    100      0      0
WIN     75      0     50
;
```

A corresponding option, omit_zero_cols, suppresses all-zero columns when set to 1, and would eliminate two columns from Trans[CLEV,*,*].

10.5 Numeric options for display

The numbers in a table or list produced by display are the result of a transformation from the computer's internal numeric representation to a string of digits and symbols. AMPL's options for adjusting this transformation are shown in Table 10-2. In this section we first consider options that affect only the appearance of numbers, and then options that affect underlying solution values as well.

Appearance of numeric values

In all of our examples so far, the display command shows each numerical value to the same number of significant digits (from Figure 6-3):

```
ampl: display {p in PROD, t in 1..T} Make[p,t]/rate[p];
Make[p,t]/rate[p] [*,*] (tr)
:      bands       coils        :=
1     29.95       10.05
2     30          10
3     20          12
4     32.1429      7.85714
;


ampl: display {p in PROD, t in 1..T} prodcost[p]*Make[p,t];
prodcost[p]*Make[p,t] [*,*] (tr)
:      bands       coils        :=
1     59900       15477
2     60000       15400
3     40000       18480
4     64285.7     12100
;
```

The default is to use six significant digits, whether the result comes out as 7.85714 or 64285.7. Some numbers seem to have fewer digits, but only because trailing zeros have been dropped; 29.95 represents the number that is exactly 29.9500 to six digits, for example, and 59900 represents 59900.0.

By changing the option display_precision to a value other than six, you can vary the number of significant digits reported:

```
ampl: option display_precision 3; display 1/3, 100000/3;
1/3 = 0.333
100000/3 = 33300


ampl: option display_precision 9; display 1/3, 100000/3;
1/3 = 0.333333333
100000/3 = 33333.3333


ampl: option display_precision 0; display 1/3, 100000/3;
1/3 = 0.333333333333333336
100000/3 = 33333.333333333333
```

In the last example, a display_precision of 0 is interpreted specially; it tells display to represent numbers as exactly as possible, using as many digits as necessary. (To be precise, the displayed number is the shortest decimal representation that, when correctly rounded to your computer's representation, gives the value exactly as stored in your computer.)

Displays to a given precision provide the same degree of useful information about each number, but they can look ragged due to the varying numbers of digits after the decimal point. To specify rounding to a fixed number of decimal places, regardless of the resulting precision, you may set the option display_round. A nonnegative value specifies the number of digits to appear after the decimal point:

```
ampl: option display_round 2;
ampl: display {p in PROD, t in 1..T} Make[p,t]/rate[p];
Make[p,t]/rate[p] [*,*] (tr)
:   bands   coils   :=
1   29.95   10.05
2   30.00   10.00
3   20.00   12.00
4   32.14    7.86
;
```

A negative value indicates rounding before the decimal point. For example, when display_round is –2 all numbers are rounded to hundreds:

```
ampl: option display_round -2;
ampl: display {p in PROD, t in 1..T} prodcost[p]*Make[p,t];
prodcost[p]*Make[p,t] [*,*] (tr)
:   bands   coils   :=
1   59900   15500
2   60000   15400
3   40000   18500
4   64300   12100
;
```

Any integer value of display_round overrides the effect of display_precision. To turn off display_round, set it to some non-integer such as the empty string ''.

Depending on the solver you employ, you may find that some of the solution values that ought to be zero do not always quite come out that way. For example, here is one solver's report of the objective function terms cost[i,j] * Trans[i,j] for the assignment problem of Section 3.3:

```
ampl: option omit_zero_rows 1;
ampl: display {i in ORIG, j in DEST} cost[i,j] * Trans[i,j];
cost[i,j]*Trans[i,j] :=
Coullard 1021      6
Coullard 2053      2.05994e-17
Hazen    1053     -1.44196e-16
Hazen    2053      4
Hopp     1055      1
Hopp     1087      4.11988e-17
Hurter   1087      3.43323e-17
Hurter   2009      1
Jones    1021      8.23976e-17
Jones    1049      2
Mehrotra 1083      3
Mehrotra 1087     -2.74659e-17
Rieders  2019      2
Spearman 1053      4
Sun      2083      1
Tamhane  1021      3.43323e-17
Tamhane  1087      3
Zazanis  2087      1
;
```

Minuscule values like 2.05994e–17 and –1.44196e–16 have no significance in the context
of this problem; they would be zeros in an exact solution, but come out slightly nonzero
as an artifact of the way that the solver's algorithm interacts with the computer's representation of numbers.

To avoid viewing these numbers in meaningless precision, you can pick a reasonable
setting for display_round — in this case 0, since there are no digits of interest after
the decimal point:

```
ampl: option display_round 0;
ampl: display {i in ORIG, j in DEST} cost[i,j] * Trans[i,j];
cost[i,j]*Trans[i,j] :=
Coullard 1021      6
Coullard 2053      0
Hazen    1053     -0
Hazen    2053      4
Hopp     1055      1
Hopp     1087      0
Hurter   1087      0
Hurter   2009      1
Jones    1021      0
Jones    1049      2
Mehrotra 1083      3
Mehrotra 1087     -0
Rieders  2019      2
Spearman 1053      4
Sun      2083      1
Tamhane  1021      0
Tamhane  1087      3
Zazanis  2087      1
;
```

The small numbers are now represented only as 0 if positive or –0 if negative. If you
want to suppress their appearance entirely, however, you must set a separate option,
display_eps:

```
ampl: option display_eps 1e-10;
ampl: display {i in ORIG, j in DEST} cost[i,j] * Trans[i,j];
cost[i,j]*Trans[i,j] :=
Coullard 1021      6
Hazen    2053      4
Hopp     1055      1
Hurter   2009      1
Jones    1049      2
Mehrotra 1083      3
Rieders  2019      2
Spearman 1053      4
Sun      2083      1
Tamhane  1087      3
Zazanis  2087      1
;
```

Any value whose magnitude is less than the value of display_eps is treated as an exact zero in all output of display.

Rounding of solution values

The options display_precision, display_round and display_eps affect only the appearance of numbers, not their actual values. You can see this if you try to display the set of all pairs of i in ORIG and j in DEST that have a positive value in the preceding table, by comparing cost[i,j]*Trans[i,j] to 0:

```
ampl: display {i in ORIG, j in DEST: cost[i,j]*Trans[i,j] > 0};
set {i in ORIG, j in DEST: cost[i,j]*Trans[i,j] > 0}  :=
(Coullard,1021)     (Hurter,2009)      (Sun,2083)
(Coullard,2053)     (Jones,1021)       (Tamhane,1021)
(Hazen,2053)        (Jones,1049)       (Tamhane,1087)
(Hopp,1055)         (Mehrotra,1083)    (Zazanis,2087)
(Hopp,1087)         (Rieders,2019)
(Hurter,1087)       (Spearman,1053);
```

Even though a value like 2.05994e–17 is treated as a zero for purposes of display, it tests greater than zero. You could fix this problem by changing > 0 above to, say, > 0.1. As an alternative, you can set the option solution_round so that AMPL rounds the solution values to a reasonable precision when they are received from the solver:

```
ampl: option solution_round 10;
ampl: solve;
MINOS 5.4: optimal solution found.
23 iterations, objective 28

ampl: display {i in ORIG, j in DEST: cost[i,j]*Trans[i,j] > 0};
set {i in ORIG, j in DEST: cost[i,j]*Trans[i,j] > 0}  :=
(Coullard,1021)     (Jones,1049)       (Sun,2083)
(Hazen,2053)        (Mehrotra,1083)    (Tamhane,1087)
(Hopp,1055)         (Rieders,2019)     (Zazanis,2087)
(Hurter,2009)       (Spearman,1053);
```

The options solution_precision and solution_round work in the same way as display_precision and display_round, except that they are applied only to solution values returned by a solver, and they permanently change these values rather than just their appearance.

Rounded values can make a difference even when they are not near zero. As an example, we first use several display options to get a compact listing of the fractional solution to the scheduling model of Figure 12-4:

```
ampl: model sched.mod;
ampl: data sched.dat;

ampl: solve;
MINOS 5.4: optimal solution found.
25 iterations, objective 265.6
```

```
ampl: option display_width 60;
ampl: option display_1col 5;

ampl: option display_eps 1e-10;

ampl: option omit_zero_rows 1;

ampl: display Work;
Work [*] :=
  7 25         32 22         81  6.8      106 20.2
 12  3.8       43 28.8       92 21.2      113  6.8
 19 10.6       55  6.8       95  3.8      116 25
 23  3         71 35.6      102 10.6      123 35.6
;
```

Each value Work[j] represents the number of workers assigned to schedule j. We can get a quick practical schedule by rounding the fractional values up to the next highest integer; using the ceil function to perform the rounding up, we see that the total number of workers needed should be:

```
ampl: display sum {j in SCHEDS} ceil(Work[j]);
sum{j in SCHEDS} ceil(Work[j]) = 272
```

If you copy the numbers from the preceding table and round them up by hand, however, you'll find that they only sum to 270. The source of the difficulty can be seen by displaying the numbers to full precision:

```
ampl: option display_eps 0;
ampl: option display_precision 0;

ampl: display Work;
Work [*] :=
  5  3.638454333654193e-14      81   6.79999999999995
  7 24.999999999999996          92  21.19999999999992
 12  3.8000000000000087         95   3.8000000000000065
 19 10.599999999999977         102  10.59999999999999
 23  3.0000000000000284        106  20.200000000000006
 32 22                         113   6.79999999999995
 43 28.79999999999998          116  25
 55  6.8000000000000025        123  35.59999999999998
 71 35.60000000000016
;
```

Half the problem is due to the minuscule positive value of Work[5], which was rounded up to 1. The other half is due to Work[23]; although it is 3 in an exact solution, it comes back from the solver with a slightly larger value of 3.0000000000000284, so it gets rounded up to 4.

The easiest way to ensure that your arithmetic works correctly in this case is again to set solution_round before solve:

```
ampl: option solution_round 10;
ampl: solve;
MINOS 5.4: optimal solution found.
17 iterations, objective 265.6

ampl: display sum {j in SCHEDS} ceil(Work[j]);
sum{j in SCHEDS} ceil(Work[j]) = 270
```

We picked a value of 10 for solution_round because we observed that the slight inaccuracies in the solver's values occurred well past the 10th decimal place.

The effect of solution_round or solution_precision applies to all values returned by the solver. If you want to modify only certain values, see the discussion of the let command in Section 10.8 below.

10.6 Other output commands

Two additional AMPL commands have much the same syntax as display, but do not automatically format their output. The print command does no formatting at all, while the printf command requires an explicit description of the desired formatting.

The print command

A print command produces a single line of output,

```
ampl: print sum {p in PROD, t in 1..T} revenue[p,t]*Sell[p,t],
ampl?           sum {p in PROD, t in 1..T} prodcost[p]*Make[p,t],
ampl?           sum {p in PROD, t in 1..T} invcost[p]*Inv[p,t];
787810 269477 3300

ampl: print {t in 1..T, p in PROD} Make[p,t];
5990 1407 6000 1400 1400 3500 2000 4200
```

or, if you follow print by an indexing expression and a colon, a line of output for each member of the index set:

```
ampl: print {t in 1..T}: {p in PROD} Make[p,t];
5990 1407
6000 1400
1400 3500
2000 4200
```

The printed entries are normally separated by a space; the option print_separator can be used to change this. For instance, you might set print_separator to a tab for data to be imported by a spreadsheet. To do this, type option print_separator "→", where → stands for the tab character.

The keyword print (with optional indexing expression and colon) is followed by a print item or comma-separated list of print items. A print item can be a value, or an indexing expression followed by a value or parenthesized list of values. Thus a print item is much like a display item, except that only individual values may appear; although you can say display rate, you must explicitly specify print {p in PROD} rate[p]. Also a set may not be an argument to print, although its members may be:

```
ampl: print PROD;
syntax error
context:  print  >>> PROD; <<<

ampl: print {p in PROD} (p,rate[p]);
bands 200 coils 140
```

Unlike display, however, print allows indexing to be nested within an indexed item:

```
ampl: print {p in PROD} (p, rate[p], {t in 1..T} Make[p,t]);
bands 200 5990 6000 1400 2000 coils 140 1407 1400 3500 4200
```

At the end of the print command, you may specify output to a file by use of > or >> followed by a filename; the rules are the same as given for display earlier in this chapter.

The representation of numbers in the output of print is governed by the print_precision and print_round options, which work exactly like the display_precision and display_round options for the display command. Initially print_precision is 0 and print_round is an empty string, so that by default print uses as many digits as necessary to represent each value as precisely as possible.

Working interactively, you may find print useful for viewing a few values on your screen in a more compact format than display produces. With output redirected to a file, print is useful for writing results in the right form for spreadsheets and other data analysis tools.

The printf command

The syntax of printf is exactly the same as that of print, except that the first print item is a character string that provides formatting instructions for the remaining items:

```
ampl: printf "Total revenue is $%6.2f.\n",
ampl?    sum {p in PROD, t in 1..T} revenue[p,t]*Sell[p,t];
Total revenue is $787810.00.
```

The format string contains two types of objects: ordinary characters, which are copied to the output, and conversion specifications, which govern the appearance of successive remaining print items. Each conversion specification begins with the character % and ends with a conversion character. For example, %6.2f specifies conversion to a decimal representation at least six characters wide with two digits after the decimal point. The complete rules are much the same as for the printf function in the C programming language; a summary appears in Appendix A.13.1.

The output from `printf` is not automatically broken into lines. A line break must be indicated explicitly by the combination \n, representing a "newline" character, in the format string. To produce a series of lines, use the indexed version of `printf`:

```
ampl: printf {t in 1..T}: "%3i%12.2f%12.2f\n", t,
ampl?     sum {p in PROD} revenue[p,t]*Sell[p,t],
ampl?     sum {p in PROD} prodcost[p]*Make[p,t];
  1   159210.00     75377.00
  2   243500.00     75400.00
  3   167300.00     52500.00
  4   217800.00     66200.00
```

This `printf` is executed once for each member of the indexing set preceding the colon; for each `t` in `1..T` the format is applied again, and the \n character generates another line break.

The `printf` command is mainly useful, in conjunction with redirection of output to a file, for printing short summary reports in a readable format. Because the number of conversion specifications in the format string must match the number of values being printed, `printf` cannot conveniently produce tables in which the number of items on a line may vary from run to run, such as a table of all `Make[p,t]` values.

10.7 Related solution values

Sets, parameters and variables are the most obvious things to look at in interpreting the solution of a linear program, but AMPL also provides ways of examining objectives, bounds, slacks, dual prices and reduced costs associated with the optimal solution.

To distinguish the different values associated with the same model component, AMPL uses "qualified" names that consist of a variable or constraint identifier, a dot (.), and a suffix. For instance, the upper bounds for the variable `Make` are called `Make.ub`, and the upper bound for `Make["coils",2]` is written `Make["coils",2].ub`. (Note that the suffix comes after the subscript.) A qualified name can be used like an unqualified one, so that `display Make.ub` prints a table of upper bounds on the `Make` variables, while `display Make, Make.ub` prints a list of the optimal values and upper bounds.

Objective functions

The name of the objective function (from a `minimize` or `maximize` declaration) refers to the objective's value computed from the current values of the variables. This name can be used to represent the optimal objective value in `display`, `print`, or `printf`:

```
ampl: print 100 * total_profit /
ampl?     sum {p in PROD, t in 1..T} revenue[p,t] * Sell[p,t];
65.37528084182735
```

If the current model declares several objective functions, you can refer to any of them, even though only one has been optimized.

Bounds and slacks

The suffixes `.lb` and `.ub` on a variable denote its lower and upper bounds; `.slack` denotes the difference of a variable's value from its nearer bound. Here's an example from Figure 2-1:

```
ampl: display Buy.lb,Buy,Buy.ub,Buy.slack;
:     Buy.lb        Buy    Buy.ub Buy.slack     :=
BEEF    2          2          10    0
CHK     2         10          10    0
FISH    2          2          10    0
HAM     2          2          10    0
MCH     2          2          10    0
MTL     2          6.23596    10    3.76404
SPG     2          5.25843    10    3.25843
TUR     2          2          10    0
;
```

The reported bounds are those that were sent to the solver. Thus they include not only the bounds specified in >= and <= phrases of `var` declarations, but also certain bounds that were deduced from the constraints by the presolve phase (Section 10.2). Other suffixes let you look at the original bounds and at additional bounds deduced by presolve; see Appendix A.11 for details.

Two equal bounds denote a fixed variable, which is normally eliminated by presolve. Thus in the planning model of Figure 4-4, the constraint `Inv[p,0] = inv0[p]` fixes the initial inventories:

```
ampl: display {p in PROD} (Inv[p,0].lb,inv0[p],Inv[p,0].ub);
:       Inv[p,0].lb inv0[p] Inv[p,0].ub    :=
bands       10        10        10
coils        0         0         0
;
```

In the production-and-transportation model of Figure 4-6, the constraint

```
sum {i in ORIG} Trans[i,j,p] = demand[j,p]
```

leads presolve to fix three variables at zero, because `demand["LAN","plate"]` is zero:

```
ampl: display {i in ORIG}
ampl?    (Trans[i,"LAN","plate"].lb,Trans[i,"LAN","plate"].ub);
:      Trans[i,'LAN','plate'].lb Trans[i,'LAN','plate'].ub     :=
CLEV              0                          0
GARY              0                          0
PITT              0                          0
;
```

As this example suggests, presolve's adjustments to the bounds may depend on the data as well as the structure of the constraints.

The concepts of bounds and slacks have an analogous interpretation for the constraints of a model. Any AMPL constraint can be put into the standard form lower-bound $\leq$ body $\leq$ upper-bound, where the body is a sum of all terms involving variables, while the lower-bound and upper-bound depend only on the data. The suffixes .lb, .body and .ub give the current values of these three parts of the constraint. For example, in the diet model of Figure 5-1 we have the declarations

```
subject to diet_min {i in MINREQ}:
    sum {j in FOOD} amt[i,j] * Buy[j] >= n_min[i];

subject to diet_max {i in MAXREQ}:
    sum {j in FOOD} amt[i,j] * Buy[j] <= n_max[i];
```

and the following constraint bounds:

```
ampl: display diet_min.lb, diet_min.body, diet_min.ub;
:    diet_min.lb diet_min.body diet_min.ub     :=
A          700         1013.98      Infinity
B1           0          605         Infinity
B2           0          492.416     Infinity
C          700          700         Infinity
CAL      16000        16000         Infinity
;

ampl: display diet_max.lb, diet_max.body, diet_max.ub;
:    diet_max.lb diet_max.body diet_max.ub     :=
A      -Infinity     1013.98       20000
CAL    -Infinity    16000          24000
NA     -Infinity    43855.9        50000
;
```

Naturally, <= constraints have no lower bounds, and >= constraints have no upper bounds; AMPL uses -Infinity and Infinity in place of a number to denote these cases. (Both the lower and the upper bound can be finite, if the constraint is specified with two <= or >= operators; see Section 8.4. For an = constraint the two bounds are the same.)

The suffix .slack refers to the difference between the body and the nearer bound:

```
ampl: display diet_min.slack;
diet_min.slack [*] :=
  A  313.978
  B1  605
  B2  492.416
  C    0
CAL    0
;
```

For constraints that have a single <= or >= operator, the slack is always the difference between the expressions to the left and right of the operator, even if there are variables on both sides. The constraints that have a slack of zero are the ones that are truly constraining at the optimal solution.

Dual values and reduced costs

Associated with each constraint in a linear program is a quantity variously known as the dual variable, marginal value or shadow price. In the AMPL command environment, these dual values are denoted by the names of the constraints, without any qualifying suffix. Thus for example in Figure 4-6 there is a collection of constraints named Demand,

```
subject to Demand {j in DEST, p in PROD}:
    sum {i in ORIG} Trans[i,j,p] = demand[j,p];
```

and a table of the dual values associated with these constraints can be viewed by

```
ampl: display Demand;
Demand [*,*]
:        bands     coils   plate     :=
DET     201       190.714   199
FRA     209       204       211
FRE     266.2     273.714   285
LAF     201.2     198.714   205
LAN     202       193.714     0
STL     206.2     207.714   216
WIN     200       190.714   198
;
```

Solvers return optimal dual values to AMPL along with the optimal values of the "primal" variables. We have space here only to summarize the most common interpretation of dual values; the extensive theory of duality and applications of dual variables can be found in any textbook on linear programming.

To start with an example, consider the constraint Demand["DET","bands"] above. If we change the value of the constant demand["DET","bands"] in this constraint, the optimal value of the objective function total_cost changes accordingly. If we were to plot the optimal value of total_cost versus all possible values of demand["DET","bands"], the result would be a cost "curve" that shows how overall cost varies with demand for bands at Detroit.

Additional computation would be necessary to determine the entire cost curve, but you can learn something about it from the optimal dual values. After you solve the linear program using a particular value of demand["DET","bands"], the dual price for the constraint tells you the *slope* of the cost curve, at the demand's current value. In our example, reading from the table above, we find that the slope of the curve at the current demand is 201. This means that total production and shipping cost is increasing at the rate of $201 for each extra ton of bands demanded at DET, or is decreasing by $201 for each reduction of one ton in the demand.

As an example of an inequality, consider the following constraint from the same model:

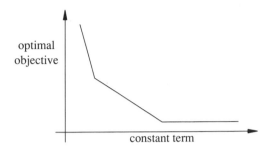

Figure 10-1: Piecewise-linear plot of objective function.

```
subject to Time {i in ORIG}:
    sum {p in PROD} (1/rate[i,p]) * Make[i,p] <= avail[i];
```

Here it is revealing to look at the dual values together with the slacks:

```
ampl: display Time,Time.slack;
:          Time    Time.slack    :=
CLEV    -1522.86    0
GARY    -3040       0
PITT        0       10.5643
;
```

Where the slack is positive, the dual value is zero. Indeed, the positive slack implies that the optimal solution does not use all of the time available at PITT; hence changing avail["PITT"] somewhat does not affect the optimum. On the other hand, where the slack is zero the dual value may be significant. In the case of GARY the value is –3040, implying that the total cost is decreasing at a rate of $3040 for each extra hour available at GARY, or is increasing at a rate of $3040 for each hour lost.

In general, if we plot the optimal objective versus a constraint's constant term, the curve will be convex piecewise-linear (Figure 10-1) for a minimization, or concave piecewise-linear (same, but upside-down) for a maximization. In terms of the standard form lower-bound ≤ body ≤ upper-bound introduced previously, the optimal dual values can be viewed as follows. If the slack of the constraint is positive, the dual value is zero. If the slack is zero, the body of the constraint must equal one (or both) of the bounds, and the dual value pertains to the equaled bound or bounds. Specifically, the dual value is the slope of the plot of the objective versus the bound, evaluated at the current value of the bound; equivalently it is the rate of change of the optimal objective with respect to the bound value.

A nearly identical analysis applies to the bounds on a variable. The role of the dual value is played by the variable's so-called reduced cost, which can be viewed from the AMPL command environment by use of the suffix .rc. As an example, here are the bounds and reduced costs for the variables in Figure 5-1:

```
ampl: display Buy.lb,Buy,Buy.ub,Buy.rc;
:      Buy.lb      Buy    Buy.ub     Buy.rc          :=
BEEF     2         2        10       1.73663
CHK      2        10        10      -0.853371
FISH     2         2        10       0.255281
HAM      2         2        10       0.698764
MCH      2         2        10       0.246573
MTL      2         6.23596  10       0
SPG      2         5.25843  10       0
TUR      2         2        10       0.343483
;
```

Since Buy["MTL"] has slack with both its bounds, its reduced cost is zero. Buy["HAM"] is at its lower bound, so its reduced cost indicates that total cost is increasing at about 70 cents per unit increase in the lower bound, or is decreasing at about 70 cents per unit decrease in the lower bound. On the other hand, Buy["CHK"] is at its upper bound, and its negative reduced cost indicates that total cost is decreasing at about 85 cents per unit increase in the upper bound, or is increasing at about 85 cents per unit decrease in the upper bound.

If the current value of a bound happens to lie right at a breakpoint in the relevant curve — one of the places where the slope changes abruptly in Figure 10-1 — the objective will change at one rate as the bound increases, but at a different rate as the bound decreases. In the extreme case either of these rates may be infinite, indicating that the linear program becomes infeasible if the bound is increased or decreased by any further amount. A solver reports only one optimal dual price or reduced cost to AMPL, however, which may be the rate in either direction, or some value between them.

In any case, moreover, a dual price or reduced cost can give you only one slope on the piecewise-linear curve of objective values. Hence these quantities should be used as only an initial guide to the objective's sensitivity to certain variable or constraint bounds. If the sensitivity is very important to your application, you can make a series of runs with different bound settings; see Section 10.8 below for ways to quickly change a small part of the data. (There do exist algorithms for finding part or all of the piecewise-linear curve, given a linear change in one or more bounds, but they are not directly supported by the current version of AMPL.)

10.8 Modifying and re-solving

Several commands are provided to help you change the current optimization problem without leaving AMPL. We first describe ones that change the model, then ones that change only the data. We conclude by explaining options for "warm starts" of a solver after data values have been changed.

Changing the model

The simplest (but most drastic) way to change the model is by typing the command `reset`, which expunges all of the current model and data. Following `reset`, you can issue new `model` and `data` commands to set up a different optimization problem; the effect is like typing `quit` and then restarting AMPL, except that options are not reset to their default values. If your operating system allows you to edit files while keeping AMPL active, `reset` is valuable for debugging and experimentation; you may make changes to the model or data files, type `reset`, then read in the modified files. The `shell` command lets you escape from AMPL to run commands; see Appendix A.13.9.

The `drop` command instructs AMPL to ignore certain constraints or objectives of the current model. As an example, the constraints of Figure 5-1 initially include

```
subject to diet_max {i in MAXREQ}:
    sum {j in FOOD} amt[i,j] * Buy[j] <= n_max[i];
```

A `drop` command can specify a particular one of these constraints to ignore:

```
drop diet_max["CAL"];
```

or it may specify all constraints or objectives indexed by some set:

```
drop {i in MAXNOT} diet_max[i];
```

where `MAXNOT` has previously been defined as a subset of `MAXREQ`. The entire collection of constraints can be ignored by

```
drop {i in MAXREQ} diet_max[i];
```

or more simply:

```
drop diet_max;
```

In general, this command consists of the keyword `drop`, an optional indexing expression, and a constraint name that may be subscripted. Successive `drop` commands have a cumulative effect.

The `restore` command reverses the effect of `drop`. It has the same syntax, except for the keyword `restore`.

The `fix` command fixes specified variables at their current values, as if there were a constraint that the variables must equal these values; the `unfix` command reverses the effect. These commands have the same syntax as `drop` and `restore`, except that they name variables rather than constraints. For example, if we follow the suggestion in Section 8.1 to initialize all the diet model's variables at their lower bounds,

```
var Buy {j in FOOD} := f_min[j], >= f_min[j], <= f_max[j];
```

the following commands force all foods that have more than 1200 mg of sodium per package to be purchased at their lower bounds and optimizes over the remaining variables:

```
ampl: fix {j in FOOD: amt["NA",j] > 1200} Buy[j];
ampl: solve;
MINOS 5.4: optimal solution found.
6 iterations, objective 86.92

ampl: display {j in FOOD} (Buy[j].lb,Buy[j],amt["NA",j]);
:     Buy[j].lb    Buy[j]    amt['NA',j]   :=
BEEF      2          2            938
CHK       2          2           2180
FISH      2         10            945
HAM       2          2            278
MCH       2          9.42857     1182
MTL       2         10            896
SPG       2          2           1329
TUR       2          2           1397
;
```

The fix and unfix commands can also be useful after the first solve, particularly in conjunction with a change in the objective or data.

Changing the data

To delete the current data for several model components, without changing the current model itself, type:

```
reset data MINREQ, MAXREQ, amt, n_min, n_max;
```

You may then use data commands to read in new values for these sets and parameters. To delete all data, type reset data.

The update data command works similarly, but does not actually delete any data until new values are assigned. Thus if you type

```
update data MINREQ, MAXREQ, amt, n_min, n_max;
```

but you only read in new values for MINREQ, amt and n_min, the previous values for MAXREQ and n_max will remain. If instead you used reset data, MAXREQ and n_max would be without values, and you would get an error message when you next tried to solve.

The let command also permits you to change particular data values while leaving the model the same, but it is more convenient for small or easy-to-describe changes than reset data or update data. You can use it, for example, to solve the diet model of Figure 5-1 for a series of upper bounds f_max["CHK"] on the purchases of food CHK:

```
ampl: model dietu.mod; data dietu.dat; solve;
MINOS 5.4: optimal solution found.
6 iterations, objective 74.27382022

ampl: let f_max["CHK"] := 11;
ampl: solve;
MINOS 5.4: optimal solution found.
3 iterations, objective 73.43818182
```

```
ampl: let f_max["CHK"] := 12;
ampl: solve;
MINOS 5.4: optimal solution found.
2 iterations, objective 73.43818182
```

Relaxing the bound to 11 reduces the cost somewhat, but further relaxation has no benefit.

An indexing expression may be given after the keyword `let`, in which case a change is made for each member of the specified indexing set. You could use this feature to change all upper bounds to 8:

```
let {f in FOOD} f_max[f] := 8;
```

or to increase all upper bounds by 10 percent:

```
let {f in FOOD} f_max[f] := 1.1 * f_max[f];
```

In general this command consists of the keyword `let`, an indexing expression if needed, and an assignment. Any set or parameter whose declaration does not have a := phrase may be specified to the left of the assignment's := operator, while to the right may appear any appropriate expression that can currently be evaluated.

Although AMPL does not impose any restrictions on what you can change using `let`, you should take care in changing any set or parameter that affects the indexing of other data. For example, after solving the multiperiod production problem of Figures 4-4 and 4-5, it might be tempting to change the number of weeks T from 4 (as given in the original data) to 3:

```
ampl: let T := 3;
ampl: solve;
error processing param avail:
        invalid subscript avail[4]
error processing param market:
        invalid subscript market[bands,4]
error processing param market:
        invalid subscript market[coils,4]
```

(additional messages omitted)

```
Ignoring solve command because of errors in model.
```

The problem here is that AMPL still has current data for 4th-week parameters such as `avail[4]`, which has become invalid with the change of T to 3. If you want to properly reduce the number of weeks in the linear program while using the same data, you must declare two parameters:

```
param Tdata integer > 0;
param T integer <= Tdata;
```

Use `1..Tdata` for indexing in the `param` declarations, while retaining `1..T` for the variables, objective and constraints; then you can use `let` to change T as you like.

You can also use the `let` command to change the current values of variables. This is sometimes a convenient feature for exploring an alternative solution. For example, here

ceil (x)	ceiling of x (next higher integer)
floor (x)	floor of x (next lower integer)
precision (x, n)	x rounded to n significant digits
round (x, n)	x rounded to n digits past the decimal point
round (x)	x rounded to the nearest integer
trunc (x, n)	x truncated to n digits past the decimal point
trunc (x)	x truncated to an integer (fractional part dropped)

Table 10-3: Rounding functions.

is what happens when we solve for the optimal diet as above, then round the optimal solution down to the next lowest integer number of packages:

```
ampl: model dietu.mod; data dietu.dat; solve;
MINOS 5.4: optimal solution found.
6 iterations, objective 74.27382022

ampl: let {j in FOOD} Buy[j] := floor(Buy[j]);
ampl: display total_cost, diet_min.slack;
total_cost = 73.29
:       n_min  diet_min.slack   :=
A       700         291
B1        0         595
B2        0         485
C       700         -20
CAL   16000        -190
;
```

Because we have used let to change the values of the variables, the objective and the slacks are automatically computed from the new, rounded values. The cost has dropped by about a dollar, but the solution is now short of the requirements for C by about 3% and for CAL by about 1%.

AMPL provides a variety of rounding functions that can be used in this way. They are summarized in Table 10-3.

Re-solving after changes

After some data values have been changed, or some constraints have been added or dropped, the current values of the variables no longer necessarily give an optimal solution. Your solver may be able to use the current solution as a starting guess, however, in its search for a new optimum. Thus when solve is issued a second or subsequent time for the same model, AMPL normally passes to the solver the current values of both the primal variables (defined by var declarations) and the dual variables (corresponding to constraints as explained above). If you want AMPL to instead pass the same starting guesses that it used the first time — as determined from the model (Section 8.1) or from the data (Section 9.3) — change the option reset_initial_guesses to 1 from its default value of 0. If you want only primal starting guesses to be sent to the solver, change the option `dual_initial_guesses` to 0 from its default value of 1. You can display both the most recent and the original initial guesses by use of appropriate suffixes, as shown in Appendix A.11.

When the current solution serves as a starting guess, a new optimum may be found much faster than the original one — particularly if only a few changes have been made to the model or data. Solvers do differ in the degree to which they can make use of starting guesses, however. In general, guesses are most directly useful to solvers for nonlinear programming, as shown by some of our examples in Chapter 13. The situation is more complicated for linear programming; details may be found in the solver-specific documentation that accompanies your AMPL software.

10.9 Batch operation

After a model has been under development for a while, you may find that you are typing the same series of commands again and again. To speed things up, you may put the commands into a file, and run them all automatically by typing `include` and the filename. For example, if the file `dietu.run` contains

```
model dietu.mod;
data dietu.dat;
solve;
option display_1col 5;
option display_round 1;
display Buy;
```

then including it will load the model and data, run the problem, and display the optimal values of the variables:

```
ampl: include dietu.run;
MINOS 5.4: optimal solution found.
6 iterations, objective 74.27382022

Buy [*] :=
BEEF  2.0    FISH  2.0    MCH  2.0    SPG  5.3
 CHK 10.0    HAM  2.0    MTL  6.2    TUR  2.0
;
```

Any line that may be typed at an AMPL prompt may appear as a line in an `include` file.

An occurrence of `include` and a filename is replaced by the contents of the file. Thus `include` may even appear in the middle of some other statement, and does not require a terminating semicolon.

The previously described `model` and `data` commands are convenient special cases of `include` that automatically put the command interpreter into model or data mode before reading the specified file. By contrast, `include` leaves the mode unchanged. To keep things simple, the examples in this book always assume that `model` reads a file of model declarations, and that `data` reads a file of data statements. You may use any of the three commands `model`, `data` and `include` to read any file, however; the only difference is the assumed mode when the reading starts. Working with a small model, for example, you might find it convenient to store in one file all the model declarations, a `data` command, and all the data statements; either a `model` or an `include` command could read this file to set up both model and data in one operation.

When an included file itself contains an `include`, `model` or `data` command, reading of the first file is suspended while the contents of the contained file are included. In the above example, the command `include dietu.run` causes the subsequent inclusion of the files `dietu.mod` and `dietu.dat`.

One particularly useful kind of `include` file contains a list of `option` commands that you want to run before any other commands, to modify the default options. You can arrange to include such a file automatically at startup; you can even have AMPL write such a file automatically at the end of a session, so that your option settings will be automatically restored the next time around. Details of this arrangement depend on your operating system; see Appendix A.13.2.

AMPL also provides options to run commands from a file in ''batch'' mode, without ever displaying a prompt on your screen. This is another procedure that is somewhat system-dependent; it generally involves typing a command such as `ampl dietu.run` at your operating system's prompt. For more information, consult the system-specific documentation that comes with your AMPL software.

Bibliography

A. L. Brearly, G. Mitra and H. P. Williams, ''Analysis of Mathematical Programming Problems Prior to Applying the Simplex Algorithm.'' Mathematical Programming **8** (1975) pp. 54–83. Origin of many of the bound-comparison tests used in AMPL's presolve phase.

11

Network Linear Programs

Models of networks have appeared in several chapters, notably in the transportation problems in Chapter 3. We now return to the formulation of these models, and AMPL's features for handling them.

Figure 11-1 shows the sort of diagram commonly used to describe a network problem. A circle represents a *node* of the network, and an arrow denotes an *arc* running from one node to another. A *flow* of some kind travels from node to node along the arcs, in the directions of the arrows.

An endless variety of models involve optimization over such networks. Many cannot be expressed in any straightforward algebraic way or are very difficult to solve. Our concern in this chapter is with a particular class of network optimization models in which the decision variables represent the amounts of flow on the arcs, and the constraints are limited to two kinds: simple bounds on the flows, and conservation of flow at the nodes. Models restricted in this way give rise to the problems known as network linear programs. They are especially easy to describe and to solve, yet retain an impressively wide range of applicability.

We begin with minimum-cost transshipment models, which are the largest and most intuitive source of network linear programs, and then proceed to other well-known cases: maximum flow, shortest path, transportation and assignment models. Examples are initially given in terms of standard AMPL variables and constraints, defined in var and subject to declarations. In the later sections of this chapter, we introduce node and arc declarations that permit models to be described more directly in terms of their network structure. A concluding section discusses the importance of formulating network models in a way that permits the resulting linear programs to be solved most efficiently.

11.1 Minimum-cost transshipment models

As a concrete example, imagine that the nodes and arcs in Figure 11-1 represent cities and intercity transportation links. A manufacturing plant at the city marked PITT will

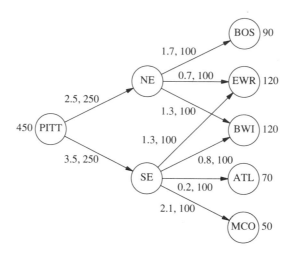

Figure 11-1: A directed network.

make 450,000 packages of a certain product in the next week, as indicated by the 450 at the left of the diagram. The cities marked NE and SE are the northeast and southeast distribution centers, which receive packages from the plant and transship them to warehouses at the cities coded as BOS, EWR, BWI, ATL and MCO. (Frequent flyers will recognize Boston, Newark, Baltimore, Atlanta, and Orlando.) These warehouses require 90, 120, 120, 70 and 50 thousand packages, respectively, as indicated by the numbers at the right. For each intercity link there is a shipping cost per thousand packages and an upper limit on the packages that can be shipped, indicated by the two numbers next to the corresponding arrow in the diagram.

The optimization problem over this network is to find the lowest-cost plan of shipments that uses only the available links, respects the specified capacities, and meets the requirements at the warehouses. We first model this as a general network flow problem, and then consider alternatives that specialize the model to the particular situation at hand. We conclude by introducing a few of the most common variations on the network flow constraints.

A general transshipment model

To write a model for any problem of shipments from city to city, we can start by defining a set of cities and a set of links. Each link is in turn defined by a start city and an end city, so we want the set of links to be a subset of the set of ordered pairs of cities:

```
set CITIES;
set LINKS within (CITIES cross CITIES);
```

Corresponding to each city there is potentially a supply of packages and a demand for packages:

```
param supply {CITIES} >= 0;
param demand {CITIES} >= 0;
```

In the case of the problem described by Figure 11-1, the only nonzero value of `supply` should be the one for PITT, where packages are manufactured and supplied to the distribution network. The only nonzero values of `demand` should be those corresponding to the five warehouses.

The costs and capacities are indexed over the links:

```
param cost {LINKS} >= 0;
param capacity {LINKS} >= 0;
```

as are the decision variables, which represent the amounts to ship over the links. These variables are nonnegative and bounded by the capacities:

```
var Ship {(i,j) in LINKS} >= 0, <= capacity[i,j];
```

The objective is

```
minimize Total_Cost:
   sum {(i,j) in LINKS} cost[i,j] * Ship[i,j];
```

which represents the sum of the shipping costs over all of the links.

It remains to describe the constraints. At each city, the packages supplied plus packages shipped in must balance the packages demanded plus packages shipped out:

```
subject to Balance {k in CITIES}:
   supply[k] + sum {(i,k) in LINKS} Ship[i,k]
      = demand[k] + sum {(k,j) in LINKS} Ship[k,j];
```

Because the expression

```
sum {(i,k) in LINKS} Ship[i,k]
```

appears within the scope of definition of the dummy index k, the summation is interpreted to run over all cities i such that (i,k) is in LINKS. That is, the summation is over all links into city k; similarly, the second summation is over all links out of k. This indexing convention, which was explained in Section 6.2, is frequently useful in describing network balance constraints algebraically. Figures 11-2a and 11-2b display the complete model and data for the particular problem depicted in Figure 11-1.

If all of the variables are moved to the left of the = sign and the constants to the right, the `Balance` constraint becomes:

```
subject to Balance {k in CITIES}:
    sum {(i,k) in LINKS} Ship[i,k]
  - sum {(k,j) in LINKS} Ship[k,j]
        = demand[k] - supply[k];
```

This variation may be interpreted as saying that, at each city k, shipments in minus shipments out must equal "net demand". If no city has both a plant and a warehouse (as in

```
set CITIES;
set LINKS within (CITIES cross CITIES);

param supply {CITIES} >= 0;    # amounts available at cities
param demand {CITIES} >= 0;    # amounts required at cities

check: sum {i in CITIES} supply[i] = sum {j in CITIES} demand[j];

param cost {LINKS} >= 0;        # shipment costs/1000 packages
param capacity {LINKS} >= 0;   # max packages that can be shipped

var Ship {(i,j) in LINKS} >= 0, <= capacity[i,j];
                                # packages to be shipped

minimize Total_Cost:
   sum {(i,j) in LINKS} cost[i,j] * Ship[i,j];

subject to Balance {k in CITIES}:
   supply[k] + sum {(i,k) in LINKS} Ship[i,k]
      = demand[k] + sum {(k,j) in LINKS} Ship[k,j];
```

Figure 11-2a: General transshipment model (net1.mod).

```
set CITIES := PITT  NE SE   BOS EWR BWI ATL MCO ;

set LINKS := (PITT,NE) (PITT,SE)
             (NE,BOS)  (NE,EWR)  (NE,BWI)
             (SE,EWR)  (SE,BWI)  (SE,ATL)  (SE,MCO);

param supply  default 0 := PITT 450 ;

param demand  default 0 :=
   BOS  90,   EWR 120,   BWI 120,   ATL 70,   MCO  50;

param:        cost  capacity  :=
   PITT NE    2.5      250
   PITT SE    3.5      250

   NE BOS     1.7      100
   NE EWR     0.7      100
   NE BWI     1.3      100

   SE EWR     1.3      100
   SE BWI     0.8      100
   SE ATL     0.2      100
   SE MCO     2.1      100 ;
```

Figure 11-2b: Data for general transshipment model (net1.dat).

our example), then positive net demand always indicates warehouse cities, negative net demand indicates plant cities, and zero net demand indicates transshipment cities. Thus we could have gotten by with just one parameter net_demand in place of demand and supply, with the sign of net_demand[k] indicating what goes on at city k. Alternative formulations of this kind are often found in descriptions of network flow models.

Specialized transshipment models

The preceding general approach has the advantage of being able to accommodate any pattern of supplies, demands, and links between cities. For example, a simple change in the data would suffice to model a plant at one of the distribution centers, or to allow shipment links between some of the warehouses.

The disadvantage of a general formulation is that it fails to show clearly what arrangement of supplies, demands and links is expected, and in fact will allow inappropriate arrangements. If we know that the situation will be like the one shown in Figure 11-1, with supply at one plant, which ships to distribution centers, which then ship to warehouses that satisfy demand, the model can be specialized to exhibit and enforce such a structure.

To show explicitly that there are three different kinds of cities in the specialized model, we can declare them separately. We use a symbolic parameter rather than a set to hold the name of the plant, in order to specify that only one plant is expected:

```
param p_city symbolic;
set D_CITY;
set W_CITY;
```

There must be a link between the plant and each distribution center, so we need a subset of pairs only to specify which links connect distribution centers to warehouses:

```
set DW_LINKS within (D_CITY cross W_CITY);
```

With the declarations organized in this way, it is impossible to specify inappropriate kinds of links, such as ones between two warehouses or from a warehouse back to the plant.

There is one parameter that represents the supply at the plant, and a collection of demand parameters indexed over the warehouses:

```
param p_supply >= 0;
param w_demand {W_CITY} >= 0;
```

These declarations allow supply and demand to be defined only where they belong.

At this juncture, we can define the sets CITIES and LINKS and the parameters supply and demand as they would be required by our previous model:

```
set CITIES := {p_city} union D_CITY union W_CITY;
set LINKS := ({p_city} cross D_CITY) union DW_LINKS;

param supply {k in CITIES} :=
   if k = p_city then p_supply else 0;

param demand {k in CITIES} :=
   if k in W_CITY then w_demand[k] else 0;
```

The rest of the model can then be exactly as in the general case, as indicated in Figures 11-3a and 11-3b.

```
param p_city symbolic;

set D_CITY;
set W_CITY;
set DW_LINKS within (D_CITY cross W_CITY);

param p_supply >= 0;               # amount available at plant
param w_demand {W_CITY} >= 0;   # amounts required at warehouses

   check: p_supply = sum {k in W_CITY} w_demand[k];

set CITIES := {p_city} union D_CITY union W_CITY;
set LINKS := ({p_city} cross D_CITY) union DW_LINKS;

param supply {k in CITIES} :=
   if k = p_city then p_supply else 0;

param demand {k in CITIES} :=
   if k in W_CITY then w_demand[k] else 0;

### Remainder same as general transshipment model ###

param cost {LINKS} >= 0;        # shipment costs/1000 packages
param capacity {LINKS} >= 0;   # max packages that can be shipped

var Ship {(i,j) in LINKS} >= 0, <= capacity[i,j];
                                 # packages to be shipped
minimize Total_Cost:
   sum {(i,j) in LINKS} cost[i,j] * Ship[i,j];

subject to Balance {k in CITIES}:
   supply[k] + sum {(i,k) in LINKS} Ship[i,k]
      = demand[k] + sum {(k,j) in LINKS} Ship[k,j];
```

Figure 11-3a: Specialized transshipment model (net2.mod).

Alternatively, we can maintain references to the different types of cities and links throughout the model. This means that we must declare two types of costs, capacities and shipments:

```
param pd_cost {D_CITY} >= 0;
param dw_cost {DW_LINKS} >= 0;

param pd_cap {D_CITY} >= 0;
param dw_cap {DW_LINKS} >= 0;

var PD_Ship {i in D_CITY} >= 0, <= pd_cap[i];
var DW_Ship {(i,j) in DW_LINKS} >= 0, <= dw_cap[i,j];
```

The "pd" quantities are associated with shipments from the plant to distribution centers; because they all relate to shipments from the same plant, they need only be indexed over D_CITY. The "dw" quantities are associated with shipments from distribution centers to warehouses, and so are naturally indexed over DW_LINKS.

The total shipment cost can now be given as the sum of two summations:

```
param p_city := PITT ;

set D_CITY := NE SE ;
set W_CITY := BOS EWR BWI ATL MCO ;

set DW_LINKS := (NE,BOS)  (NE,EWR)  (NE,BWI)
                (SE,EWR)  (SE,BWI)  (SE,ATL)  (SE,MCO);

param p_supply := 450 ;

param w_demand :=
  BOS   90,  EWR 120,  BWI 120,  ATL   70,  MCO   50;

param:      cost  capacity  :=
  PITT NE   2.5    250
  PITT SE   3.5    250

  NE BOS    1.7    100
  NE EWR    0.7    100
  NE BWI    1.3    100

  SE EWR    1.3    100
  SE BWI    0.8    100
  SE ATL    0.2    100
  SE MCO    2.1    100 ;
```

Figure 11-3b: Data for specialized transshipment model (net2.dat).

```
minimize Total_Cost:
    sum {i in D_CITY} pd_cost[i] * PD_Ship[i]
  + sum {(i,j) in DW_LINKS} dw_cost[i,j] * DW_Ship[i,j];
```

Finally, there must be three kinds of balance constraints, one for each kind of city. Shipments from the plant to the distribution centers must equal the supply at the plant:

```
    subject to P_Bal:  sum {i in D_CITY} PD_Ship[i] = p_supply;
```

At each distribution center, shipments in from the plant must equal shipments out to all the warehouses:

```
    subject to D_Bal {i in D_CITY}:
       PD_Ship[i] = sum {(i,j) in DW_LINKS} DW_Ship[i,j];
```

And at each warehouse, shipments in from all distribution centers must equal the demand:

```
    subject to W_Bal {j in W_CITY}:
       sum {(i,j) in DW_LINKS} DW_Ship[i,j] = w_demand[j];
```

The whole model, with appropriate data, is shown in Figures 11-4a and 11-4b.

The approaches shown in Figures 11-3 and 11-4 are equivalent, in the sense that they cause the same linear program to be solved. The former is more convenient for experimenting with different network structures, since any changes affect only the data for the initial declarations in the model. If the network structure is unlikely to change, however,

```
set D_CITY;
set W_CITY;
set DW_LINKS within (D_CITY cross W_CITY);

param p_supply >= 0;              # amount available at plant
param w_demand {W_CITY} >= 0;    # amounts required at warehouses

    check: p_supply = sum {j in W_CITY} w_demand[j];

param pd_cost {D_CITY} >= 0;     # shipment costs/1000 packages
param dw_cost {DW_LINKS} >= 0;

param pd_cap {D_CITY} >= 0;       # max packages that can be shipped
param dw_cap {DW_LINKS} >= 0;

var PD_Ship {i in D_CITY} >= 0, <= pd_cap[i];
var DW_Ship {(i,j) in DW_LINKS} >= 0, <= dw_cap[i,j];
                                  # packages to be shipped

minimize Total_Cost:
    sum {i in D_CITY} pd_cost[i] * PD_Ship[i] +
    sum {(i,j) in DW_LINKS} dw_cost[i,j] * DW_Ship[i,j];

subject to P_Bal:  sum {i in D_CITY} PD_Ship[i] = p_supply;

subject to D_Bal {i in D_CITY}:
    PD_Ship[i] = sum {(i,j) in DW_LINKS} DW_Ship[i,j];

subject to W_Bal {j in W_CITY}:
    sum {(i,j) in DW_LINKS} DW_Ship[i,j] = w_demand[j];
```

Figure 11-4a: Specialized transshipment model, version 2 (net3.mod).

the latter form facilitates alterations that affect only particular kinds of cities, such as the generalizations we describe next.

Variations on transshipment models

Some balance constraints in a network flow model may have to be inequalities rather than equations. In the example of Figure 11-4, if production at the plant can sometimes exceed total demand at the warehouses, we should replace = by <= in the P_Bal constraints.

A more substantial modification occurs when the quantity of flow that comes out of an arc does not necessarily equal the quantity that went in. As an example, a small fraction of the packages shipped from the plant may be damaged or stolen before they reach the distribution center. Suppose that a parameter pd_loss is introduced to represent the loss rate:

```
    param pd_loss {D_CITY} >= 0, < 1;
```

Then the balance constraints at the distribution centers must be adjusted accordingly:

```
set D_CITY := NE SE ;

set W_CITY := BOS EWR BWI ATL MCO ;

set DW_LINKS := (NE,BOS)  (NE,EWR)  (NE,BWI)
                (SE,EWR)  (SE,BWI)  (SE,ATL)  (SE,MCO);

param p_supply := 450 ;

param w_demand :=
   BOS  90,   EWR 120,   BWI 120,   ATL  70,   MCO  50;

param:  pd_cost  pd_cap :=
   NE      2.5     250
   SE      3.5     250 ;

param:      dw_cost  dw_cap :=
   NE BOS     1.7     100
   NE EWR     0.7     100
   NE BWI     1.3     100

   SE EWR     1.3     100
   SE BWI     0.8     100
   SE ATL     0.2     100
   SE MCO     2.1     100 ;
```

Figure 11-4b: Data for specialized transshipment model, version 2 (net3.dat).

```
subject to D_Bal {i in D_CITY}:
    (1-pd_loss[i]) * PD_Ship[i]
       = sum {(i,j) in DW_LINKS} DW_Ship[i,j];
```

The expression to the left of the = sign has been modified to reflect the fact that only (1-pd_loss[i]) * PD_Ship[i] packages arrive at city i when PD_Ship[i] packages are shipped from the plant.

A similar variation occurs when the flow is not measured in the same units throughout the network. If demand is reported in cartons rather than thousands of packages, for example, the model will require a parameter to represent packages per carton:

```
param ppc integer > 0;
```

Then the demand constraints at the warehouses are adjusted as follows:

```
subject to W_Bal {j in W_CITY}:
    sum {(i,j) in DW_LINKS} (1000/ppc) * DW_Ship[i,j]
       = w_demand[j];
```

The term (1000/ppc) * DW_Ship[i,j] represents the number of cartons received at warehouse j when DW_Ship[i,j] thousand packages are shipped from distribution center i.

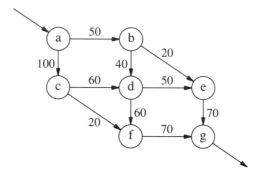

Figure 11-5: Traffic flow network.

11.2 Other network models

Not all network linear programs involve the transportation of things or the minimization of costs. We describe here three well-known model classes — maximum flow, shortest path, and transportation/assignment — that use the same kinds of variables and constraints for different purposes.

Maximum flow models

In some network design applications the concern is to send as much flow as possible through the network, rather than to send flow at lowest cost. This alternative is readily handled by dropping the balance constraints at the origins and destinations of flow, while substituting an objective that stands for total flow in some sense.

As a specific example, Figure 11-5 presents a diagram of a simple traffic network. The nodes and arcs represent intersections and roads; capacities, shown as numbers next to the roads, are in cars per hour. We want to find the maximum traffic flow that can enter the network at a and leave at g.

A model for this situation begins with a set of intersections, and symbolic parameters to indicate the intersections that serve as entrance and exit to the road network:

```
set INTER;

param entr symbolic in INTER;
param exit symbolic in INTER, <> entr;
```

The set of roads is defined as a subset of the pairs of intersections:

```
set ROADS within (INTER diff {exit}) cross (INTER diff {entr});
```

This definition ensures that no road begins at the exit or ends at the entrance.

Next, the capacity and traffic load are defined for each road:

```
set INTER;   # intersections

param entr symbolic in INTER;               # entrance to road network
param exit symbolic in INTER, <> entr;   # exit from road network

set ROADS within (INTER diff {exit}) cross (INTER diff {entr});

param cap {ROADS} >= 0;                          # capacities
var Traff {(i,j) in ROADS} >= 0, <= cap[i,j];  # traffic loads

maximize Entering_Traff: sum {(entr,j) in ROADS} Traff[entr,j];

subject to Balance {k in INTER diff {entr,exit}}:
  sum {(i,k) in ROADS} Traff[i,k] = sum {(k,j) in ROADS} Traff[k,j];

data;

set INTER := a b c d e f g ;

param entr := a ;
param exit := g ;

param:  ROADS:  cap :=
          a b      50,     a c     100
          b d      40,     b e      20
          c d      60,     c f      20
          d e      50,     d f      60
          e g      70,     f g      70 ;
```

Figure 11-6: Maximum traffic flow model and data (`netmax.mod`).

```
    param cap {ROADS} >= 0;
    var Traff {(i,j) in ROADS} >= 0, <= cap[i,j];
```

The constraints say that flow into each intersection equals flow out at all intersections except the entrance and exit:

```
    subject to Balance {k in INTER diff {entr,exit}}:
        sum {(i,k) in ROADS} Traff[i,k]
            = sum {(k,j) in ROADS} Traff[k,j];
```

Given these constraints, the flow out of the entrance must be the total flow through the network, which is to be maximized:

```
    maximize Entering_Traff: sum {(entr,j) in ROADS} Traff[entr,j];
```

We could equally well maximize the total flow into the exit. The entire model, along with data for the example shown in Figure 11-5, is presented in Figure 11-6. AMPL finds a maximum flow of 130 cars per hour.

Shortest path models

If you were to use the optimal solution to any of our models thus far, you would have to send each of the packages, cars, or whatever along some path from a supply (or entrance) node to a demand (or exit) node. The values of the decision variables do not directly say what the optimal paths are, or how much flow must go on each one. Usually it is not too hard to deduce these paths, however, especially when the network has a regular or special structure.

If a network has just one unit of supply and one unit of demand, the optimal solution assumes a quite different nature. The variable associated with each arc is either 0 or 1, and the arcs whose variables have value 1 comprise a minimum-cost path from the supply node to the demand node. Often the "costs" are in fact times or distances, so that the optimum gives a shortest path.

Only a few changes need be made to the maximum flow model of Figure 11-6 to turn it into a shortest path model. There are still a parameter and a variable associated with each road from i to j, but we call them time[i,j] and Use[i,j], and the sum of their products yields the objective:

```
param time {ROADS} >= 0;        # times to travel roads
var Use {(i,j) in ROADS} >= 0;  # 1 iff (i,j) in shortest path

minimize Total_Time: sum {(i,j) in ROADS} time[i,j] * Use[i,j];
```

Since only those variables Use[i,j] on the optimal path equal 1, while the rest are 0, this sum does correctly represent the total time to traverse the optimal path. The only other change is the addition of a constraint to ensure that exactly one unit of flow is available at the entrance to the network:

```
subject to Start:  sum {(entr,j) in ROADS} Use[entr,j] = 1;
```

The complete model is shown in Figure 11-7. If we imagine that the numbers on the arcs in Figure 11-5 are travel times in minutes rather than capacities, the data are the same; AMPL finds the solution as follows:

```
ampl: model netshort.mod;
ampl: solve;
MINOS 5.4: optimal solution found.
1 iterations, objective 140

ampl: option omit_zero_rows 1;
ampl: display Use;
Use :=
a b   1
b e   1
e g   1
;
```

The shortest path is a $\rightarrow$ b $\rightarrow$ e $\rightarrow$ g, which takes 140 minutes.

Transportation and assignment models

The best known and most widely used special network structure is the "bipartite" structure depicted in Figure 11-8. The nodes fall into two groups, one serving as origins of flow and the other as destinations. Each arc connects an origin to a destination.

```
set INTER;   # intersections

param entr symbolic in INTER;              # entrance to road network
param exit symbolic in INTER, <> entr;  # exit from road network

set ROADS within (INTER diff {exit}) cross (INTER diff {entr});

param time {ROADS} >= 0;          # times to travel roads
var Use {(i,j) in ROADS} >= 0;   # 1 iff (i,j) in shortest path

minimize Total_Time: sum {(i,j) in ROADS} time[i,j] * Use[i,j];

subject to Start:  sum {(entr,j) in ROADS} Use[entr,j] = 1;

subject to Balance {k in INTER diff {entr,exit}}:
   sum {(i,k) in ROADS} Use[i,k] = sum {(k,j) in ROADS} Use[k,j];

data;

set INTER := a b c d e f g ;

param entr := a ;
param exit := g ;

param:  ROADS:  time :=
        a b     50,    a c     100
        b d     40,    b e      20
        c d     60,    c f      20
        d e     50,    d f      60
        e g     70,    f g      70 ;
```

Figure 11-7: Shortest path model and data (netshort.mod).

The minimum-cost transshipment model on this network is known as the transportation model. The special case in which every origin is connected to every destination was introduced in Chapter 3; an AMPL model and sample data are shown in Figures 3-1a and 3-1b. A more general example analogous to the models developed earlier in this chapter, where a set LINKS specifies the arcs of the network, appeared in Figures 6-2a and 6-2b.

Every path from an origin to a destination in a bipartite network consists of one arc. Or, to say the same thing another way, the optimal flow along an arc of the transportation model gives the actual amount shipped from some origin to some destination. This property permits the transportation model to be viewed alternatively as a so-called assignment model, in which the optimal flow along an arc is the amount of something from the origin that is assigned to the destination. The meaning of assignment in this context can be broadly construed, and in particular need not involve a shipment in any sense.

One of the more common applications of the assignment model is matching people to appropriate targets, such as jobs, offices or even other people. Each origin node is associated with one person, and each destination node with one of the targets — for example, with one project. The sets might then be defined as follows:

```
    set PEOPLE;
    set PROJECTS;
    set ABILITIES within (PEOPLE cross PROJECTS);
```

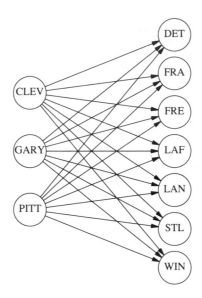

Figure 11-8: Bipartite network.

The set ABILITIES takes the role of LINKS in our earlier models; a pair (i,j) is placed in this set if and only if person i can work on project j.

As one possibility for continuing the model, the supply at node i could be the number of hours that person i is available to work, and the demand at node j could be the number of hours required for project j. Variables Assign[i,j] would represent the number of hours of person i's time assigned to project j. Also associated with each pair (i,j) would be a cost per hour, and a maximum number of hours that person i could contribute to job j. The resulting model is shown in Figure 11-9.

Another possibility is to make the assignment in terms of people rather than hours. The supply at every node i is 1 (person), and the demand at node j is the number of people required for project j. The supply constraints insure that Assign[i,j] is not greater than 1; and it will equal 1 in an optimal solution if and only if person i is assigned to project j. The coefficient cost[i,j] could be some kind of cost of assigning person i to project j, in which case the objective would still be to minimize total cost. Or the coefficient could be the ranking of person i for project j, perhaps on a scale from 1 (highest) to 10 (lowest). Then the model would produce an assignment for which the total of the rankings is as good as possible.

Finally, we can imagine an assignment model in which the demand at each node j is also 1; the problem is then to match people to projects. In the objective, cost[i,j] could be the number of hours that person i would need to complete project j, in which case the model would find the assignment that minimizes the total hours of work needed

```
set PEOPLE;
set PROJECTS;

set ABILITIES within (PEOPLE cross PROJECTS);

param supply {PEOPLE} >= 0;   # hours each person is available
param demand {PROJECTS} >= 0; # hours each project requires

check: sum {i in PEOPLE} supply[i] = sum {j in PROJECTS} demand[j];

param cost {ABILITIES} >= 0;   # cost per hour of work
param limit {ABILITIES} >= 0;  # maximum contributions to projects

var Assign {(i,j) in ABILITIES} >= 0, <= limit[i,j];

minimize total_cost:
   sum {(i,j) in ABILITIES} cost[i,j] * Assign[i,j];

subject to Supply {i in PEOPLE}:
   sum {(i,j) in ABILITIES} Assign[i,j] = supply[i];

subject to Demand {j in PROJECTS}:
   sum {(i,j) in ABILITIES} Assign[i,j] = demand[j];
```

Figure 11-9: Assignment model (`netasgn.mod`).

to finish all the projects. You can create a model of this kind by replacing all references to `supply[i]` and `demand[j]` by 1 in Figure 11-9.

11.3 Declaring network models by `node` and `arc`

AMPL's algebraic notation has great power to express a variety of network linear programs, but the resulting constraint expressions are often not as natural as we would like. While the idea of constraining "flow out minus flow in" at each node is easy to describe and understand, the corresponding algebraic constraints tend to involve terms like

```
sum {(i,k) in LINKS} Ship[i,k]
```

that are not so quickly comprehended. The more complex and realistic the network, the worse the problem. Indeed, it can be hard to tell whether a model's algebraic constraints represent a valid collection of flow balances on a network, and consequently whether specialized network optimization software (described later in this chapter) can be used.

Algebraic formulations of network flows tend to be problematical because they are constructed explicitly in terms of variables and constraints, while the nodes and arcs are merely implicit in the way that the constraints are structured. People prefer to approach network flow problems in the opposite way. They imagine giving an explicit definition of nodes and arcs, from which flow variables and balance constraints implicitly arise. To deal with this situation, AMPL provides an alternative that allows network concepts to be declared directly in a model.

The network extensions to AMPL include two new kinds of declaration, node and arc, that take the place of the subject to and var declarations in an algebraic constraint formulation. The node declarations name the nodes of a network, and characterize the flow balance constraints at the nodes. The arc declarations name and define the arcs, by specifying the nodes that arcs connect, and by providing optional information such as bounds and costs that are associated with arcs.

This section introduces node and arc by showing how they permit various examples from earlier in this chapter to be reformulated conveniently. The following section presents the rules for these declarations more systematically.

A general transshipment model

In rewriting the model of Figure 11-2a using node and arc, we can retain all of the set and param declarations and associated data. The changes affect only the three declarations — minimize, var, and subject to — that define the linear program.

There is a node in the network for every member of the set CITIES. Using a node declaration, we can say this directly:

```
node Balance {k in CITIES}: net_in = demand[k] - supply[k];
```

The keyword net_in stands for "net input", that is, the flow in minus the flow out, so this declaration says that net flow in must equal net demand at each node Balance[k]. Thus it says the same thing as the constraint named Balance[k] in the algebraic version, except that it uses the concise term net_in in place of the lengthy expression

```
sum {(i,k) in LINKS} Ship[i,k] - sum {(k,j) in LINKS} Ship[k,j]
```

Indeed, the syntax of subject to and node are practically the same except for the way that the conservation-of-flow constraint is stated. (The keyword net_out may also be used to stand for flow out minus flow in, so that we could have written net_out = supply[k] – demand[k].)

There is an arc in the network for every pair in the set LINKS. This too can be said directly, using an arc declaration:

```
arc Ship {(i,j) in LINKS} >= 0, <= capacity[i,j],
    from Balance[i], to Balance[j], obj Total_Cost cost[i,j];
```

An arc Ship[i,j] is defined for each pair in LINKS, with bounds of 0 and capacity[i,j] on its flow; to this extent, the arc and var declarations are the same. The arc declaration contains additional phrases, however, to say that the arc runs from the node named Balance[i] to the node named Balance[j], with a linear coefficient of cost[i,j] in the objective function named Total_Cost. These phrases use the keywords from, to, and obj.

Since the information about the objective function is included in the arc declaration, it is not needed in the minimize declaration, which reduces to:

```
minimize Total_Cost;
```

```
set CITIES;
set LINKS within (CITIES cross CITIES);

param supply {CITIES} >= 0;    # amounts available at cities
param demand {CITIES} >= 0;    # amounts required at cities

   check: sum {i in CITIES} supply[i] = sum {j in CITIES} demand[j];

param cost {LINKS} >= 0;         # shipment costs/1000 packages
param capacity {LINKS} >= 0;   # max packages that can be shipped

minimize Total_Cost;

node Balance {k in CITIES}: net_in = demand[k] - supply[k];

arc Ship {(i,j) in LINKS} >= 0, <= capacity[i,j],
   from Balance[i], to Balance[j], obj Total_Cost cost[i,j];
```

Figure 11-10: General transshipment model with node and arc (net1node.mod).

The whole model is shown in Figure 11-10.

As this description suggests, arc and node take the place of var and subject to, respectively. In fact AMPL treats an arc declaration as a definition of variables, so that you would still say display Ship to look at the optimal flows in the network model of Figure 11-10; it treats a node declaration as a definition of constraints. The difference is that node and arc present the model in a way that corresponds more directly to its appearance in a network diagram. The description of the nodes always comes first, followed by a description of how the arcs connect the nodes.

A specialized transshipment model

The node and arc declarations make it easy to define a linear program for a network that has several different kinds of nodes and arcs. For an example we return to the network of Figure 11-1, with the specialized sets and parameters declared as in Figure 11-4a.

The network has a plant node, a distribution center node for each member of D_CITY, and a warehouse node for each member of W_CITY. Thus the model requires three node declarations:

```
node Plant: net_out = p_supply;
node Dist {i in D_CITY};
node Whse {j in W_CITY}: net_in = w_demand[j];
```

The balance conditions say that flow out of node Plant must be p_supply, while flow into node Whse[j] is w_demand[j]. (The network has no arcs into the plant or out of the warehouses, so net_out and net_in are just the flow out and flow in, respectively.) The conditions at node Dist[i] could be written either net_in = 0 or net_out = 0, but since these are assumed by default we need not specify any condition at all.

```
set D_CITY;
set W_CITY;
set DW_LINKS within (D_CITY cross W_CITY);

param p_supply >= 0;                    # amount available at plant
param w_demand {W_CITY} >= 0;   # amounts required at warehouses

    check: p_supply = sum {j in W_CITY} w_demand[j];

param pd_cost {D_CITY} >= 0;    # shipment costs/1000 packages
param dw_cost {DW_LINKS} >= 0;

param pd_cap {D_CITY} >= 0;      # max packages that can be shipped
param dw_cap {DW_LINKS} >= 0;

minimize Total_Cost;

node Plant: net_out = p_supply;

node Dist {i in D_CITY};

node Whse {j in W_CITY}: net_in = w_demand[j];

arc PD_Ship {i in D_CITY} >= 0, <= pd_cap[i],
    from Plant, to Dist[i], obj Total_Cost pd_cost[i];

arc DW_Ship {(i,j) in DW_LINKS} >= 0, <= dw_cap[i,j],
    from Dist[i], to Whse[j], obj Total_Cost dw_cost[i,j];
```

Figure 11-11: Specialized transshipment model with node and arc (net3node.mod).

This network has two kinds of arcs. There is an arc from the plant to each member of D_CITY, which can be declared by:

```
arc PD_Ship {i in D_CITY} >= 0, <= pd_cap[i],
    from Plant, to Dist[i], obj Total_Cost pd_cost[i];
```

And there is an arc from distribution center i to warehouse j for each pair (i,j) in DW_LINKS:

```
arc DW_Ship {(i,j) in DW_LINKS} >= 0, <= dw_cap[i,j],
    from Dist[i], to Whse[j], obj Total_Cost dw_cost[i,j];
```

The arc declarations specify the relevant bounds and objective coefficients, as in our previous example. The whole model is shown in Figure 11-11.

Variations on transshipment models

The balance conditions in node declarations may be inequalities, like ordinary algebraic balance constraints. If production at the plant can sometimes exceed total demand at the warehouses, it would be appropriate to give the condition in the declaration of node Plant as net_out <= p_supply.

An arc declaration can specify losses in transit by adding a factor at the end of the to phrase:

```
    arc PD_Ship {i in D_CITY} >= 0, <= pd_cap[i],
        from Plant, to Dist[i] (1-pd_loss[i]),
        obj Total_Cost pd_cost[i];
```

This is interpreted as saying that PD_Ship[i] is the number of packages that leave node Plant, but (1-pd_loss[i]) * PD_Ship[i] is the number that enter node Dist[i].

The same option can be used to specify conversions. To use our previous example, if shipments are measured in thousands of packages but demands are measured in cartons, the arcs from distribution centers to warehouses should be declared as:

```
    arc DW_Ship {(i,j) in DW_LINKS} >= 0, <= dw_cap[i,j],
        from Dist[i], to Whse[j] (1000/ppc),
        obj Total_Cost dw_cost[i,j];
```

If the shipments to warehouses are also measured in cartons, the factor should be applied at the distribution center:

```
    arc DW_Ship {(i,j) in DW_LINKS} >= 0, <= dw_cap[i,j],
        from Dist[i] (ppc/1000), to Whse[j],
        obj Total_Cost dw_cost[i,j];
```

The loss factor could also be applied to the to phrase in these examples.

Maximum flow models

In the diagram of Figure 11-5 that we have used to illustrate the maximum flow problem, there are three kinds of intersections represented by nodes: the one where traffic enters, the one where traffic leaves, and the others where traffic flow is conserved. Thus a model of the network could have three corresponding node declarations:

```
    node Entr_Int: net_out >= 0;
    node Exit_Int: net_in >= 0;

    node Intersection {k in INTER diff {entr,exit}};
```

The condition net_out >= 0 implies that the flow out of node Entr_Int may be any amount at all; this is the proper condition, since there is no balance constraint on the entrance node. An analogous comment applies to the condition for node Exit_Int.

There is one arc in this network for each pair (i,j) in the set ROADS. Thus the declaration should look something like this:

```
    arc Traff {(i,j) in ROADS} >= 0, <= cap[i,j],  # NOT RIGHT
        from Intersection[i], to Intersection[j],
        obj Entering_Traff (if i = entr then 1);
```

Since the aim is to maximize the total traffic leaving the entrance node, the arc is given a coefficient of 1 in the objective if and only if i takes the value entr. When i does take this value, however, the arc is specified to be from Intersection[entr], a node that does not exist; the arc should rather be from node Entr_Int. Similarly, when j takes the value exit, the arc should not be to Intersection[exit], but to

Exit_Int. AMPL will catch these errors and issue a message naming one of the nonexistent nodes that has been referenced.

It might seem reasonable to use an if-then-else to get around this problem, in the following way:

```
arc Traff {(i,j) in ROADS} >= 0, <= cap[i,j],  # SYNTAX ERROR
    from (if i = entr then Entr_Int else Intersection[i]),
    to   (if j = exit then Exit_Int else Intersection[j]),
    obj Entering_Traff (if i = entr then 1);
```

However, the if-then-else construct in AMPL does not apply to model components such as Entr_Int and Intersection[i]; this version will be rejected as a syntax error. Instead you need to use from and to phrases qualified by indexing expressions:

```
arc Traff {(i,j) in ROADS} >= 0, <= cap[i,j],
    from {if i = entr} Entr_Int,
    from {if i <> entr} Intersection[i],
    to   {if j = exit} Exit_Int,
    to   {if j <> exit} Intersection[j],
    obj Entering_Traff (if i = entr then 1);
```

The special indexing expression beginning with if works much the same way here as it does for constraints (Section 8.4); the from or to phrase is processed if the condition following if is true. Thus Traff[i,j] is declared to be from Entr_Int if i equals entr, and to be from Intersection[i] if i is not equal to entr, which is what we intend.

As an alternative, we can combine the declarations of the three different kinds of nodes into one. Observing that net_out is positive or zero for Entr_Int, negative or zero for Exit_Int, and zero for all other nodes Intersection[i], we can declare:

```
node Intersection {k in INTER}:
    (if k = exit then -Infinity)
       <= net_out <= (if k = entr then Infinity);
```

The nodes that were formerly declared as Entr_Int and Exit_Int are now just Intersection[entr] and Intersection[exit], and consequently the arc declaration that we previously marked ''not right'' now works just fine. The choice between this version and the previous one is entirely a matter of convenience and taste. (Infinity is a predefined AMPL parameter that may be used to specify any ''infinitely large'' bound; its technical definition is given in Appendix A.7.2.)

Arguably the AMPL formulation that is most convenient and appealing is neither of the above, but rather comes from interpreting the network diagram of Figure 11-5 in a slightly different way. Suppose that we view the arrows into the entrance node and out of the exit node as representing additional arcs, which happen to be adjacent to only one node rather than two. Then flow in equals flow out at every intersection, and the node declaration simplifies to:

```
node Intersection {k in INTER};
```

```
set INTER;   # intersections

param entr symbolic in INTER;            # entrance to road network
param exit symbolic in INTER, <> entr;  # exit from road network

set ROADS within (INTER diff {exit}) cross (INTER diff {entr});

param cap {ROADS} >= 0;   # capacities of roads

node Intersection {k in INTER};

arc Traff_In >= 0, to Intersection[entr];
arc Traff_Out >= 0, from Intersection[exit];

arc Traff {(i,j) in ROADS} >= 0, <= cap[i,j],
   from Intersection[i], to Intersection[j];

maximize Entering_Traff: Traff_In;

data;

set INTER := a b c d e f g ;

param entr := a ;
param exit := g ;

param:  ROADS:  cap :=
          a b     50,     a c     100
          b d     40,     b e      20
          c d     60,     c f      20
          d e     50,     d f      60
          e g     70,     f g      70 ;
```

Figure 11-12: Maximum flow model with node and arc (netmax3.mod).

The two arcs "hanging" at the entrance and exit are defined in the obvious way, but include only a to or a from phrase:

```
arc Traff_In >= 0, to Intersection[entr];
arc Traff_Out >= 0, from Intersection[exit];
```

The arcs that represent roads within the network are declared as before:

```
arc Traff {(i,j) in ROADS} >= 0, <= cap[i,j],
    from Intersection[i], to Intersection[j];
```

When the model is represented in this way, the objective is to maximize Traff_In (or equivalently Traff_Out). We could do this by adding an obj phrase to the arc declaration for Traff_In, but in this case it is perhaps clearer to define the objective algebraically:

```
maximize Entering_Traff: Traff_In;
```

This version is shown in full in Figure 11-12.

11.4 Rules for `node` and `arc` declarations

Having defined `node` and `arc` by example, we now describe more comprehensively the required and optional elements of these declarations, and comment on their interaction with the conventional declarations `minimize` or `maximize`, `subject to`, and `var` when both kinds appear in the same model.

node *declarations*

A `node` declaration begins with the keyword `node`, a name, an optional indexing expression, and a colon. The expression following the colon, which describes the balance condition at the node, may have any of the following forms:

net-expr = *arith-expr*
net-expr <= *arith-expr*
net-expr >= *arith-expr*

arith-expr = *net-expr*
arith-expr <= *net-expr*
arith-expr >= *net-expr*

arith-expr <= *net-expr* <= *arith-expr*
arith-expr >= *net-expr* >= *arith-expr*

where an *arith-expr* may be any arithmetic expression that uses previously declared model components and currently defined dummy indices, while a *net-expr* is restricted to one of the following:

```
net_in                          net_out
net_in + arith-expr             net_out + arith-expr
arith-expr + net_in             arith-expr + net_out
```

Each node defined in this way induces a constraint in the resulting linear program. A node name is treated like a constraint name in the AMPL command environment, for example in a `display` statement.

For declarations that use `net_in`, AMPL generates the constraint by substituting, at the place where `net_in` appears in the balance conditions, a linear expression that represents flow into the node minus flow out of the node. Declarations that use `net_out` are handled the same way, except that AMPL substitutes flow out minus flow in. The expressions for flow in and flow out are deduced from the `arc` declarations.

arc *declarations*

An `arc` declaration consists of the keyword `arc`, a name, an optional indexing expression, and a series of optional qualifying phrases. Each arc creates a variable in the resulting linear program, whose value is the amount of flow over the arc; the arc name may be used to refer to this variable elsewhere. All of the phrases that may appear in a `var` definition have the same significance in an `arc` definition; most commonly, the `>=`

and <= phrases are used to specify values for lower and upper bounds on the flow along the arc.

The from and to phrases specify the nodes connected by an arc. Usually these consist of the keyword from or to followed by a node name. An arc is interpreted to contribute to the flow out of the from node, and to the flow into the to node; these interpretations are what permit the inference of the constraints associated with the nodes.

Typically one from and one to phrase are specified in an arc declaration. Either may be omitted, however, as in Figure 11-12. Either from or to may also be followed by an optional indexing expression, which should be one of two kinds:

- An indexing expression that specifies an empty set (in which case the from or to phrase is ignored) or a set with one member (in which case the from or to phrase is used).

- An indexing expression of the special form {if *logical-expr*}, which causes the from or to phrase to be used if and only if the *logical-expr* evaluates to true.

It is possible to specify that an arc carries flow out of or into two or more nodes, by giving more than one from or to phrase, or by using an indexing expression that specifies a set having more than one member. The result is not a network linear program, however, and AMPL displays an appropriate warning message.

At the end of a from or to phrase, you may add an arithmetic expression representing a factor to multiply the flow, as shown in our examples of shipping-loss and change-of-unit variations in Section 11.3. If the factor is in the to phrase, it multiplies the arc variable in determining the flow into the specified arc; that is, for a given flow along the arc, an amount equal to the to-factor times the flow is considered to enter the to node. A factor in the from phrase is interpreted analogously. The default factor is 1.

An optional obj phrase specifies a coefficient that will multiply the arc variable to create a linear term in a specified objective function. Such a phrase consists of the keyword obj, the name of an objective that has previously been defined in a minimize or maximize declaration, and an arithmetic expression for the coefficient value. The keyword may be followed by an indexing-expression, which is interpreted as for the from and to phrases.

Interaction with objective declarations

If all terms in the objective function are specified through obj phrases in arc declarations, the declaration of the objective is simply minimize or maximize followed by an optional indexing expression and a name. This declaration must come before the arc declarations that refer to the objective.

Alternatively, arc names may be used as variables to specify the objective function in the usual algebraic way. In this case the objective must be declared after the arcs, as in Figure 11-12.

```
set CITIES;
set LINKS within (CITIES cross CITIES);

set PRODS;

param supply {CITIES,PRODS} >= 0;   # amounts available at cities
param demand {CITIES,PRODS} >= 0;   # amounts required at cities

    check {p in PRODS}:
        sum {i in CITIES} supply[i,p] = sum {j in CITIES} demand[j,p];

param cost {LINKS,PRODS} >= 0;       # shipment costs/1000 packages
param capacity {LINKS,PRODS} >= 0;  # max packages shipped
param cap_joint {LINKS} >= 0;        # max total packages shipped/link

minimize Total_Cost;

node Balance {k in CITIES, p in PRODS}:
   net_in = demand[k,p] - supply[k,p];

arc Ship {(i,j) in LINKS, p in PRODS} >= 0, <= capacity[i,j,p],
    from Balance[i,p], to Balance[j,p], obj Total_Cost cost[i,j,p];

subject to Multi {(i,j) in LINKS}:
   sum {p in PRODS} Ship[i,j,p] <= cap_joint[i,j];
```

Figure 11-13: Multicommodity flow with side constraints (netmulti.mod).

Interaction with constraint declarations

The components defined in arc declarations may be used as variables in additional subject to declarations. The latter represent ''side constraints'' that are imposed in addition to balance of flow at the nodes.

As an example, consider how a multicommodity flow problem can be built from the node-and-arc network formulation in Figure 11-10. Following the approach in Section 4.1, we introduce a set PRODS of different products, and add it to the indexing of all parameters, nodes and arcs. The result is a separate network linear program for each product, with the objective function being the sum of the costs for all products. To tie these networks together, we provide for a joint limit on the total shipments along any link:

```
        param cap_joint {LINKS} >= 0;
        subject to Multi {(i,j) in LINKS}:
            sum {p in PRODS} Ship[p,i,j] <= cap_joint[i,j];
```

The final model, shown in Figure 11-13, is not a network linear program, but the network and non-network parts of it are cleanly separated.

Interaction with variable declarations

Just as an arc variable may be used in a subject to declaration, an ordinary var variable may be used in a node declaration. That is, the balance condition in a node declaration may contain references to variables that were defined by preceding `var` declarations. These references define ''side variables'' to the network linear program.

As an example, we again replicate the formulation of Figure 11-10 over the set PRODS. This time we tie the networks together by introducing a set of feedstocks and associated data:

```
set FEEDS;
param yield {PRODS,FEEDS} >= 0;
param limit {FEEDS,CITIES} >= 0;
```

We imagine that at city k, in addition to the amounts `supply[p,k]` of products available to be shipped, up to `limit[f,k]` of feedstock f can be converted into products; one unit of feedstock f gives rise to `yield[p,f]` units of each product p. A variable `Feed[f,k]` represents the amount of feedstock f used at city k:

```
var Feed {f in FEEDS, k in CITIES} >= 0, <= limit[f,k];
```

The balance condition for product p at city k can now say that the net flow out equals net supply plus the sum of the amounts derived from the various feedstocks:

```
node Balance {p in PRODS, k in CITIES}:
   net_out = supply[p,k] - demand[p,k]
        + sum {f in FEEDS} yield[p,f] * Feed[f,k];
```

The arcs are unchanged, leading to the model shown in Figure 11-14. At a given city k, the variables `Feed[f,k]` appear in the node balance conditions for all the different products, bringing together the product networks into a single linear program.

11.5 Solving network linear programs

All of the models that we have described in this chapter give rise to ''network'' linear programs. AMPL can send these linear programs to an LP solver and retrieve the optimal values, much as for any other class of LPs. If you use AMPL in this way, the network structure is helpful mainly as a guide to formulating the model and interpreting the results.

There is also a more restricted class of problems that (confusingly) are also known as network linear programs. In modeling terms, the variables of such a network linear program must represent flows on the arcs of a network, and the constraints must be only of two types: bounds on the flows along the arcs, and limits on flow out minus flow in at the nodes. A more technical way to say the same thing is that each variable of a network linear program must appear in at most two constraints (aside from lower or upper bounds on the variables), such that the variable has a coefficient of +1 in at most one constraint, and a coefficient of −1 in at most one constraint.

''Pure'' network linear programs of this restricted kind have some very strong properties that make their use particularly desirable. So long as the supplies, demands, and bounds are integers, a network linear program must have an optimal solution in which all

```
set CITIES;

set LINKS within (CITIES cross CITIES);

set PRODS;

param supply {PRODS,CITIES} >= 0;  # amounts available at cities

param demand {PRODS,CITIES} >= 0;  # amounts required at cities
    check {p in PRODS}:
        sum {i in CITIES} supply[p,i] = sum {j in CITIES} demand[p,j];

param cost {PRODS,LINKS} >= 0;        # shipment costs/1000 packages
param capacity {PRODS,LINKS} >= 0; # max packages shipped of product

set FEEDS;

param yield {PRODS,FEEDS} >= 0;      # amounts derived from feedstocks
param limit {FEEDS,CITIES} >= 0;    # feedstocks available at cities

minimize Total_Cost;

var Feed {f in FEEDS, k in CITIES} >= 0, <= limit[f,k];

node Balance {p in PRODS, k in CITIES}:
   net_out = supply[p,k] - demand[p,k]
       + sum {f in FEEDS} yield[p,f] * Feed[f,k];

arc Ship {p in PRODS, (i,j) in LINKS} >= 0, <= capacity[p,i,j],
   from Balance[p,i], to Balance[p,j],
   obj Total_Cost cost[p,i,j];
```

Figure 11-14: Multicommodity flow with side variables (netfeeds.mod).

flows are integers. Moreover, if the solver is of a kind that finds "extreme" solutions (such as those based on the simplex method) it will always find one of the all-integer optimal solutions. We have taken advantage of this property, without explicitly mentioning it, in assuming that the variables in the shortest path problem and in certain assignment problems come out to be either zero or one, and never some fraction in between.

Network linear programs can also be solved much faster than other linear programs of comparable size, through the use of solvers that are specialized to take advantage of the network structure. If you write your model in terms of node and arc declarations, AMPL automatically communicates the network structure to the solver, and any special network algorithms available in the solver can be applied automatically. On the other hand, a network expressed algebraically using var and subject to may or may not be recognized by the solver, and certain options may have to be set to ensure that it is recognized. For example, when using the algebraic model of Figure 11-4a, you may see the usual response from the general LP algorithm:

```
ampl: model net3.mod; data net3.dat; solve;
CPLEX 2.0: optimal solution; objective 1819
3 iterations (2 in phase I)
```

But when using the equivalent node and arc formulation of Figure 11-11, you may get a somewhat different response to reflect the application of a special network LP algorithm:

```
ampl: model net3node.mod
ampl: data net3.dat
ampl: solve;
CPLEX 2.0: optimal solution; objective 1819
6 network simplex iterations.
```

To determine how your favorite solver behaves in this situation, consult the solver-specific documentation that is supplied with your AMPL installation.

Because network linear programs are much easier to solve, especially to integrality, the success of a large-scale application may depend on whether a pure network formulation is possible. In the case of the multicommodity flow model of Figure 11-13, for example, the joint capacity constraints disrupt the network structure — they represent a third constraint in which each variable figures — but their presence cannot be avoided in a correct representation of the problem. Multicommodity flow problems thus do not necessarily have integer solutions, and are generally much harder to solve than single-commodity flow problems of comparable size.

In some cases, a judicious reformulation can turn what appears to be a more general model into a pure network model. Consider, for instance, a generalization of Figure 11-10 in which capacities are defined at the nodes as well as along the arcs:

```
param city_cap {CITIES} >= 0;
param link_cap {LINKS} >= 0;
```

The arc capacities represent, as before, upper limits on the shipments between cities. The node capacities limit the throughput, or total flow handled at a city, which may be written as the supply at the city plus the sum of the flows in, or equivalently as the demand at the city plus the sum of the flows out. Using the former, we arrive at the following constraint:

```
subject to through_limit {k in CITIES}:
    supply[k] + sum {(i,k) in LINKS} Ship[i,k] <= node_cap[k];
```

Viewed in this way, the throughput limit is another example of a "side constraint" that disrupts the network structure by adding a third coefficient for each variable. But we can achieve the same effect without a side constraint, by using two nodes to represent each city; one receives flow into a city plus any supply, and the other sends flow out of a city plus any demand:

```
node Supply {k in CITIES}: net_out = supply[k];
node Demand {k in CITIES}: net_in = demand[k];
```

A shipment link between cities i and j is represented by an arc that connects the node Demand[i] to node Supply[j]:

```
arc Ship {(i,j) in LINKS} >= 0, <= link_cap[i,j],
    from Demand[i], to Supply[j], obj Total_Cost cost[i,j];
```

```
set CITIES;
set LINKS within (CITIES cross CITIES);

param supply {CITIES} >= 0;   # amounts available at cities
param demand {CITIES} >= 0;   # amounts required at cities

   check: sum {i in CITIES} supply[i] = sum {j in CITIES} demand[j];

param cost {LINKS} >= 0;        # shipment costs per ton

param city_cap {CITIES} >= 0; # max throughput at cities
param link_cap {LINKS} >= 0;  # max shipment over links

minimize Total_Cost;

node Supply {k in CITIES}: net_out = supply[k];
node Demand {k in CITIES}: net_in = demand[k];

arc Ship {(i,j) in LINKS} >= 0, <= link_cap[i,j],
   from Demand[i], to Supply[j], obj Total_Cost cost[i,j];

arc Through {k in CITIES} >= 0, <= city_cap[k],
   from Supply[k], to Demand[k];
```

Figure 11-15: Transshipment model with node capacities (`netthru.mod`).

The throughput at city `k` is represented by a new kind of arc, from `Supply[k]` to `Demand[k]`:

```
   arc Through {k in cities} >= 0, <= city_cap[k],
      from Supply[k], to Demand[k];
```

The throughput limit is now represented by an upper bound on this arc's flow, rather than by a side constraint, and the network structure of the model is preserved. A complete listing appears in Figure 11-15.

The preceding example exhibits an additional advantage of using the `node` and `arc` declarations when developing a network model. If you use only `node` and `arc` in their simple forms — no variables in the node conditions, and no optional factors in the `from` and `to` phrases — your model is guaranteed to give rise only to pure network linear programs. By contrast, if you use `var` and `subject to`, it is your responsibility to ensure that the resulting linear program has the necessary network structure.

Some of the preceding comments can be extended to "generalized network" linear programs in which each variable still figures in at most two constraints, but not necessarily with coefficients of +1 and –1. We have seen examples of generalized networks in the cases where there is a loss of flow or change of units on the arcs. Generalized network LPs do not necessarily have integer optimal solutions, but fast algorithms for them do exist. A solver that promises a "network" algorithm may or may not have an extension to generalized networks; check the solver-specific documentation before you make any assumptions.

Bibliography

L. R. Ford, Jr. and D. R. Fulkerson, *Flows in Networks*. Princeton University Press (Princeton, NJ, 1962). A highly influential survey of network linear programming and related topics, which stimulated much subsequent study.

Walter Jacobs, "The Caterer Problem." Naval Research Logistics Quarterly **1** (1954) pp. 154–165. The origin of the network problem described in Exercise 11-8.

Katta G. Murty, *Network Programming*. Prentice-Hall (Englewood Cliffs, NJ, 1992). A categorical survey of network problems and applications — considerably longer than Ford and Fulkerson's book of 30 years earlier.

Exercises

11-1. The following diagram can be interpreted as representing a network transshipment problem:

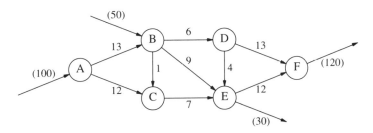

The arrows into nodes A and B represent supply in the indicated amounts, 100 and 50; the arrows out of nodes E and F similarly represent demand in the amounts 30 and 120. The remaining arrows indicate shipment possibilities, and the numbers on them are the unit shipping costs. There is a capacity of 80 on every arc.

(a) Solve this problem by developing appropriate data statements to go along with the model of Figure 11-2a.

(b) Check that you get the same solution using the node and arc formulation of Figure 11-10. Does the solver appear to be using the same algorithm as in (a)? Try this comparison with each LP solver available to you.

11-2. Reinterpret the numbers on the arcs *between* nodes, in the diagram of the preceding exercise, to solve the following problems. (Ignore the numbers on the arrows into A and B and on the arrows out of E and F.)

(a) Regarding the numbers on the arcs between nodes as lengths, use a model such as the one in Figure 11-7 to find the shortest path from A to F.

(b) Regarding the numbers on the arcs between nodes as capacities, use a model such as the one in Figure 11-6 to find the maximum flow from A to F.

(c) Generalize the model from (b) so that it can find the maximum flow from any subset of nodes to any other subset of nodes, in some meaningful sense. Use your generalization to find the maximum flow from A and B to E and F.

11-3. Section 4.2 showed how a multiperiod model could be constructed by replicating a static model over time periods, and using inventories to tie the periods together. Consider applying the same approach to the specialized transshipment model of Figure 11-4a.

(a) Construct a multi-week version of Figure 11-4a, with the inventories kept at the distribution centers (the members of D_CITY).

(b) Now consider expanding the data of Figure 11-4b for this model. Suppose that the initial inventory is 200 at NE and 75 at SE, and that the inventory carrying cost per 1000 packages is 0.15 per week at NE and 0.12 per week at SE. Let the supplies and demands over 5 weeks be as follows:

			Demand			
Week	Supply	BOS	EWR	BWI	ATL	MCO
1	450	50	90	95	50	20
2	450	65	100	105	50	20
3	400	70	100	110	50	25
4	250	70	110	120	50	40
5	325	80	115	120	55	45

Leave the cost and capacity data unchanged, the same in all weeks. Develop an appropriate data file for this situation, and solve the multi-week problem.

(c) The multi-week model in this case can be viewed as a pure network model, with arcs representing inventories as well as shipments. To show that this is the case, reformulate the model from (a) using only node and arc declarations for the constraints and variables.

11-4. For each of the following network problems, construct a data file that permits it to be solved using the general transshipment model of Figure 11-2a.

(a) The transportation problem of Figure 3-1.

(b) The assignment problem of Figure 3-2.

(c) The maximum flow problem of Figure 11-6.

(d) The shortest path problem of Figure 11-7.

11-5. Reformulate each of the following network models using node and arc declarations as much as possible:

(a) The transportation model of Figure 6-2a.

(b) The shortest path model of Figure 11-7.

(c) The production/transportation model of Figure 4-6.

(d) The multicommodity transportation model of Figure 6-5.

11-6. The professor in charge of an industrial engineering design course is faced with the problem of assigning 28 students to eight projects. Each must be assigned to one project, and each project group must have 3 or 4 students. The students have been asked to rank the projects, with 1 being the best ranking and higher numbers representing lower rankings.

(a) Formulate an algebraic assignment model, using var and subject to declarations, for this problem.

(b) Solve the assignment problem for the following table of assignments:

	A	ED	EZ	G	H1	H2	RB	SC
Allen	1	3	4	7	7	5	2	6
Black	6	4	2	5	5	7	1	3
Chung	6	2	3	1	1	7	5	4
Clark	7	6	1	2	2	3	5	4
Conners	7	6	1	3	3	4	5	2
Cumming	6	7	4	2	2	3	5	1
Demming	2	5	4	6	6	1	3	7
Eng	4	7	2	1	1	6	3	5
Farmer	7	6	5	2	2	1	3	4
Forest	6	7	2	5	5	1	3	4
Goodman	7	6	2	4	4	5	1	3
Harris	4	7	5	3	3	1	2	6
Holmes	6	7	4	2	2	3	5	1
Johnson	7	2	4	6	6	5	3	1

	A	ED	EZ	G	H1	H2	RB	SC
Knorr	7	4	1	2	2	5	6	3
Manheim	4	7	2	1	1	3	6	5
Morris	7	5	4	6	6	3	1	2
Nathan	4	7	5	6	6	3	1	2
Neuman	7	5	4	6	6	3	1	2
Patrick	1	7	5	4	4	2	3	6
Rollins	6	2	3	1	1	7	5	4
Schuman	4	7	3	5	5	1	2	6
Silver	4	7	3	1	1	2	5	6
Stein	6	4	2	5	5	7	1	3
Stock	5	2	1	6	6	7	4	3
Truman	6	3	2	7	7	5	1	4
Wolman	6	7	4	2	2	3	5	1
Young	1	3	4	7	7	6	2	5

How many students are assigned second or third choice?

(c) Some of the projects are harder than others to reach without a car. Thus it is desirable that at least a certain number of students assigned to each project must have a car; the numbers vary by project as follows:

A 1   ED 0   EZ 0   G 2   H1 2   H2 2   RB 1   SC 1

The students who have cars are:

Chung   Eng   Manheim   Nathan   Rollins
Demming   Holmes   Morris   Patrick   Young

Modify the model to add this car constraint, and solve the problem again. How many more students than before must be assigned second or third choice?

(d) Your formulation in (c) can be viewed as a transportation model with side constraints. By defining appropriate network nodes and arcs, reformulate it as a "pure" network flow model, as discussed in Section 11.5. Write the formulation in AMPL using only node and arc declarations for the constraints and variables. Solve with the same data as in (c), to show that the optimal value is the same.

11-7. To manage its excess cash over the next 12 months, a company may purchase 1-month, 2-month or 3-month certificates of deposit from any of several different banks. The current cash on hand and amounts invested are known, while the company must estimate the cash receipts and expenditures for each month, and the returns on the different certificates.

The company's problem is to determine the best investment strategy, subject to cash requirements. (As a practical matter, the company would use the first month of the optimal solution as a guide to its current purchases, and then re-solve with updated estimates at the beginning of the next month.)

(a) Draw a network diagram for this situation. Show each month as a node, and the investments, receipts and expenditures as arcs.

(b) Formulate the relevant optimization problem as an AMPL model, using node and arc declarations. Assume that any cash from previously-purchased certificates coming due in the early months is included in data for the receipts.

There is more than one way to describe the objective function for this model. Explain your choice.

(c) Suppose that the company's estimated receipts and expenses (in thousands of dollars) over the next 12 months are as follows:

	receipt	expense
1	3200	200
2	3600	200
3	3100	400
4	1000	800
5	1000	2100
6	1000	4500
7	1200	3300
8	1200	1800
9	1200	600
10	1500	200
11	1800	200
12	1900	200

The two banks competing for the business are estimating the following rates of return for the next 12 months:

CIT:	1	2	3	NBD:	1	2	3
1	0.00433	0.01067	0.01988	1	0.00425	0.01067	0.02013
2	0.00437	0.01075	0.02000	2	0.00429	0.01075	0.02025
3	0.00442	0.01083	0.02013	3	0.00433	0.01083	0.02063
4	0.00446	0.01092	0.02038	4	0.00437	0.01092	0.02088
5	0.00450	0.01100	0.02050	5	0.00442	0.01100	0.02100
6	0.00458	0.01125	0.02088	6	0.00450	0.01125	0.02138
7	0.00467	0.01142	0.02113	7	0.00458	0.01142	0.02162
8	0.00487	0.01183	0.02187	8	0.00479	0.01183	0.02212
9	0.00500	0.01217	0.02237	9	0.00492	0.01217	0.02262
10	0.00500	0.01217	0.02250	10	0.00492	0.01217	0.02275
11	0.00492	0.01217	0.02250	11	0.00483	0.01233	0.02275
12	0.00483	0.01217	0.02275	12	0.00475	0.01250	0.02312

Construct an appropriate data file, and solve the resulting linear program. Use `display` to produce a summary of the indicated purchases.

(d) Company policy prohibits investing more than 70% of its cash in new certificates of any one bank in any month. Devise a side constraint on the model from (b) to impose this restriction.

Again solve the resulting linear program, and summarize the indicated purchases. How much income is lost due to the restrictive policy?

11-8. A caterer has booked dinners for the next T days, and has as a result a requirement for a certain number of napkins each day. He has a certain initial stock of napkins, and can buy new ones each day at a certain price. In addition, used napkins can be laundered either at a slow service that takes 4 days, or at a faster but more expensive service that takes 2 days. The caterer's problem is to find the most economical combination of purchase and laundering that will meet the forthcoming demand.

(a) It is not hard to see that the decision variables for this problem should be something like the following:

Buy[t]	clean napkins bought for day t
Carry[t]	clean napkins still on hand at the end of day t
Wash2[t]	used napkins sent to the fast laundry after day t
Wash4[t]	used napkins sent to the slow laundry after day t
Trash[t]	used napkins discarded after day t

There are two collections of constraints on these variables, which can be described as follows:

– The number of clean napkins acquired through purchase, carryover and laundering on day t must equal the number sent to laundering, discarded or carried over after day t.

– The number of used napkins laundered or discarded after day t must equal the number that were required for that day's catering.

Formulate an AMPL linear programming model for this problem.

(b) Formulate an alternative network linear programming model for this problem. Write it in AMPL using node and arc declarations.

(c) The "caterer problem" was introduced in a 1954 paper by W. Jacobs of the U.S. Air Force. Although it has been presented in countless books on linear and network programming, it does not seem to have ever been used by any caterer. In what application do you suppose it really originated?

(d) Since this is an artificial problem, you might as well make up your own data for it. Use your data to check that the formulations in (a) and (b) give the same optimal value.

12

Columnwise Formulations

Because the fundamental idea of an optimization problem is to minimize or maximize a function of the decision variables, subject to constraints on them, AMPL is oriented toward explicit descriptions of variables and constraints. This is why var declarations tend to come first, followed by the minimize or maximize and subject to declarations that use the variables. A wide variety of optimization problems can be formulated with this approach, as the examples in this book demonstrate.

For certain kinds of linear programs, however, it can be preferable to declare the objective and constraints before the variables. Usually in these cases, there is a much simpler pattern or interpretation to the coefficients of a single variable down all the constraints than to the coefficients of a single constraint across all the variables. In the jargon of linear programming, it is easier to describe the matrix of constraint coefficients "columnwise" than "row-wise". As a result, the formulation is simplified by first declaring the constraints and objective, then listing the nonzero coefficients in the declarations of the variables.

One example of this phenomenon is provided by the network linear programs described in Chapter 11. Each variable has at most two nonzero coefficients in the constraints, a +1 and a −1. Rather than trying to describe the constraints algebraically, you may find it easier to specify, in each variable's declaration, the one or two constraints that the variable figures in. In fact, this is exactly what you do by using the special node and arc declarations introduced by Section 11.3. The node declarations come first to describe the nature of the constraints at the network nodes. Then the arc declarations define the network flow variables, using from and to phrases to locate their nonzero coefficients among the node constraints. This approach is particularly appealing because it corresponds directly to the way most people think about a network flow problem.

It would be impractical for AMPL to offer special declarations and phrases for every kind of linear program that you might want to declare by columns rather than by rows. Instead, additional options to the var and subject to declarations permit any linear program to be given a columnwise declaration. This chapter introduces AMPL's columnwise features with two contrasting examples — an input-output production model, and a work-shift scheduling model — and concludes with a summary of language rules.

12.1 An input-output model

In simple maximum-profit production models such as the examples in Chapter 1, the goods produced are distinct from the resources consumed, so that overall production is limited in an obvious way by resources available. In a more realistic model of a complex operation such as a steel mill or refinery, however, production is carried out at a series of units; as a result, some of a production unit's inputs may be the outputs from other units. For this situation we need a model that deals more generally with materials that may be inputs or outputs, and with production activities that may involve several inputs and outputs each.

We begin by developing an AMPL formulation in the usual row-wise (or constraint-oriented) way. Then we explain the columnwise (or variable-oriented) alternative, and discuss refinements of the model.

Formulation by constraints

The definition of our model starts with a set of materials and a set of activities:

```
set MAT;
set ACT;
```

The key data values are the input-output coefficients for all material-activity combinations:

```
param io {MAT,ACT};
```

If $io[i,j] > 0$, it is interpreted as the amount of material i produced (as an output) by a unit of activity j. On the other hand, if $io[i,j] < 0$, it represents minus the amount of material i consumed (as an input) by a unit of activity j. For example, a value of 10 represents 10 units of i produced per unit of j, while a value of -10 represents 10 units consumed per unit of j. Of course, we can expect that for many combinations of i and j we will have $io[i,j] = 0$, signifying that material i does not figure in activity j at all.

To see why we want to interpret $io[i,j]$ in this manner, suppose we define $X[j]$ to be the level at which we operate activity j:

```
param act_min {ACT} >= 0;
param act_max {j in ACT} >= act_min[j];

var X {j in ACT} >= act_min[j], <= act_max[j];
```

Then $io[i,j] * X[j]$ is the amount of material i produced (if $io[i,j] > 0$) or minus the amount of material i consumed (if $io[i,j] < 0$) by activity j. Summing over all activities, we see that

```
sum {j in ACT} io[i,j] * X[j]
```

represents the amount of material i produced in the operation minus the amount consumed. These amounts must balance, as expressed by the following constraint:

```
set MAT;               # materials
set ACT;               # activities
param io {MAT,ACT};   # input-output coefficients

param revenue {ACT};
param act_min {ACT} >= 0;
param act_max {j in ACT} >= act_min[j];

var X {j in ACT} >= act_min[j], <= act_max[j];

maximize Net_Profit:  sum {j in ACT} revenue[j] * X[j];

subject to Balance {i in MAT}: sum {j in ACT} io[i,j] * X[j] = 0;
```

Figure 12-1: Input-output model by rows (`iorow.mod`).

```
subject to Balance {i in MAT}:
   sum {j in ACT} io[i,j] * X[j] = 0;
```

What about the availability of resources, or the requirements for finished goods? These are readily modeled through additional activities that represent the purchase or sale of materials. A purchase activity for material i has no inputs and just i as an output; the upper limit on X[i] represents the amount of this resource available. Similarly, a sale activity for material i has no outputs and just i as an input, and the lower limit on X[i] represents the amount of this good that must be produced for sale.

We complete the model by associating unit revenues with the activities. Sale activities necessarily have positive revenues, while purchase and production activities have negative revenues — that is, costs. The sum of unit revenues times activity levels gives the total net profit of the operation:

```
param revenue {ACT};
maximize Net_Profit: sum {j in ACT} revenue[j] * X[j];
```

The completed model is shown in Figure 12-1.

A columnwise formulation

As our discussion of purchase and sale activities suggests, everything in this model can be organized by activity. Specifically, for each activity j we have a decision variable X[j], a cost or income represented by revenue[j], limits act_min[j] and act_max[j], and a collection of input-output coefficients io[i,j]. Changes such as improving the yield of a unit, or acquiring a new source of supply, are accommodated by adding an activity or by modifying the data for an activity.

In the formulation by rows, the activities' importance to this model is somewhat hidden. While act_min[j] and act_max[j] appear in the declaration of the variables, revenue[j] is in the objective, and the io[i,j] values are in the constraint declaration. The columnwise alternative brings all of this information together, by adding obj and coeff phrases to the var declaration:

```
set MAT;               # materials
set ACT;               # activities
param io {MAT,ACT};    # input-output coefficients

param revenue {ACT};
param act_min {ACT} >= 0;
param act_max {j in ACT} >= act_min[j];

maximize Net_Profit;
subject to Balance {i in MAT}: to_come = 0;

var X {j in ACT} >= act_min[j], <= act_max[j],
    obj Net_Profit revenue[j],
    coeff {i in MAT} Balance[i] io[i,j];
```

Figure 12-2: Columnwise formulation (`iocol1.mod`).

```
var X {j in ACT} >= act_min[j], <= act_max[j],
    obj Net_Profit revenue[j],
    coeff {i in MAT} Balance[i] io[i,j];
```

The `obj` phrase says that in the objective function named `Net_Profit`, the variable `X[j]` has the coefficient `revenue[j]`; that is, the term `revenue[j] * X[j]` should be added in. The `coeff` phrase is a bit more complicated, because it is indexed over a set. It says that for each material `i`, in the constraint `Balance[i]` the variable `X[j]` should have the coefficient `io[i,j]`, so that the term `io[i,j] * X[j]` is added in. Together, these phrases describe all the coefficients of all the variables in the linear program.

Since we have placed all the coefficients in the `var` declaration, we must remove them from the other declarations:

```
maximize Net_Profit;
subject to Balance {i in MAT}: to_come = 0;
```

The keyword `to_come` indicates where the terms `io[i,j] * X[j]` generated by the `var` declaration are to be "added in." You can think of `to_come = 0` as a template for the constraint, which will be filled out as the coefficients are declared. No template is needed for the objective in this example, however, since it is exclusively the sum of the terms `revenue[j] * X[j]`. Templates may be written in a limited variety of ways, as shown in Section 12.3 below.

Because the `obj` and `coeff` phrases refer to `Net_Profit` and `Balance`, the `var` declaration must come after the `maximize` and `subject to` declarations in the columnwise formulation. The complete model is shown in Figure 12-2.

Refinements of the columnwise formulation

The advantages of a columnwise approach become more evident as the model becomes more complicated. As one example, consider what happens if we want to have separate variables to represent sales of finished materials. We declare a subset of materials that can be sold, and use it to index new collections of bounds, revenues and variables:

```
set MATF within MAT;   # finished materials

param revenue {MATF} >= 0;

param sell_min {MATF} >= 0;
param sell_max {i in MATF} >= sell_min[i];

var Sell {i in MATF} >= sell_min[i], <= sell_max[i];
```

We may now dispense with the special sale activities previously described. Since the remaining members of ACT represent purchase or production activities, we can introduce a nonnegative parameter cost associated with them:

```
param cost {ACT} >= 0;
```

In the row-wise approach, the new objective is written as

```
maximize Net_Profit:
    sum {i in MATF} revenue[i] * Sell[i]
    - sum {j in ACT} cost[j] * X[j];
```

to represent total sales revenue minus total raw material and production costs.

So far we seem to have improved upon the model in Figure 12-1. The composition of net profit is more clearly modeled, and sales are restricted to explicitly designated finished materials; also the optimal amounts sold are more easily examined apart from the other variables, by a command such as display Sell. It remains to fix up the constraints. We would like to say that the net output of material i from all activities, represented as

```
sum {j in ACT} io[i,j] * X[j]
```

in Figure 12-1, must balance the amount sold — either Sell[i] if i is a finished material, or zero. Thus the constraint declaration must be written:

```
subject to Balance {i in MAT}:
    sum {j in ACT} io[i,j] * X[j]
       = if i in MATF then Sell[i] else 0;
```

Unfortunately this constraint seems less clear than our original one, due to the complication introduced by the if-then-else expression.

In the columnwise alternative, the objective and constraints are the same as in Figure 12-2, while all the changes are reflected in the declarations of the variables:

```
var X {j in ACT} >= act_min[j], <= act_max[j],
    obj Net_Profit -cost[j],
    coeff {i in MAT} Balance[i] io[i,j];

var Sell {i in MATF} >= sell_min[i], <= sell_max[i],
    obj Net_Profit revenue[i],
    coeff Balance[i] -1;
```

```
set MAT;                # materials
set ACT;                # activities
param io {MAT,ACT};     # input-output coefficients

set MATF within MAT;    # finished materials

param revenue {MATF} >= 0;
param sell_min {MATF} >= 0;
param sell_max {i in MATF} >= sell_min[i];

param cost {ACT} >= 0;
param act_min {ACT} >= 0;
param act_max {j in ACT} >= act_min[j];

maximize Net_Profit;
subject to Balance {i in MAT}: to_come = 0;

var X {j in ACT} >= act_min[j], <= act_max[j],
    obj Net_Profit -cost[j],
    coeff {i in MAT} Balance[i] io[i,j];

var Sell {i in MATF} >= sell_min[i], <= sell_max[i],
    obj Net_Profit revenue[i],
    coeff Balance[i] -1;
```

Figure 12-3: Columnwise formulation, with sales activities (iocol2.mod).

In this view, the variable Sell[i] represents the kind of sale activity that we previously described, with only material i as input and no materials as output — hence the single coefficient of −1 in constraint Balance[i]. We need not specify all the zero coefficients for Sell[i]; a zero is assumed for any constraint not explicitly cited in a coeff phrase in the declaration. The whole model is shown in Figure 12-3.

This example suggests that a columnwise approach is particularly suited to refinements of the input-output model that distinguish different kinds of activities. It would be easy to add another group of variables that represent purchases of raw materials, for instance. On the other hand, versions of the input-output model that involve numerous specialized constraints would lend themselves more to a formulation by rows.

12.2 A scheduling model

In Section 2.4 we observed that the general form of a blending model was applicable to certain scheduling applications. Here we describe a related scheduling model for which the columnwise approach is particularly attractive.

Suppose that a factory's production for the next week is divided into fixed time periods, or shifts. You want to assign employees to shifts, so that the required number of people are working on each shift. You cannot fill each shift independently of the others, however, because only certain weekly schedules are allowed; for example, a person cannot work five shifts in a row. Your problem is thus more properly viewed as one of assigning employees to schedules, so that each shift is covered and the overall assignment is the most economical.

We can conveniently represent the schedules for this problem by an indexed collection of subsets of shifts:

```
set SHIFTS;                 # shifts

param Nsched;               # number of schedules;
set SCHEDS := 1..Nsched;    # set of schedules

set SHIFT_LIST {SCHEDS} within SHIFTS;
```

For each schedule j in the set SCHEDS, the shifts that a person works on schedule j are contained in the set SHIFT_LIST[j]. We also specify a pay rate per person on each schedule, and the number of people required on each shift:

```
param rate {SCHEDS} >= 0;
param required {SHIFTS} >= 0;
```

We let the variable Work[j] represent the number of people assigned to work each schedule j, and minimize the sum of rate[j] * Work[j] over all schedules:

```
var Work {SCHEDS} >= 0;
minimize Total_Cost: sum {j in SCHEDS} rate[j] * Work[j];
```

Finally, our constraints say that the total of employees assigned to each shift i must be at least the number required:

```
subject to Shift_Needs {i in SHIFTS}:
    sum {j in SCHEDS: i in SHIFT_LIST[j]} Work[j]
        >= required[i];
```

On the left we take the sum of Work[j] over all schedules j such that i is in SHIFT_LIST[j]. This sum represents the total employees who are assigned to schedules that contain shift i, and hence equals the total employees covering shift i.

The somewhat awkward description of the constraint in this formulation motivates us to try a columnwise formulation. As in our previous examples, we declare the objective and constraints first, but with the variables left out:

```
minimize Total_Cost;
subject to Shift_Needs {i in SHIFTS}: to_come >= required[i];
```

The coefficients of Work[j] appear instead in its var declaration. In the objective, it has a coefficient of rate[j]. In the constraints, the membership of SHIFT_LIST[j] tells us exactly what we need to know: Work[j] has a coefficient of 1 in constraint Shift_Needs[i] for each i in SHIFT_LIST[j], and a coefficient of 0 in the other constraints. This leads us to the following concise declaration:

```
var Work {j in SCHEDS} >= 0,
    obj Total_Cost rate[j],
    coeff {i in SHIFT_LIST[j]} Shift_Needs[i] 1;
```

The full model is shown in Figure 12-4.

```
set SHIFTS;                    # shifts

param Nsched;                  # number of schedules;
set SCHEDS := 1..Nsched;       # set of schedules

set SHIFT_LIST {SCHEDS} within SHIFTS;

param rate {SCHEDS} >= 0;
param required {SHIFTS} >= 0;

minimize Total_Cost;
subject to Shift_Needs {i in SHIFTS}: to_come >= required[i];

var Work {j in SCHEDS} >= 0,
   obj Total_Cost rate[j],
   coeff {i in SHIFT_LIST[j]} Shift_Needs[i] 1;
```

Figure 12-4: Columnwise scheduling model (`sched.mod`).

```
set SHIFTS := Mon1 Tue1 Wed1 Thu1 Fri1 Sat1
              Mon2 Tue2 Wed2 Thu2 Fri2 Sat2
              Mon3 Tue3 Wed3 Thu3 Fri3 ;

param Nsched := 126 ;

set SHIFT_LIST[  1] := Mon1 Tue1 Wed1 Thu1 Fri1 ;
set SHIFT_LIST[  2] := Mon1 Tue1 Wed1 Thu1 Fri2 ;
set SHIFT_LIST[  3] := Mon1 Tue1 Wed1 Thu1 Fri3 ;
set SHIFT_LIST[  4] := Mon1 Tue1 Wed1 Thu1 Sat1 ;
set SHIFT_LIST[  5] := Mon1 Tue1 Wed1 Thu1 Sat2 ;
```

(117 lines omitted)

```
set SHIFT_LIST[123] := Tue1 Wed1 Thu1 Fri2 Sat2 ;
set SHIFT_LIST[124] := Tue1 Wed1 Thu2 Fri2 Sat2 ;
set SHIFT_LIST[125] := Tue1 Wed2 Thu2 Fri2 Sat2 ;
set SHIFT_LIST[126] := Tue2 Wed2 Thu2 Fri2 Sat2 ;

param rate  default 1 ;

param required :=  Mon1 100   Mon2 78   Mon3 52
                   Tue1 100   Tue2 78   Tue3 52
                   Wed1 100   Wed2 78   Wed3 52
                   Thu1 100   Thu2 78   Thu3 52
                   Fri1 100   Fri2 78   Fri3 52
                   Sat1 100   Sat2 78 ;
```

Figure 12-5: Partial data for scheduling model (`sched.dat`).

As a specific instance of this model, imagine that you have three shifts a day on Monday through Friday, and two shifts on Saturday. Each day you need 100, 78 and 52 employees on the first, second and third shifts, respectively. To keep things simple, suppose that the cost per person is the same regardless of schedule, so that you may just minimize the total number of employees by setting `rate[j]` to 1 for all `j`.

As for the schedules, a reasonable scheduling rule might be that each employee works five shifts in a week, but never more than one shift in any 24-hour period. Part of the data file is shown in Figure 12-5; we don't show the whole file, because there are 126 schedules that satisfy the rule! The resulting 126-variable linear program is not hard to solve, however:

```
ampl: model sched.mod; data sched.dat; solve;
MINOS 5.4: optimal solution found.
25 iterations, objective 265.6

ampl: option display_eps .000001;
ampl: option omit_zero_rows 1;
ampl: option display_1col 0;

ampl: display Work;
Work [*] :=
    7 25         32 22         81   6.8     106 20.2
   12  3.8       43 28.8       92 21.2      113  6.8
   19 10.6       55  6.8       95  3.8      116 25
   23  3         71 35.6      102 10.6      123 35.6
   ;
```

As you can see, this optimal solution makes use of 16 of the schedules, some in fractional amounts. (There exist many other optimal solutions to this problem, so the results you get may differ.) If you round each fraction in this solution up to the next highest value, you get a pretty good feasible solution using 270 employees. To determine whether this is the best whole-number solution, however, it is necessary to use integer programming techniques, which are the subject of Chapter 15.

The columnwise formulation works well in this case as a direct result of how we have chosen to represent the data. We imagine that the modeler will be thinking in terms of schedules, and will want to try adding, dropping or modifying different schedules to see what solutions can be obtained. The subsets SHIFT_LIST[j] provide a convenient and concise way of maintaining the schedules in the data. Since the data are then organized by schedules, and there is also a variable for each schedule, it proves to be simpler — and for larger examples, more efficient — to specify the coefficients by variable.

Models of this kind are used for a variety of scheduling problems. As a convenience, the keyword cover may be used (in the manner of from and to for networks) to specify a coefficient of 1:

```
var Work {j in SCHEDS} >= 0,
   obj Total_Cost rate[j],
   cover {i in SHIFT_LIST[j]} Shift_Needs[i];
```

Some of the best known and largest examples are in airline crew scheduling, where the variables may represent the assignment of crews rather than individuals, the shifts become flights, and the requirement is one crew for each flight. We then have what is known as a set covering problem, in which the objective is to most economically cover the set of all flights with subsets representing crew schedules.

12.3 General rules

The rules for a constraint declaration (Section 8.4) say that the algebraic description of the constraint can be written in any of the following ways:

arith-expr <= *arith-expr*
arith-expr = *arith-expr*
arith-expr >= *arith-expr*

const-expr <= *arith-expr* <= *const-expr*
const-expr >= *arith-expr* >= *const-expr*

Each *const-expr* must be an arithmetic expression not containing variables, while an *arith-expr* may be any valid arithmetic expression — though it must be linear in the variables (Section 8.2) if the result is to be a linear program. To permit a columnwise formulation, one of the *arith-expr*s may have one of the following forms:

```
to_come
to_come + arith-expr
arith-expr + to_come
```

Most often a "template" constraint of this kind consists, as in our examples, of `to_come`, a relational operator and a *const-expr*; the constraint's linear terms are all provided in subsequent `var` declarations, and `to_come` shows where they should go. If the template constraint does contain variables, they must be from previous `var` declarations, and the model becomes a sort of hybrid between row-wise and columnwise forms.

The expression for an objective function may also incorporate `to_come` in one of the ways shown above. If the objective is a sum of linear terms specified entirely by subsequent `var` declarations, as in our examples, the expression for the objective is just `to_come` and may be omitted.

In a `var` declaration, constraint coefficients may be specified by one or more phrases consisting of the keyword `coeff`, an optional indexing expression, a constraint name, and an *arith-expr*. If an indexing expression is present, a coefficient is generated for each member of the indexing set; otherwise, one coefficient is generated. The indexing expression may also take the special form { `if` *logical-expr* } as seen in Section 8.4 or 11.3, in which case a coefficient is generated only if the *logical-expr* evaluates to true. Our simple examples have required just one `coeff` phrase in each `var` declaration, but in general a separate `coeff` phrase is needed for each different indexed collection of constraints in which a variable appears.

Objective function coefficients may be specified in the same way, except that the keyword `obj` is used instead of `coeff`.

The `obj` phrase in a `var` declaration is the same as the `obj` phrase used in `arc` declarations for networks (Section 11.4). The constraint coefficients for the variables defined by an `arc` declaration are normally given in `from` and `to` phrases, but `coeff` phrases may be present as well; they can be useful if you want to give a columnwise description of "side" constraints that apply in addition to the balance-of-flow constraints. As an example, Figure 12-6 shows how a `coeff` phrase can be used to rewrite the multi-commodity flow model of Figure 11-13 in an entirely columnwise manner.

```
set CITIES;
set LINKS within (CITIES cross CITIES);

set PRODS;

param supply {CITIES,PRODS} >= 0;  # amounts available at cities
param demand {CITIES,PRODS} >= 0;  # amounts required at cities

    check {p in PRODS}:
        sum {i in CITIES} supply[i,p] = sum {j in CITIES} demand[j,p];

param cost {LINKS,PRODS} >= 0;      # shipment costs/1000 packages
param capacity {LINKS,PRODS} >= 0;  # max packages shipped
param cap_joint {LINKS} >= 0;       # max total packages shipped/link

minimize Total_Cost;

node Balance {k in CITIES, p in PRODS}:
   net_in = demand[k,p] - supply[k,p];

subject to Multi {(i,j) in LINKS}:
   to_come <= cap_joint[i,j];

arc Ship {(i,j) in LINKS, p in PRODS} >= 0, <= capacity[i,j,p],
   from Balance[i,p], to Balance[j,p],
   coeff Multi[i,j] 1.0,
   obj Total_Cost cost[i,j,p];
```

Figure 12-6: Columnwise formulation of Figure 11-13 (netmcol.mod).

Bibliography

Gerald Kahan, ''Walking Through a Columnar Approach to Linear Programming of a Business.'' Interfaces **12**, 3 (1982) pp. 32–39. A brief for the columnwise approach to linear programming, with a small example.

Exercises

12-1. (a) Construct a columnwise formulation of the diet model of Figure 2-1.

(b) Construct a columnwise formulation of the diet model of Figure 5-1. Since there are two separate collections of constraints in this diet model, you will need two coeff phrases in the var declaration.

12-2. Expand the columnwise production model of Figure 12-3 to incorporate variables Buy[i] that represent purchases of raw materials.

12-3. Formulate a multiperiod version of the model of Figure 12-3, using the approach introduced in Section 4.2: first replicate the model over a set of weeks, then introduce inventory variables to tie the weeks together. Use only columnwise declarations for all of the variables, including those for inventory.

12-4. The "roll trim" or "cutting stock" problem has much in common with the scheduling problem described in this chapter. Review the description of the roll trim problem in Exercise 2-6, and use it to answer the following questions.

(a) What is the relationship between the available cutting patterns in the roll trim problem, and the coefficients of the variables in the linear programming formulation?

(b) Formulate an AMPL model for the roll trim problem that uses only columnwise declarations of the variables.

(c) Solve the roll trim problem for the data given in Exercise 2-6. As a test of your formulation, show that it gives the same optimal value as a row-wise formulation of the model.

12-5. The set covering problem, mentioned at the end of Section 12.2, can be stated in a general way as follows. You are given a set S, and certain subsets $T_1, T_2, \ldots, T_n$ of S; a cost is associated with each of the subsets. A selection of some of the subsets T_j is said to *cover* S if every member of S is also a member of at least one of the selected subsets. For example, if

$$S = \{1,2,3,4\}$$

and

$$T1 = \{1,2,4\} \quad T2 = \{2,3\} \quad T3 = \{1\} \quad T4 = \{3,4\} \quad T5 = \{1,3\}$$

the selections (T1, T2) and (T2, T4, T5) cover S, but the selection (T3, T4, T5) does not.

The goal of the set covering problem is to find the least costly selection of subsets that covers S. Formulate a columnwise linear program that solves this problem for any given set and subsets.

13

Nonlinear Programs

Although any model that violates the linearity rules of Chapter 8 is "not linear", the term "nonlinear program" is traditionally used in a more narrow sense. For our purposes in this chapter, a nonlinear program, like a linear program, has continuous (rather than integer or discrete) variables; the expressions in its objective and constraints need not be linear, but they must represent "smooth" functions. Intuitively, a smooth function of one variable has a graph like that in Figure 13-1a, for which there is a well-defined slope at every point; there are no jumps as in Figure 13-1b, or kinks as in Figure 13-1c. Mathematically, a smooth function of any number of variables must be continuous and must have a well-defined gradient (vector of first derivatives) at every point; Figures 13-1b and 13-1c exhibit points at which a function is discontinuous and nondifferentiable, respectively.

Optimization problems in functions of this kind have been singled out because they are readily distinguished from other "not linear" problems, because they have many and varied applications, and because they are amenable to solution by well-established types of algorithms. Indeed, most solvers for nonlinear programming use methods that rely on the assumptions of continuity and differentiability. Even with these assumptions, nonlinear programs are typically a lot harder to formulate and solve than comparable linear ones.

This chapter begins with an introduction to sources of nonlinearity in mathematical programs. We will not try to cover the variety of nonlinear models systematically, but instead will give a few examples to indicate why and how nonlinearities occur. Subsequent sections discuss the implications of nonlinearity for AMPL variables and expressions. Finally, we point out some of the difficulties that you are likely to confront in trying to solve nonlinear programs.

While the focus of this chapter is on nonlinear optimization, keep in mind that AMPL can also express systems of nonlinear equations or inequalities, even if there is no objective to optimize. There exist solvers specialized to this case, and many solvers for nonlinear optimization can also do a decent job of finding a feasible solution to an equation or inequality system.

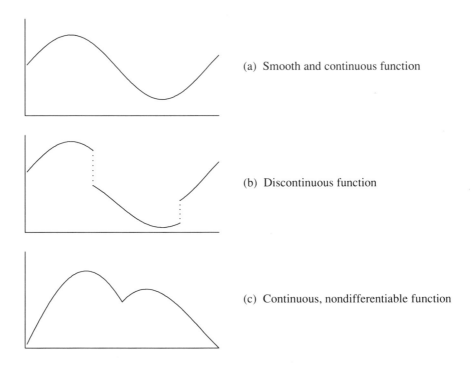

(a) Smooth and continuous function

(b) Discontinuous function

(c) Continuous, nondifferentiable function

Figure 13-1: Classes of nonlinear functions.

13.1 Sources of nonlinearity

Many of our examples of linear programs have depended on some kind of linearity assumption in the formulation. As an example of how nonlinearities can appear in a model, we consider here how one particular linearity assumption might be relaxed. At the end of this section we comment briefly on other ways in which nonlinear programs arise.

Examples of nonlinear costs

In the transportation model of Chapter 3, the objective function is defined in terms of sets ORIG and DEST representing the origins and destinations of shipments:

```
param cost {ORIG,DEST} >= 0;    # shipment costs per unit
var Trans {ORIG,DEST} >= 0;     # units to be shipped

minimize total_cost:
   sum {i in ORIG, j in DEST} cost[i,j] * Trans[i,j];
```

For a particular origin `i` and destination `j`, the variable `Trans[i,j]` contributes an amount `cost[i,j] * Trans[i,j]` to the total cost of the shipments. Figure 13-2a shows a typical plot of shipping cost versus amount shipped. The plot is a line with slope `cost[i,j]` (hence the term linear). The other plots in Figure 13-2 show a variety of other ways, none of them linear, in which shipping cost could depend on the shipment amount.

In Figure 13-2b the cost also tends to increase linearly with the amount shipped, but at certain critical amounts the cost per unit (that is, the slope of the line) makes an abrupt change. This kind of function is called piecewise-linear. It is not linear, strictly speaking, but it is also not smoothly nonlinear. The use of piecewise-linear objectives is the topic of Chapter 14.

In Figure 13-2c the function itself jumps abruptly. When nothing is shipped, the shipping cost is zero; but when there is any shipment at all, the cost is linear starting from a value much greater than zero. In this case there is a fixed cost for using the link from `i` to `j`, plus a variable cost per unit shipped. Again, this is not a function that can be handled by linear programming techniques, but it is also not a smooth nonlinear function. Fixed costs are most commonly handled with integer variables, which are the topic of Chapter 15.

The remaining plots illustrate the sorts of smooth nonlinear functions that we want to consider in this chapter. Figure 13-2d shows a kind of concave cost function. The incremental cost for each additional unit shipped (that is, the slope of the plot) is great at first, but becomes less as more units are shipped; after a certain point, the cost is nearly linear. This is a continuous alternative to the fixed cost function of Figure 13-2c. It could also be used to approximate the cost for a situation in which volume discounts become available as the amount shipped increases.

Figure 13-2e shows a kind of convex cost function. The cost is more or less linear for smaller shipments, but rises steeply toward infinity as shipments approach some critical amount. This sort of function would be used to model a situation in which the lowest cost shippers are used first, while shipping becomes progressively more expensive as the number of units increases. The critical amount represents, in effect, an upper bound on the shipments.

These are some of the simplest functional forms. The functions that you consider will depend on the kind of situation that you are trying to model. Figure 13-2f shows a possibility that is neither concave nor convex, combining features of the previous two examples.

Whereas linear functions are essentially all the same except for the choice of coefficients (or slopes), nonlinear functions can be defined by an infinite variety of different formulas. Thus in building a nonlinear programming model, it is up to you to derive or specify nonlinear functions that properly represent the situation at hand. In the objective of the transportation example, for instance, one possibility would be to replace the product `cost[i,j] * Trans[i,j]` by

```
rate[i,j] * Trans[i,j] / (1 - Trans[i,j]/limit[i,j])
```

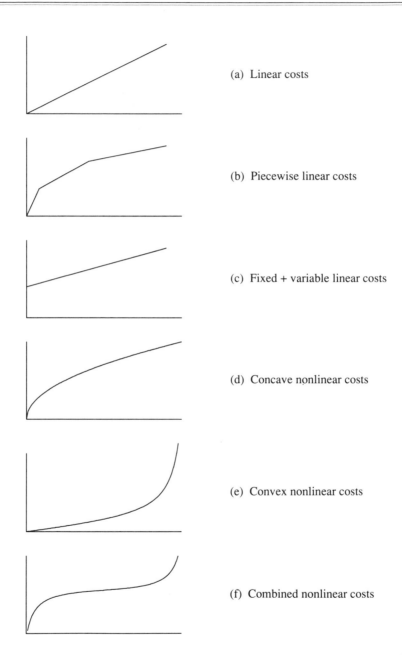

(a) Linear costs

(b) Piecewise linear costs

(c) Fixed + variable linear costs

(d) Concave nonlinear costs

(e) Convex nonlinear costs

(f) Combined nonlinear costs

Figure 13-2: Nonlinear cost functions.

where `rate[i,j]` and `limit[i,j]` are declared as parameters. This function is essentially linear with slope `rate[i,j]` when `Trans[i,j]` is small, and goes to infinity as `Trans[i,j]` approaches `limit[i,j]`, so it represents one way to implement the curve shown in Figure 13-2e.

Another way to approach the specification of a nonlinear objective function is to focus on the slopes of the plots in Figure 13-2. In the linear case of Figure 13-2a, the slope of the plot is constant; that is why we can use a single parameter `cost[i,j]` to represent the cost per unit shipped. In the piecewise-linear case of Figure 13-2b, the slope is constant within each interval; we will use this property to specify piecewise-linear functions in Chapter 14. In the nonlinear case, however, the slope varies continuously with the amount shipped. This suggests that we go back to our original linear formulation of the transportation problem, and turn the parameter `cost[i,j]` into a variable `Cost[i,j]`:

```
var Cost {ORIG,DEST};        # shipment costs per unit
var Trans {ORIG,DEST} >= 0;  # units to ship

minimize total_cost:
    sum {i in ORIG, j in DEST} Cost[i,j] * Trans[i,j];
```

This is no longer a linear objective, because it multiplies a variable by another variable. We add some equations to specify how the cost relates to the amount shipped:

```
subject to cost_relation {i in ORIG, j in DEST}:
    Cost[i,j] = rate[i,j] / (1 - Trans[i,j]/limit[i,j]);
```

These equations are also nonlinear, because they involve division by an expression that contains a variable. As noted above, `Cost[i,j]` is practically constant at `rate[i,j]` when `Trans[i,j]` is small, but grows rapidly as `Trans[i,j]` approaches `limit[i,j]`.

Other sources of nonlinearity

Although our examples have concentrated on the costs, there is a potential for nonlinearity in any aspect of a model that assumes linearity. In the production model of Figure 1-4, for instance, the constraint

```
subject to Time:
    sum {p in PROD} (1/rate[p]) * Make[p] <= avail;
```

embodies an assumption that the number of hours used in making product p grows linearly with the level of production. In the transportation model of Figure 3-1a, on the other hand, the constraint

```
subject to Supply {i in ORIG}:
    sum {j in DEST} Trans[i,j] = supply[i];
```

embodies only a weak linearity assumption, to the effect that the total shipped out of origin i is the sum of the shipments to the different destinations.

In addition to these examples based on models of economic phenomena, there are many sources of nonlinearity in models of physical activities such as oil refining, power transmission, and structural design. More often than not, the nonlinearities in these models cannot be traced to the relaxation of any linearity assumptions, but are a consequence of inherently nonlinear relationships that govern forces, volumes, currents and so forth. The forms of the nonlinear functions in physical models may be easier to determine, because they are given by the laws of physics rather than of economics. On the other hand, the nonlinearities in physical models tend to involve more complicated functional forms and interactions among the variables.

If you know the algebraic form of a nonlinear expression that you want to include in your model, you can probably see a way to write it in AMPL. The next two sections of this chapter consider some of the specific issues and features relevant to declaring variables for nonlinear programs and to writing nonlinear expressions. Lest you get carried away by the ease of writing nonlinear expressions, however, the last section offers some cautionary advice on solving nonlinear programs.

13.2 Nonlinear variables

Although AMPL variables are declared for nonlinear programs in the same way as for linear programs, two features of variables — initial values and automatic substitution — are particularly useful in working with nonlinear models.

Initial values of variables

You may specify values for AMPL variables, much as you do for parameters. Prior to optimization, these ''initial'' values can be displayed and manipulated by AMPL commands. When you type `solve`, they are passed to the solver, which may use them as a starting guess at the optimum. After the solver has finished its work, the initial values are replaced by the computed optimal ones.

All of the AMPL features for assigning values to parameters are also available for variables. A `var` declaration may specify initial values in an optional `:=` phrase; for the transportation example, you can write

```
var Trans {ORIG,DEST} >= 0, := 1;
```

to set every `Trans[i,j]` initially to 1, or

```
var Trans {i in ORIG, j in DEST} >= 0, := limit[i,j] - 1;
```

to initialize each `Trans[i,j]` to 1 less than `limit[i,j]`. Alternatively, initial values may be given in a data statement along with the other data for a model:

```
var Trans:  FRA  DET  LAN  WIN  STL  FRE  LAF :=
     GARY   800  400  400  200  400  200  200
     CLEV   800  800  800  600  600  500  600
     PITT   800  800  800  200  300  800  500  ;
```

Any of the data statements for parameters can also be used for variables, as explained in Section 9.3.

You can often speed up the work of the solver by suggesting good initial values. This can be so even for linear programs, but the effect is much stronger in the nonlinear case. The choice of an initial guess may determine what value of the objective is found to be "optimal" by the solver, or even whether the solver finds any optimal solution at all. These possibilities are discussed further in the last section of this chapter.

If you don't give any initial value for a variable, then AMPL will tentatively set it to zero. If the solver incorporates a routine for determining initial values, then it may re-set the values of any uninitialized variables, while making use of the values of variables that have been initialized. Otherwise, uninitialized variables will be left at zero. Although zero is an obvious starting point, it has no special significance; for some of the examples that we will give in Section 13.4, the solver cannot optimize successfully unless the initial values are reset away from zero.

Automatic substitution of variables

The issue of substituting variables has already arisen in an example of the previous section, where we declared variables to represent the shipping costs, and then defined them in terms of other variables by use of a constraint:

```
subject to cost_relation {i in ORIG, j in DEST}:
   Cost[i,j] = rate[i,j] / (1 - Trans[i,j]/limit[i,j]);
```

If the expression to the right of the = sign is substituted for every appearance of Cost[i,j], the Cost variables can be eliminated from the model, and these constraints need not be passed to the solver. There are two ways in which you can tell AMPL to make such substitutions automatically.

First, by typing the command

```
ampl: option substout 1;
```

you can tell AMPL to look for all "defining" constraints that have the form shown above: a single variable to the left of an = sign. AMPL tries to use as many of these constraints as possible to substitute variables out of the model.

When this alternative is employed, AMPL looks for substitutions as follows. After you have typed solve and a nonlinear program has been generated from a model and data, the constraints are scanned in the order that they appeared in the model. A constraint is identified as ''defining'' provided that

- It has just one variable to the left of an = sign;
- The left-hand variable's declaration did not specify any restrictions, such as integrality or bounds; and
- The left-hand variable has not already appeared in a constraint identified as defining.

The expression to the right of the = sign is then substituted for every appearance of the left-hand variable in the other constraints, and the defining constraint is dropped. These rules give AMPL an easy way to avoid circular substitutions, but they do imply that the nature and number of substitutions may depend on the ordering of the constraints.

As an alternative, if you want to specify explicitly that a certain collection of variables is to be substituted out, use an = phrase in the declarations of the variables. For the preceding example, you could write:

```
var Cost {i in ORIG, j in DEST}
    = rate[i,j] / (1 - Trans[i,j]/limit[i,j]);
```

Then the variables Cost [i, j] would be replaced everywhere by the expression following the = sign. Declarations of this kind can appear in any order, subject to the usual requirement that every variable appearing in an = phrase must be previously defined.

Whenever variables can be substituted out, they are not mathematically necessary to the optimization problem. Nevertheless, they often serve an essential descriptive purpose; by associating names with nonlinear expressions, they permit more complicated expressions to be written understandably. Moreover, even though these variables have been removed from the problem sent to the solver, their names remain available for use in browsing through the results of optimization.

When the same nonlinear expression appears more than once in the objective and constraints, substituting a variable for it may make the model more concise as well as more readable. AMPL also processes such a substitution efficiently; when it generates a representation of the nonlinear program for the solver, it doesn't literally substitute the whole defining expression for each occurrence of the variable, but rather saves one copy of the expression and an indication of where its value is to be substituted. Thus the expression's evaluation is never repeated unnecessarily.

From the solver's standpoint, substitutions reduce the number of constraints and variables, but tend to make the constraint and objective expressions more complex. As a result, there are circumstances in which a solver will perform better if defined variables are not substituted out. When developing a new model, you may have to experiment to determine which substitutions give the best results.

13.3 Nonlinear expressions

Any of AMPL's arithmetic operators (Table 7-1) and arithmetic functions (Table 7-2) may be applied to variables as well as parameters. If any resulting objective or constraint

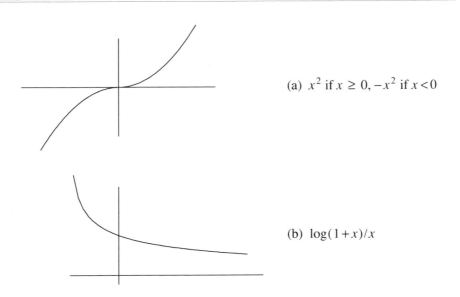

(a) x^2 if $x \geq 0$, $-x^2$ if $x < 0$

(b) $\log(1+x)/x$

Figure 13-3: Smooth nonlinear functions.

does not satisfy the rules for linearity (Chapter 8) or piecewise-linearity (Chapter 14), AMPL treats it as "not linear". When you type `solve`, AMPL tries to pass along instructions that are sufficient for your solver to evaluate every expression in every objective and constraint, together with derivatives if appropriate.

If you are using a typical nonlinear solver, it is up to you to define your objective and constraints in terms of the "smooth" functions that the solver requires. The generality of AMPL's expression syntax can be misleading in this regard. For example, if you are trying to use variables `Flow[i,j]` representing flow between points `i` and `j`, it is tempting to write expressions like

        cost[i,j] * abs(Flow[i,j])

or

        if Flow[i,j] = 0 then 0 else base[i,j] + cost[i,j]*Flow[i,j]

These are certainly not linear, but the first is not smooth (its slope changes abruptly at zero) and the second is not even continuous (its value jumps suddenly at zero). If you try to use such expressions, AMPL will not complain, and your solver may even return what it claims to be an optimal solution — but the results are likely to be wrong.

Expressions that apply nonsmooth functions (such as `abs`, `min`, and `max`) to variables generally produce nonsmooth results; the same is true of `if-then-else` expressions in which a condition involving variables follows `if`. Nevertheless, there are useful exceptions where a carefully written expression can preserve smoothness. For example, the function graphed in Figure 13-3a can be written in AMPL as

```
if X[j] >= 0 then X[j]^2 else - X[j]^2
```

but is smooth because the functions following `then` and `else` are smooth and have the same slope where they meet at $X[j] = 0$. As another example, the function in Figure 13-3b is most easily written $\log(1+X[j]) / X[j]$, but unfortunately if $X[j]$ is zero this simplifies to 0/0, which would be reported as an error. In fact, this expression does not evaluate accurately if $X[j]$ is even close to zero. If instead we write

```
if abs(X[j]) > 0.00001 then log(1+X[j])/X[j] else 1 - X[j]/2
```

a highly accurate linear approximation is substituted at small magnitudes of $X[j]$. This alternative is not smooth in a literal mathematical sense, but it is numerically close enough to being smooth to suffice for use with some solvers.

Although AMPL expressions are general enough to describe most nonlinear functions of interest, they are limited to functions with a known mathematical form. In some situations it is desirable to incorporate nonlinear functions defined by an existing procedure (a "black box"), computed by an algorithm, or determined by a simulation. For these purposes, AMPL provides a `function` declaration to define a new function that will be subsequently recognized in the model.

A `function` declaration specifies a new function's name and required arguments, and describes how some program outside of AMPL is invoked to compute the function's value. Declared functions can have any number of arguments, and even iterated collections of arguments. The connection between a declared function and another program must be set up in a way that depends in part on the environment in which AMPL is being run. As a result, the rules for a function declaration are somewhat complex and technical. If you need to use this feature, you will find an outline of the rules in Appendix A.14 and full details in the system-specific documentation that accompanies your AMPL software.

13.4 Pitfalls of nonlinear programming

While AMPL gives you the power to formulate diverse nonlinear optimization models, no solver can guarantee an acceptable solution every time you type `solve`. The algorithms used by solvers are susceptible to a variety of difficulties inherent in the complexities of nonlinear functions. This is in unhappy contrast to the linear case, where a well-designed solver can be relied upon to solve almost any linear program.

This section offers a brief introduction to pitfalls of nonlinear programming, illustrated with the nonlinear transportation example developed earlier in this chapter. We focus on two common kinds of difficulties, function range violations and multiple local optima; at the end we mention some other traps.

```
set ORIG;    # origins
set DEST;    # destinations

param supply {ORIG} >= 0;    # amounts available at origins
param demand {DEST} >= 0;    # amounts required at destinations

    check: sum {i in ORIG} supply[i] = sum {j in DEST} demand[j];

param rate {ORIG,DEST} >= 0;    # base shipment costs per unit
param limit {ORIG,DEST} > 0;    # limit on units shipped

var Trans {i in ORIG, j in DEST} >= 0, := 0; # units to ship

minimize total_cost:
   sum {i in ORIG, j in DEST}
      rate[i,j] * Trans[i,j] / (1 - Trans[i,j]/limit[i,j]);

subject to Supply {i in ORIG}:
   sum {j in DEST} Trans[i,j] = supply[i];

subject to Demand {j in DEST}:
   sum {i in ORIG} Trans[i,j] = demand[j];
```

Figure 13-4: Nonlinear transportation model (nltrans.mod).

```
param: ORIG:  supply :=
       GARY   1400    CLEV   2600     PITT    2900 ;

param: DEST:  demand :=
       FRA      900
       DET     1200
       LAN      600
       WIN      400
       STL     1700
       FRE     1100
       LAF     1000 ;

param rate :  FRA  DET  LAN  WIN  STL  FRE  LAF :=
       GARY    39   14   11   14   16   82    8
       CLEV    27    9   12    9   26   95   17
       PITT    24   14   17   13   28   99   20 ;

param limit : FRA  DET  LAN  WIN  STL  FRE  LAF :=
       GARY   500 1000 1000 1000  800  500 1000
       CLEV   500  800  800  800  500  500 1000
       PITT   800  600  600  600  500  500  900 ;
```

Figure 13-5: Data for nonlinear transportation model (nltrans.dat).

Function range violations

The complete transportation model is shown in Figure 13-4; associated data values, similar to those used for the linear transportation example in Chapter 3, are given in Figure 13-5. AMPL reads this model and its data as for a linear program, and invokes a solver in the same way when we type solve:

```
ampl: model nltrans.mod;
ampl: data nltrans.dat;

ampl: solve;
MINOS 5.4 Error evaluating objective total_cost
can't compute 8000/0
MINOS 5.4: solution aborted.
8 iterations, objective 0
```

The solver's message is cryptic, but strongly suggests a division by zero while evaluating the objective. That could only happen if the expression

```
1 - Trans[i,j]/limit[i,j]
```

is zero at some point. If we use `display` to print the pairs where `trans[i,j]` equals `limit[i,j]`:

```
ampl: display {i in ORIG, j in DEST: Trans[i,j] = limit[i,j]};
set {i in ORIG, j in DEST: Trans[i,j] == limit[i,j]}
        := (GARY,LAF) (PITT,LAN);

ampl: display Trans['GARY','LAF'], limit['GARY','LAF'];
Trans['GARY','LAF'] = 1000
limit['GARY','LAF'] = 1000
```

we can see the problem. The solver has allowed `Trans[GARY,LAF]` to have the value 1000, which equals `limit[GARY,LAF]`. As a result, the objective function term

```
rate[GARY,LAF] * Trans[GARY,LAF]
   / (1 - Trans[GARY,LAF]/limit[GARY,LAF])
```

evaluates to 8000/0. Since the solver is unable to evaluate the objective function, it gives up without finding an optimal solution.

Because the behavior of a nonlinear optimization algorithm can be sensitive to the choice of starting guess, we might hope that the solver will have greater success from a different start. AMPL's `let` command may be used, for example, to suggest a new initial value for each `Trans[i,j]` that is half of `limit[i,j]`:

```
ampl: let {i in ORIG, j in DEST} Trans[i,j] := limit[i,j]/2

ampl: solve;
MINOS 5.4: the current point cannot be improved.
32 iterations, objective -7.385903389e+18
```

This time the solver runs to completion, but there is still something wrong. The objective is less than -10^{18}, or $-\infty$ for all practical purposes, and the solution is described as "cannot be improved" rather than optimal.

Examining the values of `Trans[i,j]/limit[i,j]` in the solution that the solver has returned gives a clue to the difficulty:

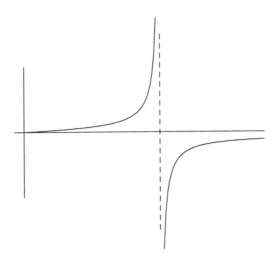

Figure 13-6: Singularity in cost function $y = x/(1 - x/c)$.

```
ampl: display {i in ORIG, j in DEST} Trans[i,j]/limit[i,j];
Trans[i,j]/limit[i,j] [*,*] (tr)
:          CLEV          GARY         PITT      :=
DET     -6.125e-14    -4.9e-14        2
FRA      0             1.5            0.1875
FRE      0.7           1              0.5
LAF      0.4           0.15           0.5
LAN      0.375         7.03288e-15    0.5
STL      2.9           0              0.5
WIN      0.125         0              0.5
;
```

These ratios show that the shipments for several pairs, such as Trans[CLEV,STL],
significantly exceed their limits. More seriously, Trans[GARY,FRE] seems to be right
at limit[GARY,FRE], since their ratio is given as 1; if we display them to full precision, however, we see:

```
ampl: option display_precision 0;
ampl: display Trans['GARY','FRE'],limit['GARY','FRE'];
Trans['GARY','FRE'] = 500.0000000000028
limit['GARY','FRE'] = 500
```

The variable is just slightly larger than the limit, so the cost term has a huge negative
value. If we graph the entire cost function, as in Figure 13-6, we see that indeed the cost
function goes off to $-\infty$ to the right of the singularity at limit[GARY,FRE].

The source of error in both of the runs above is our assumption that, since the objective goes to $+\infty$ as Trans[i,j] approaches limit[i,j] from below, the solver will
keep Trans[i,j] between 0 and limit[i,j]. At least for this solver, we must

enforce such an assumption by giving each `Trans[i,j]` an explicit upper bound that is slightly less than `limit[i,j]`, but close enough not to affect the eventual optimal solution:

```
var Trans {i in ORIG, j in DEST} >= 0, <= .9999 * limit[i,j];
```

With this modification, the solver readily finds an optimum:

```
ampl: model nltransb.mod;
ampl: data nltrans.dat;
ampl: solve;
MINOS 5.4: optimal solution found.
81 iterations, objective 1212117

ampl: display Trans;
Trans [*,*] (tr)
:         CLEV       GARY       PITT        :=
DET     586.372    187.385    426.242
FRA     294.993     81.2205   523.787
FRE     365.5      369.722    364.778
LAF     490.537      0        509.463
LAN     294.148      0        305.852
STL     469.691    761.672    468.637
WIN      98.7595     0        301.241
;
```

These values of the variables are well away from any singularity, with `Trans[i,j]/limit[i,j]` being less than 0.96 in every case. (If you change the starting guess to be `limit[i,j]/2` as before, you should find that the solution is the same but the solver needs only about half as many iterations to find it.)

The immediate lesson here is that nonlinear functions can behave quite badly outside the intended range of the variables. The incomplete graph in Figure 13-2e made this cost function look misleadingly well-behaved, whereas Figure 13-6 shows the need for a bound to keep the variable away from the singularity. A more general lesson is that difficulties posed by a nonlinear function may lead the solver to fail in a variety of ways. When developing a nonlinear model, you need to be alert to bizarre results from the solver, and you may have to do some detective work to trace the problem back to a flaw in the model.

Multiple local optima

To illustrate a different kind of difficulty in nonlinear optimization, we consider a slightly modified objective function that has the following formula:

```
minimize total_cost:
   sum {i in ORIG, j in DEST}
      rate[i,j] * Trans[i,j]^0.8 / (1 - Trans[i,j]/limit[i,j]);
```

By raising the amount shipped to the power 0.8, we cause the cost function to be concave at lower shipment amounts and convex at higher amounts, in the manner of Figure 13-2f.

Upon attempting to solve this new model, we again initially run into technical difficulties:

```
ampl: model nltransc.mod;
ampl: data nltrans.dat;
ampl: solve;
MINOS 5.4: Error evaluating objective total_cost:
                        can't evaluate pow'(0,0.8)
MINOS 5.4: solution aborted.
0 iterations, objective 0
```

This time our suspicion naturally centers upon `Trans[i,j]^0.8`, the only expression that we have changed in the model. A further clue is provided by the error message's reference to `pow'(0,0.8)`, which denotes the derivative of the exponential (power) function at zero. When `Trans[i,j]` is zero, this function has a well-defined value, but its derivative with respect to the variable is infinite. As a result, the partial derivative of the total cost with respect to any variable at zero cannot be returned to the solver; since the solver has requested all the partial derivatives in order to apply its optimization algorithm, it gives up.

This is another variation on the range violation problem, and again it can be remedied by imposing some bounds to keep the solution away from troublesome points. In this case, we move the lower bound from zero to a very small positive number:

```
var Trans {i in ORIG, j in DEST}
    >= 1e-10, <= .9999 * limit[i,j], := 0;
```

We might also move the starting guess away from zero, but in this example the solver takes care of that automatically, since the initial values only suggest a starting point. With the bounds adjusted, the solver runs normally and reports a solution:

```
ampl: model nltransd.mod;
ampl: data nltrans.dat;
ampl: solve;
MINOS 5.4: optimal solution found.
65 iterations, objective 427568.1225

ampl: display Trans;
Trans [*,*] (tr)
:         CLEV        GARY        PITT        :=
DET       689.091     1e-10       510.909
FRA       1e-10       199.005     700.995
FRE       385.326     326.135     388.54
LAF       885.965     114.035     1e-10
LAN       169.662     1e-10       430.338
STL       469.956     760.826     469.218
WIN       1e-10       1e-10       400
;
```

We can regard each `1e-10` as a zero, since such a small value is negligible in comparison with the rest of the solution.

Next we again try a starting guess at `limit[i,j]/2`, in the hope of speeding things up. This is the result:

```
ampl: let {i in ORIG, j in DEST} Trans[i,j] := limit[i,j]/2
ampl: solve;
MINOS 5.4: optimal solution found.
40 iterations, objective 355438.2006

ampl: display Trans;
Trans [*,*] (tr)
:        CLEV        GARY       PITT       :=
DET    540.601     265.509    393.89
FRA    328.599     1e-10      571.401
FRE    364.639     371.628    363.732
LAF    491.262     1e-10      508.738
LAN    301.741     1e-10      298.259
STL    469.108     762.863    468.029
WIN    104.049     1e-10      295.951
;
```

Not only is the solution completely different, but the optimal value is 17% lower! The first solution could not truly have minimized the objective over all solutions that are feasible in the constraints.

Actually both solutions can be considered correct, in the sense that each is *locally* optimal. That is, each solution is less costly than any other nearby solutions. All of the classical methods of nonlinear optimization, which are the methods commonly implemented in solvers, are designed to seek a local optimum. Started from one specified initial guess, these methods are not guaranteed to find a solution that is *globally* optimal, in the sense of giving the best objective value among all solutions that satisfy the constraints. In general, finding a global optimum is much harder than finding a local one.

Fortunately, there are many cases in which a local optimum is entirely satisfactory. When the objective and constraints satisfy certain properties, any local optimum is also global; the model considered at the beginning of this section is one example, where the convexity of the objective, together with the linearity of the constraints, guarantees that the solver will find a global optimum. (Linear programming is an even more special case with this property; that's why in previous chapters we never encountered local optima that were not global.) Even when there is more than one local optimum, a knowledge of the situation being modeled may help you to identify the global one. Perhaps you can choose an initial solution near to the global optimum, or you can add some constraints that rule out regions known to contain local optima.

Finally, you may be content to find a very good local optimum, even if you don't have a guarantee that it is global. One straightforward approach is to try a series of starting points systematically, and take the best among the solutions. As a simple illustration, suppose that we declare the variables in our example as follows:

```
param alpha >=0, <= 1;

var Trans {i in ORIG, j in DEST}
   >= 1e-10, <= .9999 * limit[i,j], := alpha * limit[i,j];
```

For each choice of `alpha` we get a different starting guess, and potentially a different solution. Here are some resulting objective values for `alpha` ranging from 0 to 1:

alpha	total_cost
0.0	427568.1
0.1	366791.2
0.2	366791.2
0.3	366791.2
0.4	366791.2
0.5	355438.2
0.6	356531.5
0.7	376043.3
0.8	367014.4
0.9	402795.3
1.0	365827.2

The solution that we previously found for an `alpha` of 0.5 is still the best, but in light of these results we are now more inclined to believe that it is a very good solution. We might also observe that, although the reported objective value varies somewhat erratically with the choice of starting point — a feature of nonlinear programs generally — the second-best value of `total_cost` was found by setting `alpha` to 0.6. This suggests that a closer search of `alpha` values around 0.5 might be worthwhile. Some of the more sophisticated methods for global optimization attempt to search through starting points in this way, but with a more elaborate and systematic procedure for deciding which starting points to try next.

Other pitfalls

Many other factors can influence the efficiency and success of a nonlinear solver, including the way that the model is formulated and the choice of units (or scaling) for the variables. As a rule, nonlinearities are more easily handled when they appear in the objective function rather than in the constraints. AMPL's option to substitute variables automatically, described earlier in this chapter, may help in this regard. Another rule of thumb is that the values of the variables should differ by at most a few orders of magnitude; solvers can be misled when some variables are, say, in millions and others are in thousandths. Some solvers automatically scale a problem to try to avoid such a situation, but you can help them considerably by judiciously picking the units in which the variables are expressed.

Nonlinear solvers also have many modes of failure besides the ones we have discussed. Some methods of nonlinear optimization can get stuck at ``stationary'' points that are not optimal in any sense; can identify a maximum when a minimum is desired (or vice-versa); and can falsely give an indication that there is no feasible solution to the constraints. In these cases your only recourse may be to try a different starting guess; it can sometimes help to specify a start that is feasible for many of the nonlinear constraints. You may also improve the solver's chances of success by placing realistic bounds on the variables. If you know, for instance, that an optimal value of 80 is plausible for some variables, but a value of 800 is not, you may want to give them a bound of 400. (Once an indicated optimum is at hand, you should be sure to check whether any of these ''safety'' bounds has been reached; if so, the bounds should be relaxed and the problem re-solved.)

The intent of this section has been to illustrate that extra caution is advisable in working with nonlinear models. If you encounter a difficulty that cannot be resolved by any of the simple devices described here, you may need to consult a textbook in nonlinear programming, the documentation for the particular solver that you are using, or a numerical analyst versed in nonlinear optimization techniques.

Bibliography

Roger Fletcher, *Practical Methods of Optimization.* John Wiley & Sons (New York, NY, 1987). A concise survey of theory and methods.

Philip E. Gill, Walter Murray and Margaret H. Wright, *Practical Optimization.* Academic Press (New York, NY, 1981). Theory, algorithms and practical advice.

Judith Liebman, Leon Lasdon, Linus Schrage and Allan Waren, *Modeling and Optimization with GINO.* The Scientific Press (South San Francisco, CA, 1986). A collection of many small nonlinear programming examples, and some advice on solving them.

Exercises

13-1. In the last example of Section 13.4, try some more starting points to see if you can find an even better locally optimal solution. What is the best solution you can find?

13-2. The following little model determines the dimensions of a rectangular field of maximum area that can be surrounded by a fence of given length:

```
param fence > 0;

var Xfield >= 0;
var Yfield >= 0;

maximize Area:  Xfield * Yfield;
subject to Enclose:  2*Xfield + 2*Yfield <= fence;
```

It's well known that the optimum field is a square.

(a) When we tried to solve this problem for a fence of 40 meters, with the default initial guess of zero for the variables, we got the following result:

```
ampl: solve;
MINOS 5.4: optimal solution found.
0 iterations, objective 0

ampl: display Xfield,Yfield;
Xfield = 0
Yfield = 0
```

What could explain this unexpected outcome? Try the same problem on any nonlinear solver available to you, and report the behavior that you observe.

(b) Using a different starting point if necessary, run your solver to confirm that the optimal dimensions for 40 meters of fence are indeed 10×10.

(c) Experiment with an analogous model for determining the dimensions of a box of maximum volume that can be wrapped by paper of a given area.

(d) Solve the same problem as in (c), but for wrapping a cylinder rather than a box.

13-3. A falling object on a nameless planet has been observed to have approximately the following heights h_j at (mostly) one-second intervals t_j:

t_j	0.0	0.5	1.5	2.5	3.5	4.5	5.5	6.5	7.5	8.5	9.5	10.0
h_j	100	95	87	76	66	56	47	38	26	15	6	0

According to the laws of physics on this planet, the height of the object at any time should be given by the formula

$$h_j = a_0 - a_1 t_j - \tfrac{1}{2} a_2 t_j^2,$$

where a_0 is the initial height, a_1 is the initial velocity, and a_2 is the acceleration due to gravity. But since the observations were not made exactly, there exists no choice of a_0, a_1, and a_2 that will cause all of the data to fit this formula exactly. Instead, we wish to estimate these three values by choosing them so as to minimize the "sum of squares"

$$\sum_{j=1}^{n} [h_j - (a_0 - a_1 t_j - \tfrac{1}{2} a_2 t_j^2)]^2.$$

where t_j and h_j are the observations from the jth entry of the table, and n is the number of observations. This expression measures the error between the ideal formula and the observed behavior.

(a) Create an AMPL model that minimizes the sum of squares for any number n of observations t_j and h_j. This model should have three variables and an objective function, but no constraints.

(b) Use AMPL data statements to represent the sample observations given above, and solve the resulting nonlinear program to determine the estimates of a_0, a_1, and a_2.

13-4. This problem involves a very simple "traffic flow" network:

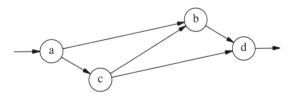

Traffic proceeds in the direction of the arrows, entering at intersection a, exiting at d, and passing through b or c or both. These data values are given for the roads connecting the intersections:

From	To	Time	Capacity	Sensitivity
a	b	5.0	10	0.1
a	c	1.0	30	0.9
c	b	2.0	10	0.9
b	d	1.0	30	0.9
c	d	5.0	10	0.1

To be specific, we imagine that the times are in minutes, the capacities are in cars per minute, and the sensitivities are in minutes per (car per minute).

The following AMPL statements can be used to represent networks of this kind:

```
set inters;    # road intersections

param EN symbolic in inters;   # entrance to network
param EX symbolic in inters;   # exit from network

set roads within {i in inters, j in inters: i <> EX and j <> EN};

param time {roads} > 0;
param cap {roads} > 0;
param sens {roads} > 0;
```

(a) What is the shortest path, in minutes, from the entrance to the exit of this network? Construct a shortest path model, along the lines of Figure 11-7, that verifies your answer.

(b) What is the maximum traffic flow from entrance to exit, in cars per minute, that the network can sustain? Construct a maximum flow model, along the lines of Figure 11-6, that verifies your answer.

(c) Question (a) above was concerned only with the speed of traffic, and question (b) only with the volume of traffic. In reality, these quantities are interrelated. As the traffic volume on a road increases from zero, the time required to travel the road also increases.

Travel time increases only moderately when there are just a few cars, but it increases very rapidly as the traffic approaches capacity. Thus a nonlinear function is needed to model this phenomenon. Let us define $T[i,j]$, the travel time on road (i,j), by the following constraints:

```
var X {roads} >= 0;   # cars per minute entering road (i,j)
var T {roads};        # travel time for road (i,j)

subject to Travel_Time {(i,j) in roads}:
   T[i,j] = time[i,j] + (sens[i,j]*X[i,j]) / (1-X[i,j]/cap[i,j]);
```

You can confirm that the travel time on (i,j) is close to $time[i,j]$ when the road is lightly used, but goes to infinity as the use approaches $cap[i,j]$ cars per minute. The magnitude of $sens[i,j]$ controls the rate at which travel time increases, as more cars try to enter the road.

Suppose that we want to analyze how the network can best handle some number of cars per minute. The objective is to minimize average travel time from entrance to exit:

```
param through > 0;   # cars per minute using the network

minimize avg_time:
   (sum {(i,j) in roads} T[i,j] * X[i,j]) / through;
```

The nonlinear expression $T[i,j] * X[i,j]$ is travel minutes on road (i,j) times cars per minute entering the road — hence, the number of cars on road (i,j). The summation in the objective thus gives the total cars in the entire system. Dividing by the number of cars per minute getting through, we have the average number of minutes for each car.

Complete the model by adding the following:

– A constraint that total cars per minute in equals total cars per minute out at each intersection, except the entrance and exit.

– A constraint that total cars per minute leaving the entrance equals the total per minute (represented by through) that are using the network.

– Appropriate bounds on the variables. (The example in Section 13.4 should suggest what bounds are needed.)

Use AMPL to show that, for the example given above, a throughput of 4.0 cars per minute is optimally managed by sending half the cars along $a{\rightarrow}b{\rightarrow}d$ and half along $a{\rightarrow}c{\rightarrow}d$, giving an average travel time of about 8.18 minutes.

(d) By trying values of parameter `through` both greater than and less than 4.0, develop a graph of minimum average travel time as a function of throughput. Also, keep a record of which travel routes are used in the optimal solutions for different throughputs, and summarize this information on your graph.

What is the relationship between the information in your graph and the solutions from parts (a) and (b)?

(e) The model in (c) assumes that you can make the cars' drivers take certain routes. For example, in the optimal solution for a throughput of 4.0, no drivers are allowed to "cut through" from c to b.

What would happen if instead all drivers could take whatever route they pleased? Observation has shown that, in such a case, the traffic tends to reach a *stable* solution in which no route has a travel time less than the average. The optimal solution for a throughput of 4.0 is not stable, since — as you can verify — the travel time on $a{\rightarrow}c{\rightarrow}b{\rightarrow}d$ is only about 7.86 minutes; some drivers would try to cut through if they were permitted.

To find a stable solution using AMPL, we have to add some data specifying the possible routes from entrance to exit:

```
param choices integer > 0;   # number of routes
set route {1..choices} within roads;
```

Here `route` is an indexed collection of sets; for each `r` in `1..choices`, the expression `route[r]` denotes a different subset of roads that together form a route from EN to EX. For our network, `choices` should be 3, and the `route` sets should be `{(a,b),(b,d)}`, `{(a,c),(c,d)}` and `{(a,c),(c,b),(b,d)}`. Using these data values, the stability conditions may be ensured by one additional collection of constraints, which say that the time to travel any route is no less than the average travel time for all routes:

```
subject to stability {r in 1..choices}:
    sum {(i,j) in route[r]} T[i,j] >=
        (sum {(i,j) in roads} T[i,j] * X[i,j]) / through;
```

Show that, in the stable solution for a throughput of 4.0, a little more than 5% of the drivers cut through, and the average travel time increases to about 8.27 minutes. Thus traffic would have been faster if the road from c to b had never been built! (This phenomenon is known as Braess's paradox, in honor of a traffic analyst who noticed that when a certain link was added to Munich's road system, traffic seemed to get worse.)

(f) By trying throughput values both greater than and less than 4.0, develop a graph of the stable travel time as a function of throughput. Indicate, on the graph, for which throughputs the stable time is greater than the optimal time.

(g) Suppose now that you have been hired to analyze the possibility of opening an additional winding road, directly from a to d, with travel time 5 minutes, capacity 10, and sensitivity 1.5. Working with the models developed above, write an analysis of the consequences of making this change, as a function of the throughput value.

13-5. Return to the model constructed in (e) of the previous exercise. This exercise asks about reducing the number of variables by substituting some out, as explained in Section 13.2.

(a) Set the option `show_stats` to 1, and solve the problem. From the extra output you get, verify that there are 10 variables.

Next repeat the session with option `substout` set to 1. Verify from the resulting messages that some of the variables are eliminated by substitution. Which of the variables must these be?

(b) Rather than setting `substout`, you can specify that a variable is to be substituted out by placing an appropriate = phrase in its `var` declaration. Modify your model from (a) to use this feature, and verify that the results are the same.

(c) There is a long expression for average travel time that appears twice in this model. Define a new variable `Avg` to stand for this expression. Verify that AMPL also substitutes this variable out when you solve the resulting model, and that the result is the same as before.

13-6. Liebman, Lasdon, Schrage and Waren describe the following problem in designing a steel tank to hold ammonia. The decision variables are

T the temperature inside the tank
I the thickness of the insulation

The pressure inside the tank is a function of the temperature,

$$P = e^{-3950/(T+460)+11.86}$$

It is desired to minimize the cost of the tank, which has three components: insulation cost, which depends on the thickness; the cost of the steel vessel, which depends on the pressure; and the cost of a recondensation process for cooling the ammonia, which depends on both temperature and insulation thickness. A combination of engineering and economic considerations has yielded the following formulas:

$$C_I = 400 I^{0.9}$$
$$C_V = 1000 + 22(P - 14.7)^{1.2}$$
$$C_R = 144(80 - T)/I$$

(a) Formulate this problem in AMPL as a two-variable problem, and alternatively as a six-variable problem in which four of the variables can be substituted out. Which formulation would you prefer to work with?

(b) Using your preferred formulation, determine the parameters of the least-cost tank.

(c) Increasing the factor 144 in C_R causes a proportional increase in the recondensation cost. Try several larger values, and describe in general terms how the total cost increases as a function of the increase in this factor.

13-7. A social accounting matrix is a table that shows the flows from each sector in an economy to each other sector. Here is simple five-sector example, with blank entries indicating flows known to be zero:

	LAB	H1	H2	P1	P2	total
LAB		15	3	130	80	220
H1	?					?
H2	?					?
P1		15	130		20	190
P2		25	40	55		105

If the matrix were estimated perfectly, it would be *balanced:* each row sum (the flow out of a sector) would equal the corresponding column sum (the flow in). As a practical matter, however, there are several difficulties:

– Some entries, marked ? above, have no reliable estimates.

– In the estimated table, the row sums do not necessarily equal the column sums.

– We have separate estimates of the total flows into (or out of) each sector, shown to the right of the rows in our table. These do not necessarily equal the sums of the estimated rows or columns.

Nonlinear programming can be used to adjust this matrix by finding the balanced matrix that is closest, in some sense, to the one given.

For a set S of sectors, let $E_T \subseteq S$ be the subset of sectors for which we have estimated total flows, and let $E_A \subseteq S \times S$ contain all sector pairs (i, j) for which there are known estimates. The given data values are:

$$t_i \quad \text{estimated row/column sums, } i \in E_T$$
$$a_{ij} \quad \text{estimated social accounting matrix entries, } (i, j) \in E_A$$

Let $S_A \subseteq S \times S$ contain all row-column pairs (i, j) for which there should be entries in the matrix — this includes entries that contain ? instead of a number. We want to determine adjusted entries A_{ij}, for each $(i, j) \in S_A$, that are truly balanced:

$$\sum_{j \in S:(i,j) \in S_A} A_{ij} = \sum_{j \in S:(j,i) \in S_A} A_{ji}$$

for all $i \in S$. You can think of these equations as the constraints on the variables A_{ij}.

There is no best function for measuring "close", but one popular choice is the sum of squared differences between the estimates and the adjusted values — for both the matrix and the row and column sums — scaled by the estimated values. For convenience, we write the adjusted sums as defined variables,

$$T_i = \sum_{j \in S:(i,j) \in S_A} A_{ij}$$

Then the objective is to minimize

$$\sum_{(i,j) \in E_A} (a_{ij} - A_{ij})^2 / a_{ij} + \sum_{i \in E_T} (t_i - T_i)^2 / t_i$$

Formulate an AMPL model for this problem, and determine an optimal adjusted matrix.

13-8. A network of pipes has the following layout:

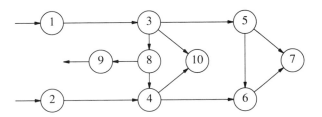

The circles represent joints, and the arrows are pipes. Joints 1 and 2 are sources of flow, and joint 9 is a sink or destination for flow, but flow through a pipe can be in either direction. Associated with each joint i is an amount w_i to be withdrawn from the flow at that joint, and an elevation e_i:

	1	2	3	4	5	6	7	8	9	10
w_i	0	0	200	0	0	200	150	0	0	150
e_i	50	40	20	20	0	0	0	20	20	20

Our decision variables are the flows F_{ij} through the pipes from i to j, with a positive value representing flow in the direction of the arrow, and a negative value representing flow in the opposite direction. Naturally, flow in must equal flow out plus the amount withdrawn at every joint, except for the sources and the sink.

The "head loss" of a pipe is a measure of the energy required to move a flow through it. In our situation, the head loss for the pipe from i to j is proportional to the square of the flow rate,

$$H_{ij} = Kc_{ij}F_{ij}^2 \text{ if } F_{ij} > 0,$$

$$H_{ij} = -Kc_{ij}F_{ij}^2 \text{ if } F_{ij} < 0,$$

where $K = 4.96407 \times 10^{-6}$ is a conversion constant, and c_{ij} is a factor computed from the diameter, friction, and length of the pipe:

from	to	c_{ij}
1	3	6.36685
2	4	28.8937
3	10	28.8937
3	5	6.36685
3	8	43.3406
4	10	28.8937
4	6	28.8937
5	6	57.7874
5	7	43.3406
6	7	28.8937
8	4	28.8937
8	9	705.251

For two joints i and j at the same elevation, the pressure drop for flow from i to j is equal to the head loss. Both pressure and head loss are measured in feet, so that after correcting for differences in elevation between the joints we have the relation:

$$H_{ij} = (P_i + e_i) - (P_j + e_j)$$

Finally, we wish to maintain the pressure at both the sources and the sink at 200 feet.

(a) Formulate a general AMPL model for this situation, and put together data statements for the data given above.

(b) There is no objective function here, but you can still employ a nonlinear solver to seek a feasible solution. By setting the option show_stats to 1, confirm that the number of variables equals the number of equations, so that there are no "degrees of freedom" in the solution. (This does not guarantee that there is just one solution, however.)

Check that your solver finds a solution to the equations, and display the results.

14

Piecewise-Linear Programs

Several kinds of linear programming problems use functions that are not really linear, but are pieced together from connected linear segments:

These "piecewise-linear" terms are easy to imagine, but can be hard to describe in conventional algebraic notation. Hence AMPL provides a special, concise way of writing them.

This chapter introduces AMPL's piecewise-linear notation through examples of piecewise-linear objective functions. In Section 14.1, terms of potentially many pieces are used to describe costs more accurately than a single linear relationship. Section 14.2 shows how terms of two or three pieces can be valuable for such purposes as penalizing deviations from constraints, dealing with infeasibilities, and modeling "reversible" activities. Finally, Section 14.3 describes piecewise-linear functions that can be written with other AMPL operators and functions; some are most effectively handled by converting them to the piecewise-linear notation, while others can be accommodated only through more extensive transformations.

Although the piecewise-linear examples in this chapter are all easy to solve, seemingly similar examples can be much more difficult. The last section of this chapter thus offers guidelines for forming and using piecewise-linear terms. We show that the distinction between easy and hard depends mainly on the convexity or concavity of the piecewise-linear terms.

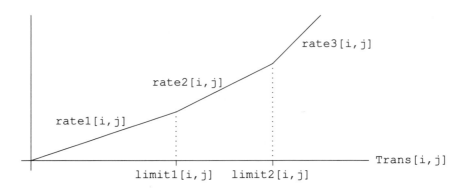

Figure 14-1: Piecewise-linear function, with three slopes.

14.1 Cost terms

Piecewise-linearities are often employed to give a more realistic description of costs than can be achieved by linear terms alone. In this kind of application, piecewise-linear terms serve much the same purpose as nonlinear ones, but without some of the difficulties described in Chapter 13.

To make the comparison explicit, we use the same transportation example as in Chapter 13. We introduce AMPL's notation for piecewise-linear terms with a simple example in which there are a fixed number of cost levels (and linear pieces) for each shipping link. Then we show how an extension of the notation can use indexing expressions to specify a varying number of pieces that is controlled by the data.

Fixed numbers of pieces

In a linear transportation model like Figure 3-1a, any number of units can be shipped from a given origin to a given destination at the same cost per unit. More realistically, however, the most favorable rate may be available for only a limited number of units; shipments beyond this limit pay higher rates. As an example, imagine that three cost rate levels are specified for each origin-destination pair. Then the total cost of shipments along a link increases with the amount shipped in a piecewise-linear fashion, with three pieces as shown in Figure 14-1.

To model the three-piece costs, we replace the parameter cost of Figure 3-1a by three rates and two limits:

```
param rate1 {i in ORIG, j in DEST} >= 0;
param rate2 {i in ORIG, j in DEST} >= rate1[i,j];
param rate3 {i in ORIG, j in DEST} >= rate2[i,j];

param limit1 {i in ORIG, j in DEST} > 0;
param limit2 {i in ORIG, j in DEST} > limit1[i,j];
```

Shipments from `i` to `j` are charged at `rate1[i,j]` per unit up to `limit1[i,j]` units, then at `rate2[i,j]` per unit up to `limit2[i,j]`, and then at `rate3[i,j]`. Normally `rate2[i,j]` would be greater than `rate1[i,j]` and `rate3[i,j]` would be greater than `rate2[i,j]`, but they may be equal if the link from `i` to `j` does not have three distinct rates.

In the linear transportation model, the variables and objective use the parameter `cost` as follows:

```
var Trans {ORIG,DEST} >= 0;

minimize total_cost:
    sum {i in ORIG, j in DEST} cost[i,j] * Trans[i,j];
```

We could express the piecewise-linear transportation model analogously, by introducing three collections of variables, one to represent the amount shipped at each rate:

```
var Trans1 {i in ORIG, j in DEST} >= 0,
                                  <= limit1[i,j];
var Trans2 {i in ORIG, j in DEST} >= 0,
                                  <= limit2[i,j] - limit1[i,j];
var Trans3 {i in ORIG, j in DEST} >= 0;

minimize total_cost:
    sum {i in ORIG, j in DEST} (rate1[i,j] * Trans1[i,j]
        + rate2[i,j] * Trans2[i,j] + rate3[i,j] * Trans3[i,j]);
```

But then the new variables would have to be introduced into all the constraints, and we would also have to deal with these variables whenever we wanted to display the optimal results. Rather than go to all this trouble, we would much prefer to describe the piecewise-linear cost function explicitly in terms of the original variables. Since there is no standard way to describe piecewise-linear functions in algebraic notation, AMPL supplies its own syntax for this purpose.

The piecewise-linear function depicted in Figure 14-1 is written in AMPL as follows:

```
<<limit1[i,j], limit2[i,j];
    rate1[i,j], rate2[i,j], rate3[i,j]>> Trans[i,j]
```

The expression between `<<` and `>>` describes the piecewise-linear function, and is followed by the name of the variable to which it applies. (You can think of it as "multiplying" `Trans[i,j]`, but by a series of coefficients rather than just one.) There are two parts to the expression, a list of breakpoints where the slope of the function changes, and a list of the slopes — which in this case are the cost rates. The lists are separated by a semicolon, and members of each list are separated by commas. Since the first slope applies to values before the first breakpoint, and the last slope to values after the last breakpoint, the number of slopes is always one more than the number of breakpoints.

Although the lists of breakpoints and slopes are sufficient to describe the piecewise-linear cost function for optimization, they do not quite specify the function uniquely. If we added, say, 10 to the cost at every point, we would have a different cost function even though all the breakpoints and slopes would be the same. To resolve this ambiguity,

```
set ORIG;   # origins
set DEST;   # destinations

param supply {ORIG} >= 0;    # amounts available at origins
param demand {DEST} >= 0;    # amounts required at destinations

   check: sum {i in ORIG} supply[i] = sum {j in DEST} demand[j];

param rate1 {i in ORIG, j in DEST} >= 0;
param rate2 {i in ORIG, j in DEST} >= rate1[i,j];
param rate3 {i in ORIG, j in DEST} >= rate2[i,j];

param limit1 {i in ORIG, j in DEST} > 0;
param limit2 {i in ORIG, j in DEST} > limit1[i,j];

var Trans {ORIG,DEST} >= 0;      # units to be shipped

minimize total_cost:
   sum {i in ORIG, j in DEST}
      <<limit1[i,j], limit2[i,j];
         rate1[i,j], rate2[i,j], rate3[i,j]>> Trans[i,j];

subject to Supply {i in ORIG}:
   sum {j in DEST} Trans[i,j] = supply[i];

subject to Demand {j in DEST}:
   sum {i in ORIG} Trans[i,j] = demand[j];
```

Figure 14-2: Piecewise-linear model with three slopes (transpll.mod).

AMPL assumes that a piecewise-linear function evaluates to zero at zero, as in Figure 14-1. Options for other possibilities are discussed later in this chapter.

Summing the cost over all links, the piecewise-linear objective function is now written

```
minimize total_cost:
   sum {i in ORIG, j in DEST}
      <<limit1[i,j], limit2[i,j];
         rate1[i,j], rate2[i,j], rate3[i,j]>> Trans[i,j];
```

The declarations of the variables and constraints stay the same as before; the complete model is shown in Figure 14-2.

Varying numbers of pieces

The approach taken in the preceding example is most useful when there are only a few linear pieces for each term. If there were, for example, 12 pieces instead of three, a model defining rate1[i,j] through rate12[i,j] and limit1[i,j] through limit11[i,j] would be unwieldy. Realistically, moreover, there would more likely be up to 12 pieces, rather than exactly 12, for each term; a term with fewer than 12 pieces could be handled by making some rates equal, but for large numbers of pieces this would be a cumbersome device that would require many unnecessary data values and would obscure the actual number of pieces in each term.

A much better approach is to let the number of pieces (that is, the number of shipping rates) itself be a parameter of the model, indexed over the links:

```
param npiece {ORIG,DEST} integer >= 1;
```

We can then index the rates and limits over all combinations of links and pieces:

```
param rate {i in ORIG, j in DEST, p in 1..npiece[i,j]}
   >= if p = 1 then 0 else rate[i,j,p-1];

param limit {i in ORIG, j in DEST, p in 1..npiece[i,j]-1}
   > if p = 1 then 0 else limit[i,j,p-1];
```

For any particular origin i and destination j, the number of linear pieces in the cost term is given by npiece[i,j]. The slopes are rate[i,j,p] for p ranging from 1 to npiece[i,j], and the intervening breakpoints are limit[i,j,p] for p from 1 to npiece[i,j]-1. As before, there is one more slope than breakpoint.

To use AMPL's piecewise-linear function notation with these data values, we have to give indexed lists of breakpoints and slopes, rather than the explicit lists of the previous example. This is done by placing indexing expressions in front of the slope and breakpoint values:

```
minimize total_cost:
    sum {i in ORIG, j in DEST}
        <<{p in 1..npiece[i,j]-1} limit[i,j,p];
          {p in 1..npiece[i,j]} rate[i,j,p]>> Trans[i,j];
```

Once again, the rest of the model is the same. Figure 14-3a shows the whole model and Figure 14-3b illustrates how the data would be specified. Notice that since npiece["PITT","STL"] is 1, Trans["PITT","STL"] has only one slope and no breakpoints; this implies a one-piece linear term for Trans["PITT","STL"] in the objective function.

14.2 Common two-piece and three-piece terms

Simple piecewise-linear terms have a variety of uses in otherwise linear models. In this section we present three cases: penalizing deviations from constraint levels, concentrating infeasibility in one area of a model, and representing costs for variables that are meaningful at negative as well as positive levels.

Penalty terms for "soft" constraints

Linear programs most easily express "hard" constraints: that production must be at least at a certain level, for example, or that resources used must not exceed those available. Real situations are often not nearly so definite. Production and resource use may

```
set ORIG;   # origins
set DEST;   # destinations

param supply {ORIG} >= 0;   # amounts available at origins
param demand {DEST} >= 0;   # amounts required at destinations

   check: sum {i in ORIG} supply[i] = sum {j in DEST} demand[j];

param npiece {ORIG,DEST} integer >= 1;

param rate {i in ORIG, j in DEST, p in 1..npiece[i,j]}
  >= if p = 1 then 0 else rate[i,j,p-1];

param limit {i in ORIG, j in DEST, p in 1..npiece[i,j]-1}
  > if p = 1 then 0 else limit[i,j,p-1];

var Trans {ORIG,DEST} >= 0;     # units to be shipped

minimize total_cost:
   sum {i in ORIG, j in DEST}
      <<{p in 1..npiece[i,j]-1} limit[i,j,p];
        {p in 1..npiece[i,j]} rate[i,j,p]>> Trans[i,j];

subject to Supply {i in ORIG}:
   sum {j in DEST} Trans[i,j] = supply[i];

subject to Demand {j in DEST}:
   sum {i in ORIG} Trans[i,j] = demand[j];
```

Figure 14-3a: Piecewise-linear model with indexed slopes (transp12.mod).

have certain preferred levels, yet we may be allowed to violate these levels by accepting some extra costs or reduced profits. The resulting "soft" constraints can be modeled by adding piecewise-linear "penalty" terms to the objective function.

For an example, we return to the multi-week production model developed in Chapter 4. As seen in Figure 4-4, the constraints say that, in each of weeks 1 through T, total hours used to make all products may not exceed hours available:

```
subject to time {t in 1..T}:
   sum {p in PROD} (1/rate[p]) * Make[p,t] <= avail[t];
```

Suppose that, in reality, a larger number of hours may be used in each week, but at some penalty per hour to the total profit. Specifically, we replace the parameter avail[t] by two availability levels and an hourly penalty rate:

```
param avail_min {1..T} >= 0;
param avail_max {t in 1..T} >= avail_min[t];

param time_penalty {1..T} > 0;
```

Up to avail_min[t] hours are available without penalty in week t, and up to avail_max[t] hours are available at a loss of time_penalty[t] in profit for each hour above avail_min[t].

To model this situation, we introduce a new variable Use[t] to represent the hours used by production. Clearly Use[t] may not be less than zero, or greater than

```
param: ORIG:   supply :=
        GARY 1400   CLEV 2600   PITT 2900 ;

param: DEST:   demand :=
        FRA   900   DET 1200   LAN  600   WIN  400
        STL 1700   FRE 1100   LAF 1000 ;

param npiece:   FRA DET LAN WIN STL FRE LAF :=
        GARY     3   3   3   2   3   2   3
        CLEV     3   3   3   3   3   3   3
        PITT     2   2   2   2   1   2   1 ;

param rate :=
  [GARY,FRA,*] 1 39  2 50  3 70   [GARY,DET,*] 1 14  2  17  3 33
  [GARY,LAN,*] 1 11  2 12  3 23   [GARY,WIN,*] 1 14  2  17
  [GARY,STL,*] 1 16  2 23  3 40   [GARY,FRE,*] 1 82  2  98
  [GARY,LAF,*] 1  8  2 16  3 24

  [CLEV,FRA,*] 1 27  2 37  3 47   [CLEV,DET,*] 1  9  2  19  3  24
  [CLEV,LAN,*] 1 12  2 32  3 39   [CLEV,WIN,*] 1  9  2  14  3  21
  [CLEV,STL,*] 1 26  2 36  3 47   [CLEV,FRE,*] 1 95  2 105  3 129
  [CLEV,LAF,*] 1  8  2 16  3 24

  [PITT,FRA,*] 1 24  2 34         [PITT,DET,*] 1 14  2  24
  [PITT,LAN,*] 1 17  2 27         [PITT,WIN,*] 1 13  2  23
  [PITT,STL,*] 1 28              [PITT,FRE,*] 1 99  2 140
  [PITT,LAF,*] 1 20 ;

param limit :=
  [GARY,*,*]  FRA 1  500   FRA 2 1000   DET 1  500   DET 2 1000
              LAN 1  500   LAN 2 1000   WIN 1 1000
              STL 1  500   STL 2 1000   FRE 1 1000
              LAF 1  500   LAF 2 1000

  [CLEV,*,*]  FRA 1  500   FRA 2 1000   DET 1  500   DET 2 1000
              LAN 1  500   LAN 2 1000   WIN 1  500   WIN 2 1000
              STL 1  500   STL 2 1000   FRE 1  500   FRE 2 1000
              LAF 1  500   LAF 2 1000

  [PITT,*,*]  FRA 1 1000   DET 1 1000   LAN 1 1000   WIN 1 1000
              FRE 1 1000 ;
```

Figure 14-3b: Data for piecewise-linear model (`transp12.dat`).

`avail_max[t]`. In place of our previous constraint, we say that the total hours used to make all products must equal `Use[t]`:

```
var Use {t in 1..T} >= 0, <= avail_max[t];

subject to time {t in 1..T}:
   sum {p in PROD} (1/rate[p]) * Make[p,t] = Use[t];
```

We can now describe the hourly penalty in terms of this new variable. If `Use[t]` is between 0 and `avail_min[t]`, there is no penalty; if `Use[t]` is between `avail_min[t]` and `avail_max[t]`, the penalty is `time_penalty[t]` per hour

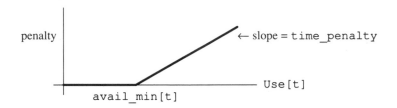

Figure 14-4: Piecewise-linear penalty function for hours used.

that it exceeds `avail_min[t]`. That is, the penalty is a piecewise-linear function of `Use[t]` as shown in Figure 14-4, with slopes of 0 and `time_penalty[t]` surrounding a breakpoint at `avail_min[t]`. Using the syntax previously introduced, we can rewrite the expression for the objective function as:

```
maximize net_profit:
   sum {p in PROD, t in 1..T} (revenue[p,t]*Sell[p,t] -
      prodcost[p]*Make[p,t] - invcost[p]*Inv[p,t])
 - sum {t in 1..T} <<avail_min[t]; 0,time_penalty[t]>> Use[t];
```

The first summation is the same expression for total profit as before, while the second is the sum of the piecewise-linear penalty functions over all weeks. Between << and >> are the breakpoint `avail_min[t]` and a list of the surrounding slopes 0 and `time_penalty[t]`; this is followed by the argument `Use[t]`.

The complete revised model is shown in Figure 14-5a, and our small data set from Chapter 4 is expanded with the new availabilities and penalties in Figure 14-5b. In the optimal solution, we find that the hours used are as follows:

```
ampl: model steelpll.mod; data steelpll.dat; solve;
MINOS 5.4: optimal solution found.
27 iterations, objective 457572.8571

ampl: display avail_min,Use,avail_max;
: avail_min  Use avail_max     :=
1     35       35     42
2     35       42     42
3     30       30     40
4     35       42     42
;
```

In weeks 1 and 3 we use only the unpenalized hours available, while in weeks 2 and 4 we also use the penalized hours. Solutions to piecewise-linear programs usually display this sort of solution, in which many (though not necessarily all) of the variables "stick" at one of the breakpoints.

```
set PROD;        # products
param T > 0;     # number of weeks

param rate {PROD} > 0;               # tons per hour produced
param inv0 {PROD} >= 0;              # initial inventory
param commit {PROD,1..T} >= 0;  # minimum tons sold in week
param market {PROD,1..T} >= 0;  # limit on tons sold in week

param avail_min {1..T} >= 0;         # unpenalized hours available
param avail_max {t in 1..T} >= avail_min[t]; # total hours avail
param time_penalty {1..T} > 0;

param prodcost {PROD} >= 0;          # cost/ton produced
param invcost {PROD} >= 0;           # carrying cost/ton of inventory
param revenue {PROD,1..T} >= 0;  # revenue/ton sold

var Make {PROD,1..T} >= 0;              # tons produced
var Inv {PROD,0..T} >= 0;               # tons inventoried
var Sell {p in PROD, t in 1..T}
   >= commit[p,t], <= market[p,t];   # tons sold

var Use {t in 1..T} >= 0, <= avail_max[t];  # hours used

maximize total_profit:
   sum {p in PROD, t in 1..T} (revenue[p,t]*Sell[p,t] -
      prodcost[p]*Make[p,t] - invcost[p]*Inv[p,t])
 - sum {t in 1..T} <<avail_min[t]; 0,time_penalty[t]>> Use[t];

                 # Objective: total revenue less costs in all weeks
subject to time {t in 1..T}:
   sum {p in PROD} (1/rate[p]) * Make[p,t] = Use[t];

                 # Total of hours used by all products
                 # may not exceed hours available, in each week
subject to initial {p in PROD}:  Inv[p,0] = inv0[p];

                 # Initial inventory must equal given value

subject to balance {p in PROD, t in 1..T}:
   Make[p,t] + Inv[p,t-1] = Sell[p,t] + Inv[p,t];

                 # Tons produced and taken from inventory
                 # must equal tons sold and put into inventory
```

Figure 14-5a: Piecewise-linear objective with penalty function (steelpl1.mod).

Dealing with infeasibility

The parameters commit[p,t] in Figure 14-5b represent the minimum production amounts for each product in each week. If we change the data to raise these commitments:

```
param commit:      1      2      3      4  :=
       bands    3500   5900   3900   6400
       coils    2500   2400   3400   4100  ;
```

```
param T := 4;
set PROD := bands coils;

param:      rate  inv0  prodcost  invcost :=
   bands     200   10     10        2.5
   coils     140    0     11        3 ;

param: avail_min  avail_max  time_penalty :=
   1        35        42          3100
   2        35        42          3000
   3        30        40          3700
   4        35        42          3100 ;

param revenue:    1     2     3     4 :=
         bands   25    26    27    27
         coils   30    35    37    39 ;

param commit:     1     2     3     4 :=
         bands 3000  3000  3000  3000
         coils 2000  2000  2000  2000 ;

param market:     1     2     3     4 :=
         bands 6000  6000  4000  6500
         coils 4000  2500  3500  4200 ;
```

Figure 14-5b: Data for Figure 14-5a (`steelpl1.dat`).

then there are not enough hours to produce even the minimum amounts, and the solver reports that the problem is infeasible:

```
ampl: model steelpl1.mod;
ampl: data steelpl2.dat;

ampl: solve;
MINOS 5.4: infeasible problem.
13 iterations
```

In the solution that is returned, the inventory of coils in the last period is negative:

```
ampl: option display_1col 0;
ampl: display Inv;
Inv [*,*] (tr)
: bands    coils     :=
0    10        0
1    0       937
2    0       287
3    0         0
4    0     -2700
;
```

and production of coils in several periods is below the minimum:

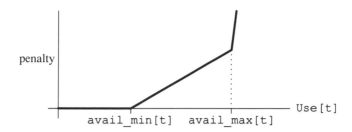

Figure 14-6: Penalty function for hours used, with two breakpoints.

```
ampl:  display commit,Make,market;
:         commit    Make market      :=
bands 1    3500     3490    6000
bands 2    5900     5900    6000
bands 3    3900     3900    4000
bands 4    6400     6400    6500
coils 1.   2500     3437    4000
coils 2    2400     1750    2500
coils 3    3400     2870    3500
coils 4    4100     1400    4200
;
```

These are typical of the infeasible results that solvers return. The infeasibilities are scattered around the solution, so that it is hard to tell what changes might be necessary to achieve feasibility. By extending the idea of penalties, we can better concentrate the infeasibility where it can be understood.

Suppose that we want to view the infeasibility in terms of a shortage of hours. Imagine that we extend the piecewise-linear penalty function of Figure 14-4 to the one shown in Figure 14-6. Now `Use[t]` is allowed to increase past `avail_max[t]`, but only with an extremely steep penalty per hour — so that the solution will use hours above `avail_max[t]` only to the extent absolutely necessary.

In AMPL, the new penalty function is introduced through the following changes:

```
var Use {t in 1..T} >= 0;

maximize total_profit:
   sum {p in PROD, t in 1..T} (revenue[p,t]*Sell[p,t] -
      prodcost[p]*Make[p,t] - invcost[p]*Inv[p,t])
   - sum {t in 1..T} <<avail_min[t],avail_max[t];
                      0,time_penalty[t],100000>> Use[t];
```

The former bound `avail_max[t]` has become a breakpoint, and to its right an arbitrarily large slope of 100,000 has been introduced. Now we get a feasible solution, which uses hours as follows:

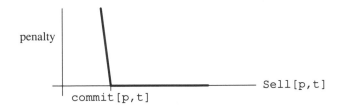

Figure 14-7: Penalty function for sales.

```
ampl: model steelp12.mod; data steelp12.dat; solve;
MINOS 5.4: optimal solution found.
19 iterations, objective -1576814.857

ampl: display avail_max,Use;
: avail_max  Use    :=
1     42       42
2     42       42
3     40       41.7357
4     42       61.2857
;
```

This table implies that the commitments can be met only by adding about 21 hours, mostly in the last week.

Alternatively, we may view the infeasibility in terms of an excess of commitments. For this purpose we subtract a very large penalty from the objective for each unit that Sell[p,t] falls below commit[p,t]; the penalty as a function of Sell[p,t] is depicted in Figure 14-7.

Since this function has a breakpoint at commit[p,t], with a slope of 0 to the right and a very negative value to the left, it would seem that the AMPL representation could be

```
<<commit[p,t]; -100000,0>> Sell[p,t]
```

Recall, however, AMPL's convention that such a function takes the value zero at zero. Figure 14-7 clearly shows that we want our penalty function to take a positive value at zero, so that it will fall to zero at commit[p,t] and beyond. In fact we want the function to take a value of 100000 * commit[p,t] at zero, and we could express the function properly by adding this constant to the penalty expression:

```
<<commit[p,t]; -100000,0>> Sell[p,t] + 100000*commit[p,t]
```

The same thing may be said more concisely by using a second argument that states explicitly where the piecewise-linear function should evaluate to zero:

```
<<commit[p,t]; -100000,0>> (Sell[p,t],commit[p,t])
```

This says that the function should be zero at commit[p,t], as Figure 14-7 shows. In the completed model, we have:

```
var Sell {p in PROD, t in 1..T} >= 0, <= market[p,t];

maximize total_profit:
    sum {p in PROD, t in 1..T} (revenue[p,t]*Sell[p,t] -
       prodcost[p]*Make[p,t] - invcost[p]*Inv[p,t])
  - sum {t in 1..T} <<avail_min[t]; 0,time_penalty[t]>> Use[t]
  - sum {p in PROD, t in 1..T}
        <<commit[p,t]; -100000,0>> (Sell[p,t],commit[p,t]);
```

The rest of the model is the same as in Figure 14-5a. Notice that Sell[p,t] appears in both a linear and a piecewise-linear term within the objective function; AMPL automatically recognizes that the sum of these terms is also piecewise-linear.

This version, using the same data, produces a solution in which the amounts sold are as follows:

```
ampl: model steelp13.mod; data steelp12.dat; solve;
MINOS 5.4: optimal solution found.
28 iterations, objective -293856347

ampl: display Sell,commit;
:           Sell commit     :=
bands 1     3500    3500
bands 2     5900    5900
bands 3     3900    3900
bands 4     6400    6400
coils 1        0    2500
coils 2     2400    2400
coils 3     3400    3400
coils 4     3657    4100
;
```

To get by with the given number of hours, commitments to deliver coils are cut by 2500 tons in the first week and 443 tons in the fourth week.

Reversible activities

Almost all of the linear programs in this book are formulated in terms of nonnegative variables. Sometimes a variable makes sense at negative as well as positive values, however, and in many such cases the associated cost is piecewise-linear with a breakpoint at zero.

One example is provided by the inventory variables in Figure 14-5a. We have defined Inv[p,t] to represent the tons of product p inventoried at the end of week t. That is, after week t there are Inv[p,t] tons of product p that have been made but not sold. A negative value of Inv[p,t] could thus reasonably be interpreted as representing tons of product p that have been sold but not made — tons backordered, in effect. The material balance constraints,

```
subject to balance {p in PROD, t in 1..T}:
    Make[p,t] + Inv[p,t-1] = Sell[p,t] + Inv[p,t];
```

remain valid under this interpretation.

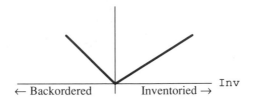

Figure 14-8: Inventory cost function.

This analysis suggests that we remove the >= 0 from the declaration of Inv in our model. Then backordering might be especially attractive if the sales price were expected to drop in later weeks, like this:

```
param revenue:      1      2      3      4 :=
          bands     25     26     23     20
          coils     30     35     31     25 ;
```

When we re-solve with appropriately revised model and data files, however, the results are not what we expect:

```
ampl: model steelp14.mod; data steelp14.dat; solve;
MINOS 5.4: optimal solution found.
25 iterations, objective 1194250

ampl: display Make,Inv,Sell;
:         Make      Inv      Sell     :=
bands 0     .        10        .
bands 1     0      -5990     6000
bands 2     0     -11990     6000
bands 3     0     -15990     4000
bands 4     0     -22490     6500
coils 0     .         0        .
coils 1     0      -4000     4000
coils 2     0      -6500     2500
coils 3     0     -10000     3500
coils 4     0     -14200     4200
;
```

The source of difficulty is in the objective function, where invcost[p] * Inv[p,t] is subtracted from the sales revenue. When Inv[p,t] is negative, a negative amount is subtracted, increasing the apparent total profit. The greater the amount backordered, the more the total profit is increased — hence the odd solution in which the maximum possible sales are backordered, while nothing is produced!

A proper inventory cost function for this model should look like the one graphed in Figure 14-8. It increases both as Inv[p,t] becomes more positive (greater inventories) and as Inv[p,t] becomes more negative (greater backorders). We represent this piecewise-linear function in AMPL by declaring a backorder cost to go with the inventory cost:

```
param invcost {PROD} >= 0;
param backcost {PROD} >= 0;
```

Then the slopes for the `Inv[p,t]` term in the objective are `-backcost[p]` and `invcost[p]`, with the breakpoint at zero, and the correct objective function is:

```
maximize total_profit:
   sum {p in PROD, t in 1..T}
       (revenue[p,t]*Sell[p,t] - prodcost[p]*Make[p,t]
        - <<0; -backcost[p],invcost[p]>> Inv[p,t])
     - sum {t in 1..T} <<avail_min[t]; 0,time_penalty[t]>> Use[t];
```

In contrast to our first example, the piecewise-linear function is subtracted rather than added. The result is still piecewise-linear, though; it's the same as if we had added the expression `<<0; backcost[p], -invcost[p]>> Inv[p,t]`.

When we make this change, and add some backorder costs to the data, we get a more reasonable-looking solution. Nevertheless, there remains a tendency to make nothing and backorder everything in the later periods; this is an ''end effect'' that occurs because the model does not account for the eventual production cost of items backordered past the last period. As a quick fix, we can rule out any remaining backorders at the end, by adding a constraint that final-week inventory must be nonnegative:

```
subject to final {p in PROD}:  Inv[p,T] >= 0;
```

Solving with this constraint, and with `backcost` values of 1.5 for band and 2 for coils:

```
ampl: model steelp15.mod; data steelp15.dat; solve;
MINOS 5.4: optimal solution found.
20 iterations, objective 370752.8571

ampl: display Make,Inv,Sell;
:            Make      Inv     Sell        :=
bands 0        .         10        .
bands 1     4142.86       0     4152.86
bands 2     6000          0     6000
bands 3     3000          0     3000
bands 4     3000          0     3000
coils 0        .          0        .
coils 1     2000          0     2000
coils 2     1680       -820     2500
coils 3     2100       -800     2080
coils 4     2800          0     2000
;
```

About 800 tons of coils for weeks 2 and 3 will be delivered a week late under this plan.

14.3 Other piecewise-linear functions

Many simple piecewise-linear functions can be modeled in several equivalent ways in AMPL. The function of Figure 14-4, for example, could be written as

```
if Use[t] > avail_min[t]
   then time_penalty[t] * (Use[t] - avail_min[t]) else 0
```

or more concisely as

```
max(0, time_penalty[t] * (Use[t] - avail_min[t]))
```

The current version of AMPL does not detect that these expressions are piecewise-linear, so you are unlikely to get satisfactory results if you try to solve a model that has expressions like these in its objective. To take advantage of linear programming techniques that can be applied for piecewise-linear terms, you need to use the piecewise-linear terminology

```
<<avail_min[t]; 0,time_penalty[t]>> Use[t]
```

so the structure can be noted and passed to a solver.

The same advice applies to the function abs. Imagine that we would like to encourage the number of hours used to be close to avail_min[t]. Then we would want the penalty term to equal time_penalty[t] times the amount that Use[t] deviates from avail_min[t], either above or below. Such a term can be written as

```
time_penalty[t] * abs(Use[t] - avail_min[t])
```

To express it in an explicitly piecewise-linear fashion, however, you should write it as

```
time_penalty[t] * <<avail_min[t]; -1,1>> Use[t]
```

or equivalently,

```
<<avail_min[t]; -time_penalty[t],time_penalty[t]>> Use[t]
```

As this example shows, multiplying a piecewise-linear function by a constant is the same as multiplying each of its slopes individually.

As a final example of a common piecewise-linearity in the objective, we return to the kind of assignment model that was discussed in Chapter 11. Recall that, for i in the set PEOPLE and j in the set PROJECTS, cost[i,j] is the cost for person i to work an hour on project j, and the decision variable Assign[i,j] is the number of hours that person i is assigned to work on project j:

```
set PEOPLE;
set PROJECTS;

param cost {PEOPLE,PROJECTS} >= 0;
var Assign {PEOPLE,PROJECTS} >= 0;
```

We originally formulated the objective as the total cost of all assignments,

```
sum {i in PEOPLE, j in PROJECTS} cost[i,j] * Assign[i,j]
```

What if we want the fairest assignment instead of the cheapest? Then we might minimize the maximum cost of any one person's assignments:

```
set PEOPLE;
set PROJECTS;

param supply {PEOPLE} >= 0;    # hours each person is available
param demand {PROJECTS} >= 0; # hours each project requires
    check: sum {i in PEOPLE} supply[i]
           = sum {j in PROJECTS} demand[j];
param cost {PEOPLE,PROJECTS} >= 0;    # cost per hour of work
param limit {PEOPLE,PROJECTS} >= 0;   # maximum contributions
                                      # to projects

var M;
var Assign {i in PEOPLE, j in PROJECTS} >= 0, <= limit[i,j];

minimize max_cost: M;
subject to M_def {i in PEOPLE}:
    M >= sum {j in PROJECTS} cost[i,j] * Assign[i,j];

subject to Supply {i in PEOPLE}:
    sum {j in PROJECTS} Assign[i,j] = supply[i];

subject to Demand {j in PROJECTS}:
    sum {i in PEOPLE} Assign[i,j] = demand[j];
```

Figure 14-9: Min-max assignment model (`minmax.mod`).

```
    minimize max_cost:
        max {i in PEOPLE}
            sum {j in PROJECTS} cost[i,j] * Assign[i,j];
```

This function is also piecewise-linear, in a sense; it is pieced together from the linear functions `sum {j in PROJECTS} cost[i,j] * Assign[i,j]` for different people `i`. However, it is not piecewise-linear in the individual variables — in mathematical jargon, it is not separable — and hence it cannot be written using the << ... >> notation.

This is a case in which piecewise-linearity can only be handled by rewriting the model as a linear program. We introduce a new variable M to represent the maximum, and then write constraints to guarantee that M is greater than or equal to each cost of which it is the maximum:

```
    var M;
    minimize max_cost: M;

    subject to M_def {i in PEOPLE}:
        M >= sum {j in PROJECTS} cost[i,j] * Assign[i,j];
```

Since M is being minimized, at the optimal solution it will in fact equal the maximum of `sum {j in PROJECTS} cost[i,j] * Assign[i,j]` over all `i in PEOPLE`. The other constraints are the same as in any assignment problem, as shown in Figure 14-9.

This kind of reformulation can be applied to any problem that has a "min-max" objective. The same idea works for the analogous "max-min" objective, with `maximize` instead of `minimize` and with `M <= ...` in the constraints.

14.4 Guidelines for piecewise-linear optimization

AMPL's piecewise-linear notation has the power to specify a variety of useful functions. We summarize below the general rules, most of which have been illustrated earlier in this chapter.

Because this notation is so general, it can be used to specify many functions that are too complicated for standard algorithms to optimize. We conclude by describing the kinds of piecewise-linear functions that are most likely to appear in tractable models, with particular emphasis on the property of convexity or concavity.

Rules for piecewise-linear expressions

An AMPL piecewise-linear term has the general form

>   <<breakpoint-list; slope-list>> pl-argument

where *breakpoint-list* and *slope-list* each consist of a comma-separated list of one or more items. An item may be an individual arithmetic expression, or an indexing expression followed by an arithmetic expression. In the latter case, the indexing expression must be an ordered set; the item is expanded to a list by evaluating the arithmetic expression once for each set member (as in the example of Figure 14-3a).

After any indexed items are expanded, the number of slopes must be one more than the number of breakpoints, and the breakpoints must be nondecreasing. The resulting piecewise-linear function is constructed by interleaving the slopes and breakpoints in the order given, with the first slope to the left of the first breakpoint, and the last slope to the right of the last breakpoint. By indexing breakpoints over an empty set, it is possible to specify no breakpoints and one slope, in which case the function is linear.

The *pl-argument* may have one of the forms

>   *var-ref*
>   (*arg-expr*)
>   (*arg-expr*, *zero-expr*)

The *var-ref* (a reference to a previously declared variable) or the *arg-expr* (an arithmetic expression) specifies the point where the piecewise-linear function is to be evaluated. The *zero-expr* is an arithmetic expression that specifies a place where the function is zero; when the *zero-expr* is omitted, the function is assumed to be zero at zero.

Suggestions for piecewise-linear models

The major practical use of piecewise-linear notation, as seen in all of our examples, is to describe piecewise-linear functions of individual variables. For these cases the expressions in the *breakpoint-list* and *slope-list* must contain no variables, while any *arg-expr* in the *pl-argument* must be a reference to an individual variable.

A piecewise-linear function of an individual variable remains such a function when it is multiplied or divided by an arithmetic expression that does not contain variables. Furthermore, any sum or difference of piecewise-linear and linear functions of the same variable is treated by AMPL as representing one piecewise-linear function of that variable.

A "separable" piecewise-linear function of a model's variables consists of a sum or difference (using +, − or sum) of piecewise-linear or linear functions of the individual variables. These separable functions are the ones that appear in our examples, and that optimizers can handle effectively.

A piecewise-linear function is convex if successive slopes are nondecreasing (along with the breakpoints), and is concave if the slopes are nonincreasing. The two kinds of piecewise-linear optimization most easily handled by solvers are minimizing a separable convex piecewise-linear function, and maximizing a separable concave piecewise-linear function, subject to linear constraints. You can easily check that all of this chapter's examples are of these kinds. AMPL can obtain solutions in these cases by translating to an equivalent linear program, applying any LP solver, and then translating the solution back; the whole sequence occurs automatically when you type `solve`.

Outside of these two cases, optimizing a separable piecewise-linear function must be viewed as an application of integer programming — the topic of Chapter 15 — and AMPL must translate piecewise-linear terms to equivalent integer programming forms. This, too, is done automatically, if an appropriate solver is available. Because integer programs are usually much harder to solve than similar linear programs of comparable size, however, you should not assume that just any separable piecewise-linear function can be readily optimized; a degree of experimentation may be necessary to determine how large an instance your solver can handle. The best results are likely to be obtained by solvers that accept an option known (mysteriously) as "special ordered sets of type 2"; check the solver-specific documentation for details.

The situation for the constraints can be described in a similar way. However, a separable piecewise-linear function in a constraint can be handled through linear programming only under a restrictive set of circumstances:

- If it is convex and on the left-hand side of a ≤ constraint (or equivalently, the right-hand side of a ≥ constraint);
- If it is concave and on the left-hand side of a ≥ constraint (or equivalently, the right-hand side of a ≤ constraint).

Other piecewise-linearities in the constraints must be dealt with through integer programming techniques, and the preceding comments for the case of the objective apply.

If you have access to a solver that can handle piecewise-linearities directly, you can turn off AMPL's translation to the linear or integer programming form by resetting the option `pl_linearize` to 0. The case of minimizing a convex or maximizing a concave separable piecewise-linear function can in particular be handled very efficiently by piecewise-linear generalizations of LP techniques. A solver intended for nonlinear programming may also accept piecewise-linear functions, but it is unlikely to handle them reliably unless it has been specially designed for "nondifferentiable" optimization.

The differences between hard and easy piecewise-linear cases can be slight. This chapter's transportation example is easy, in particular because the shipping rates increase along with shipping volume. The same example would be hard if shipping rates decreased with volume, since then we would be minimizing a concave rather than a convex function. We cannot say definitively that shipping rates ought to be one way or the other; their behavior depends upon the specifics of the situation being modeled.

In all cases, the difficulty of piecewise-linear optimization gradually increases with the total number of pieces. Thus piecewise-linear cost functions are most effective when the costs can be described or approximated by relatively few pieces. If you need more than about a dozen pieces to describe a cost accurately, you may be better off using a nonlinear function as described in Chapter 13.

Bibliography

Robert Fourer, ''A Simplex Algorithm for Piecewise-Linear Programming III: Computational Analysis and Applications.'' Mathematical Programming **53** (1992) pp. 213–235. A survey of conversions from piecewise-linear to linear programs, and of applications.

Robert Fourer and Roy E. Marsten, ''Solving Piecewise-Linear Programs: Experiments with a Simplex Approach.'' ORSA Journal on Computing **4** (1992) pp. 16–31. Descriptions of varied applications and of experience in solving them.

Exercises

14-1. (a) Reformulate the model of Figure 13-4 so that it approximates each nonlinear term

```
Trans[i,j] / (1 - Trans[i,j]/limit[i,j])
```

by a piecewise-linear term having three pieces. Set the breakpoints at (1/3) * limit[i,j] and (2/3) * limit[i,j]. Pick the slopes so that the approximation equals the original nonlinear term when Trans[i,j] is 0, 1/3 * limit[i,j], 2/3 * limit[i,j], or 11/12 * limit[i,j]; you should find that the three slopes are 3/2, 9/2 and 36 in every term, regardless of the size of limit[i,j]. Finally, place an explicit upper limit of 0.99 * limit[i,j] on Trans[i,j].

(b) Solve the approximation with the data given in Figure 13-5, and compare the optimal shipment amounts to the amounts recommended by the nonlinear model.

(c) Formulate a more sophisticated version in which the number of linear pieces for each term is given by a parameter nsl. Pick the breakpoints to be at (k/nsl) * limit[i,j] for k from 1 to nsl-1. Pick the slopes so that the piecewise-linear function equals the original nonlinear function when Trans[i,j] is (k/nsl) * limit[i,j] for any k from 0 to nsl-1, or when Trans[i,j] is (nsl-1/4)/nsl * limit[i,j].

Check your model by showing that you get the same results as in (b) when nsl is 3. Then, by trying higher values of nsl, determine how many linear pieces the approximation requires in order to determine all shipment amounts to within about 10% of the amounts recommended by the original nonlinear model.

14-2. This exercise asks how you might convert the demand constraints in the transportation model of Figure 3-1a into the kind of ''soft'' constraints described in Section 14.2.

Suppose that instead of a single parameter called demand[j] at each destination j, you are given the following four parameters that describe a more complicated situation:

dem_min_abs[j]	absolute minimum that must be shipped to j
dem_min_ask[j]	preferred minimum amount shipped to j
dem_max_ask[j]	preferred maximum amount shipped to
dem_max_abs[j]	absolute maximum that may be shipped to j

There are also two penalty costs for shipment amounts outside of the preferred limits:

dem_min_pen	penalty per unit that shipments fall below dem_min_ask[j]
dem_max_pen	penalty per unit that shipments exceed dem_max_ask[j]

Because the total shipped to j is no longer fixed, a new variable Receive[j] is introduced to represent the amount received at j.

(a) Modify the model of Figure 3-1a to use this new information. The modifications will involve declaring Receive[j] with the appropriate lower and upper bounds, adding a three-piece piecewise-linear penalty term to the objective function, and substituting Receive[j] for demand[j] in the constraints.

(b) Add the following demand information to the data of Figure 3-1b,:

	dem_min_abs	dem_min_ask	dem_max_ask	dem_max_abs
FRA	800	850	950	1100
DET	975	1100	1225	1250
LAN	600	600	625	625
WIN	350	375	450	500
STL	1200	1500	1800	2000
FRE	1100	1100	1100	1125
LAF	800	900	1050	1175

Let dem_min_pen and dem_max_pen be 2 and 4, respectively. Find the new optimal solution. In the solution, which destinations receive shipments that are outside the preferred levels?

14-3. When the diet model of Figure 2-1 is run with the data of Figure 2-3, there is no feasible solution. This exercise asks you to use the ideas of Section 14.2 to find some good near-feasible solutions.

(a) Modify the model so that it is possible, at a very high penalty, to purchase more than the specified maximum of a food. In the resulting solution, which maximums are exceeded?

(b) Modify the model so that it is possible, at a very high penalty, to supply more than the specified maximum of a nutrient. In the resulting solution, which maximums are exceeded?

(c) Using extremely large penalties, such as 10^{20} may give the solver numerical difficulties. Experiment to see how the available solvers behave when you use such penalty terms as 10^{20} and 10^{30}.

14-4. In the model of Exercise 4-4(b), the change in crews from one period to the next is limited to some number M. As an alternative to imposing this limit, suppose that we introduce a new variable D_t that represents the change in number of crews (in all shifts) at period t. This variable may be positive, indicating an increase in crews over the previous period, or negative, indicating a decrease in crews.

To make use of this variable, we introduce a defining constraint,

$$D_t = \sum_{s \in S} (Y_{st} - Y_{s,t-1}),$$

for each $t = 1, \ldots, T$. We then estimate costs of c^+ per crew added from period to period, and c^- per crew dropped from period to period; as a result, the following cost must be included in the objective for each month t:

$$c^- D_t, \qquad \text{if } D_t < 0;$$
$$c^+ D_t, \qquad \text{if } D_t > 0.$$

Reformulate the model in AMPL accordingly, using a piecewise-linear function to represent this extra cost.

Solve using $c^- = -20000$ and $c^+ = 100000$, together with the data previously given. How does this solution compare to the one from Exercise 4-4(b)?

14-5. The following "credit scoring" problem appears in many contexts, including the processing of credit card applications. A set APPL of people apply for a card, each answering a set QUES of questions on the application. The response of person i to question j is converted to a number, ans[i,j]; typical numbers are years at current address, monthly income, and a home ownership indicator (say, 1 if a home is owned and 0 otherwise).

To summarize the responses, the card issuer chooses weights Wt[j], from which a score for each person i in APPL is computed by the linear formula

```
sum {j in QUES} ans[i,j] * Wt[j]
```

The issuer also chooses a cutoff, Cut; credit is granted when an applicant's score is greater than or equal to the cutoff, and is denied when the score is less than the cutoff. In this way the decision can be made objectively (if not always most wisely).

To choose the weights and the cutoff, the card issuer collects a sample of previously accepted applications, some from people who turned out to be good customers, and some from people who never paid their bills. If we denote these two collections of people by sets GOOD and BAD, then the ideal weights and cutoff (for this data) would satisfy

```
sum {j in QUES} ans[i,j] * Wt[j] >= Cut    for each i in GOOD
sum {j in QUES} ans[i,j] * Wt[j] < Cut     for each i in BAD
```

Since the relationship between answers to an application and creditworthiness is imprecise at best, however, no values of Wt[j] and Cut can be found to satisfy all of these inequalities. Instead, the issuer has to choose values that are merely as good as possible, in some sense. There are any number of ways to make such a choice; here, naturally, we consider an optimization approach.

(a) Suppose that we define a new variable Diff[i] that equals the difference between person i's score and the cutoff:

```
Diff[i] = sum {j in QUES} ans[i,j] * Wt[j] - Cut
```

Clearly the undesirable cases are where Diff[i] is negative for i in GOOD, and where it is nonnegative for i in BAD. To discourage these cases, we can advise the issuer to minimize the following function:

```
sum {i in GOOD} max(0,-Diff[i]) + sum {i in BAD} max(0,Diff[i])
```

Explain why minimizing this function tends to produce a desirable choice of weights and cutoff.

(b) The expression above is a piecewise-linear function of the variables Diff[i]. Rewrite it using AMPL's notation for piecewise-linear functions.

(c) Incorporate the expression from (b) into an AMPL model for determining the weights and cutoff.

(d) Given this approach, any positive value for Cut is as good as any other. We can fix it at a convenient round number — say, 100. Explain why this is the case.

(e) Using a Cut of 100, apply the model to the following imaginary credit data:

```
set GOOD := _17 _18 _19 _22 _24 _26 _28 _29 ;
set BAD  := _15 _16 _20 _21 _23 _25 _27 _30 ;

set QUES := Q1 Q2 R1 R2 R3 S2 T4 ;

param ans:   Q1     Q2     R1     R2     R3     S2     T4 :=
     _15    1.0    10     15     20     10      8     10
     _16    0.0     5     15     40      8     10      8
     _17    0.5    10     25     35      8     10     10
     _18    1.5    10     25     30      8      6     10
     _19    1.5     5     20     25      8      8      8
     _20    1.0     5      5     30      8      8      6
     _21    1.0    10     20     30      8     10     10
     _22    0.5    10     25     40      8      8     10
     _23    0.5    10     25     25      8      8     14
     _24    1.0    10     15     40      8     10     10
     _25    0.0     5     15     15     10     12     10
     _26    0.5    10     15     20      8     10     10
     _27    1.0     5     10     25     10      8      6
     _28    0.0     5     15     40      8     10      8
     _29    1.0     5     15     40      8      8     10
     _30    1.5     5     20     25     10     10     14 ;
```

What are the chosen weights? Using these weights, how many of the good customers would be denied a card, and how many of the bad risks would be granted one?

You should find that a lot of the bad risks have scores right at the cutoff. Why does this happen in the solution? How might you adjust the cutoff to deal with it?

(f) To force scores further away from the cutoff (in the desired direction), it might be preferable to use the following objective,

```
sum {i in GOOD} max(0,-Diff[i]+offset) +
sum {i in BAD}  max(0,Diff[i]+offset)
```

where offset is a positive parameter whose value is supplied. Explain why this change has the desired effect. Try offset values of 2 and 10 and compare the results with those in (e).

(g) Suppose that giving a card to a bad credit risk is considered much more undesirable than refusing a card to a good credit risk. How would you change the model to take this into account?

(h) Suppose that when someone's application is accepted, his or her score is also used to suggest an initial credit limit on their card. Thus it is particularly important that bad credit risks not receive very large scores. How would you add pieces to the piecewise-linear objective function terms to account for this concern?

14-6. In Exercise 13-3, we suggested a way to estimate position, velocity and acceleration values from imprecise data, by minimizing a nonlinear function that was a "sum of squares",

$$\sum_{j=1}^{n} [h_j - (a_0 - a_1 t_j - \tfrac{1}{2} a_2 t_j^2)]^2.$$

We now turn to an alternative approach that minimizes instead a sum of absolute values,

$$\sum_{j=1}^{n} |h_j - (a_0 - a_1 t_j - \tfrac{1}{2} a_2 t_j^2)| .$$

(a) Substitute the sum of absolute values directly for the sum of squares in the model from Exercise 13-3, first with the abs function, and then with AMPL's explicit piecewise-linear notation.

Explain why neither of these formulations is likely to be handled effectively by any solver.

(b) To model this situation effectively, we introduce variables e_j to represent the individual formulas $h_j - (a_0 - a_1 t_j - \frac{1}{2} a_2 t_j^2)$ whose absolute values are being taken. Then we can express the minimization of the sum of absolute values as the following constrained optimization problem:

$$\text{Minimize} \sum_{j=1}^{n} |e_j|$$

$$\text{Subject to} \quad e_j = h_j - (a_0 - a_1 t_j - \frac{1}{2} a_2 t_j^2), \quad j = 1, \dots, n$$

Write an AMPL model for this formulation, using the piecewise-linear notation for the terms $|e_j|$.

(c) Solve for a_0, a_1, and a_2 using the data from Exercise 13-3. How much difference is there between this estimate and the least-squares one?

Use display to print the e_j values for both the least-squares and the least-absolute-values solutions. What is the most obvious qualitative difference?

(d) Yet another possibility is to focus on the greatest absolute deviation, rather than the sum:

$$\max_{j=1,\dots,n} |h_j - (a_0 - a_1 t_j - \frac{1}{2} a_2 t_j^2)|.$$

Formulate an AMPL linear program that will minimize this quantity, and test it on the same data as before. Compare the resulting estimates and e_j values. Which of the three estimates would you choose in this case?

14-7. A planar *structure* consists of a set of *joints* connected by *bars*. For example, in the following diagram, the joints are represented by circles, and the bars by lines between two circles:

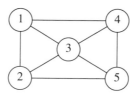

Consider the problem of finding a minimum-weight structure to meet certain external forces. We let J be the set of joints, and $B \subseteq J \times J$ be the set of *admissible* bars; for the diagram above, we could take $J = \{1,2,3,4,5\}$, and

$$B = \{(1,2), (1,3), (1,4), (2,3), (2,5), (3,4), (3,5), (4,5)\}.$$

The "origin" and "destination" of a bar are arbitrary. The bar between joints 1 and 2, for example, could be represented in B by either (1,2) or (2,1), but it need not be represented by both.

We can use two-dimensional Euclidean coordinates to specify the position of each joint in the plane, taking some arbitrary point as the origin:

a_i^x horizontal position of joint i relative to the origin

a_i^y vertical position of joint i relative to the origin

For the example, if the origin lies exactly at joint 2, we might have

$$(a_1^x, a_1^y) = (0, 2), \quad (a_2^x, a_2^y) = (0, 0), \quad (a_3^x, a_3^y) = (2, 1),$$
$$(a_4^x, a_4^y) = (4, 2), \quad (a_5^x, a_5^y) = (4, 0).$$

The remaining data consist of the external forces on the joints:

f_i^x horizontal component of the external force on joint i

f_i^y vertical component of the external force on joint i

To resist this force, a subset $F \subseteq J$ of joints is fixed in position. (It can be proved that fixing two joints is sufficient to guarantee a solution.)

The external forces induce stresses on the bars, which we can represent as

F_{ij} if > 0, tension on bar (i,j)

    if < 0, compression of bar (i,j)

A set of stresses is in *equilibrium* if the external forces, tensions and compressions balance at all joints, in both the horizontal and vertical components — except at the fixed joints. That is, for each joint $k \notin F$,

$$\sum_{i \in J:(i,k) \in B} c_{ik}^x F_{ik} - \sum_{j \in J:(k,j) \in B} c_{kj}^x F_{kj} = f_k^x$$

$$\sum_{i \in J:(i,k) \in B} c_{ik}^y F_{ik} - \sum_{j \in J:(k,j) \in B} c_{kj}^y F_{kj} = f_k^y,$$

where c_{st}^x and c_{st}^y are the cosines of the direction from joint s to joint t with the horizontal and vertical axes,

$$c_{st}^x = (a_t^x - a_s^x)/l_{st},$$
$$c_{st}^y = (a_t^y - a_s^y)/l_{st},$$

and l_{st} is the length of the bar (s,t):

$$l_{st} = \sqrt{(a_t^x - a_s^x)^2 + (a_t^y - a_s^y)^2}.$$

In general, there are infinitely many different sets of equilibrium stresses. However, it can be shown that a given system of stresses will be realized in the structure if and only if the cross-sectional areas of the bars are proportional to the absolute values of the stresses. Since the weight of a bar is proportional to the cross section times length, we can take the (suitably scaled) weight of bar (i,j) to be

$$w_{ij} = l_{ij} \cdot |F_{ij}|.$$

The problem is then to find a system of stresses F_{ij} that meet the equilibrium conditions, and that minimize the sum of the weights w_{ij} over all bars $(i,j) \in B$.

(a) The indexing sets for this linear program can be declared in AMPL as:

```
set joints;
set fixed within joints;
set bars within {i in joints, j in joints: i <> j};
```

Using these set declarations, formulate an AMPL model for the minimum-weight structural design problem. Use the piecewise-linear notation of this chapter to represent the absolute-value terms in the objective function.

(b) Now consider in particular a structure that has the following joints:

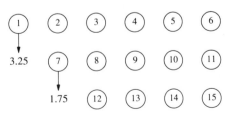

Assume that there is one unit horizontally and vertically between joints, and that the origin is at the lower left; thus $(a_1^x, a_1^y) = (0, 2)$ and $(a_{15}^x, a_{15}^y) = (5, 0)$.

Let there be external forces of 3.25 and 1.75 units straight downward on joints 1 and 7, so that $f_1^y = -3.25$, $f_7^y = -1.75$, and otherwise all $f_i^x = 0$ and $f_i^y = 0$. Let $F = \{6, 15\}$. Finally, let the admissible bars consist of all possible bars that do not go directly through a joint; for example, $(1, 2)$ or $(1, 9)$ or $(1, 13)$ would be admissible, but not $(1, 3)$ or $(1, 12)$ or $(1, 14)$.

Determine all the data for the problem that is needed by the linear program, and represent it as AMPL data statements.

(c) Use AMPL to solve the linear program and to examine the minimum-weight structure that is determined.

Draw a diagram of the optimal structure, indicating the cross sections of the bars and the nature of the stresses. If there is zero force on a bar, it has a cross section of zero, and may be left out of your diagram.

(d) Repeat parts (b) and (c) for the case in which all possible bars are admissible. Is the resulting structure different? Is it any lighter?

15

Integer Linear Programs

Many linear programming problems require whole number, or integer, values of the variables in order to be properly solved. Such a requirement arises naturally when the variables represent entities like packages or people that can not be fractionally divided — at least, not in a meaningful way for the situation being modeled. Integer variables also play a role in formulating equation systems that model logical conditions, as we will show later in this chapter.

In some situations, the optimization techniques described in previous chapters are sufficient to find an integer solution. An integer optimal solution is guaranteed for certain network linear programs, as explained in Section 11.5. Even where there is no guarantee, a linear programming solver may happen to find an integer optimal solution for the particular instances of a model in which you are interested. This happened in the solution of the multicommodity transportation model (Figure 4-1) for the particular data that we specified (Figure 4-2).

Even if you do not obtain an integer solution from the solver, chances are good that you'll get a solution in which most of the variables lie at integer values. Specifically, many solvers are able to return an "extreme" solution in which the number of variables not lying at their bounds is at most the total number of variables minus the number of constraints. If the bounds are integral, all of the variables at their bounds will have integer values; and if the rest of the data is integral, many of the remaining variables may turn out to be integers, too. You may then be able to adjust the relatively few non-integer variables to produce a completely integer solution that is close enough to feasible and optimal for practical purposes.

An example is provided by the scheduling linear program of Figures 12-4 and 12-5. Since the number of variables greatly exceeds the number of constraints, most of the variables end up at their bound of 0 in the optimal solution, and some other variables come out at integer values as well. The remaining variables can be rounded up to a nearby integer solution that is a little more costly but still satisfies the constraints.

Despite all these possibilities, there remain many circumstances in which the restriction to integrality must be enforced explicitly by the solver. Integer programming solvers face a much more difficult problem than their linear programming counterparts, however;

they generally require more computer time and memory, and often demand more help from the user in formulation and in choice of options. As a result, the size of problem that you can solve will be more limited for integer programs than for linear ones.

This chapter first describes AMPL declarations of ordinary integer variables, and then introduces the use of zero-one (or binary) integer variables for modeling logical conditions. A concluding section offers some advice on formulating and solving integer programs effectively.

15.1 Integer variables

By adding the keyword `integer` to the qualifying phrases of a `var` declaration, you restrict the declared variables to integer values.

As an example, in analyzing the diet model in Section 2.3, we arrived at the following optimal solution:

```
ampl: model diet.mod;
ampl: data diet2a.dat;

ampl: solve;
MINOS 5.4: optimal solution found.
12 iterations, objective 118.0594032

ampl: display Buy;
Buy [*] :=
BEEF   5.36061
 CHK   2
FISH   2
 HAM  10
 MCH  10
 MTL  10
 SPG   9.30605
 TUR   2
;
```

If we want the foods to be purchased in integral amounts, we add `integer` to the model's `var` declaration (Figure 2-1) as follows:

```
var Buy {j in FOOD} integer >= f_min[j], <= f_max[j];
```

We can then try to re-solve:

```
ampl: model dieti.mod; data diet2a.dat;
ampl: solve;
MINOS 5.4: ignoring integrality of 8 variables
MINOS 5.4: optimal solution found.
12 iterations, objective 118.0594032
```

As you can see, the current solver does not handle integrality constraints. It has ignored them and returned the same optimal value as before.

To get the integer optimum, we switch to a solver that does accommodate integrality:

```
ampl: option solver osl;
ampl: solve;
OSL 1.2: optimal solution; objective 119.3
85 simplex iterations
3 integer solutions found

ampl: display Buy;
Buy [*] :=
BEEF    9
 CHK    2
FISH    2
 HAM    8
 MCH   10
 MTL   10
 SPG    7
 TUR    2
;
```

Comparing this solution to the previous one, we see a few features typical of integer programming. The minimum cost has increased from $118.06 to $119.30; because integrality is an additional constraint on the values of the variables, it can only make the objective less favorable. The amounts of the different foods in the diet have also changed, but in unpredictable ways. The two foods that had fractional amounts in the original optimum, BEEF and SPG, have increased from 5.36061 to 9 and decreased from 9.30605 to 7, respectively; also, HAM has dropped from the upper limit of 10 to 8. Clearly, you cannot always deduce the integer optimum by rounding the non-integer optimum to the closest integer values.

15.2 Zero-one variables and logical conditions

Variables that can take only the values zero and one are a special case of integer variables. By cleverly incorporating these "zero-one" or "binary" variables into objectives and constraints, integer linear programs can specify a variety of logical conditions that cannot be described in any practical way by linear constraints alone.

To introduce the use of zero-one variables in AMPL, we return to the multicommodity transportation model of Figure 4-1. The decision variables Trans[i,j,p] in this model represent the tons of product p in set PROD to be shipped from originating city i in ORIG to destination city j in DEST. In the small example of data given by Figure 4-2, the products are bands, coils and plate; the origins are GARY, CLEV and PITT, and there are seven destinations.

The cost that this model associates with shipment of product p from i to j is cost[i,j,p] * Trans[i,j,p], regardless of the amount shipped. This "variable cost" is typical of purely linear programs, and in this case allows small shipments between many origin-destination pairs. In the following examples we describe ways to use zero-one variables to discourage shipments of small amounts; the same techniques can be adapted to many other logical conditions as well.

To provide a convenient basis for comparison, we focus on the tons shipped from each origin to each destination, summed over all products. The optimal values of these total shipments are determined by a linear programming solver as follows:

```
ampl: model multi.mod;
ampl: data multi.dat;
ampl: solve;
MINOS 5.4: optimal solution found.
45 iterations, objective 199500

ampl: option display_eps .000001;
ampl: option display_transpose -10;

ampl: display {i in ORIG, j in DEST} sum {p in PROD} Trans[i,j,p];
sum{p in PROD} Trans[i,j,p] [*,*]
:     DET   FRA   FRE   LAF   LAN   STL   WIN      :=
CLEV  625   275   150   225   400   550   375
GARY    0     0   625   150     0   625     0
PITT  525   625   400   625   100   625     0
;
```

The quantity 625 appears often in this solution, as a consequence of the multicommodity constraints:

```
subject to Multi {i in ORIG, j in DEST}:
    sum {p in PROD} Trans[i,j,p] <= limit[i,j];
```

In the data for our example, `limit[i,j]` is 625 for all `i` and `j`; its six appearances in the solution correspond to the six routes on which the multicommodity limit constraint is tight. Other routes have positive shipments as low as 100; the five instances of 0 indicate routes that are not used.

Even though all of the shipment amounts happen to be integers in this solution, we would be willing to ship fractional amounts. Thus we will not declare the `Trans` variables to be integer, but will instead extend the model by using other integer variables.

Fixed costs

One way to discourage small shipments is to add a fixed cost for each origin-destination route that is actually used. For this purpose we rename the `cost` parameter `vcost`, and declare another parameter `fcost` to represent the fixed assessment for using the route from `i` to `j`:

```
param vcost {ORIG,DEST,PROD} >= 0; # variable cost on routes
param fcost {ORIG,DEST} > 0;       # fixed cost on routes
```

We want `fcost[i,j]` to be added to the objective function if the total shipment of products from `i` to `j` — that is, sum {p in PROD} Trans[i,j,p] — is positive; we want nothing to be added if the total shipment is zero. Using AMPL expressions, we could write the objective function most directly as follows:

```
minimize total_cost:    # NOT PRACTICAL
    sum {i in ORIG, j in DEST, p in PROD}
        vcost[i,j,p] * Trans[i,j,p]
  + sum {i in ORIG, j in DEST}
        if sum {p in PROD} Trans[i,j,p] > 0 then fcost[i,j];
```

AMPL accepts this objective, but treats it as merely "not linear" in the sense of Chapter 13, so that you are unlikely to get acceptable results trying to minimize it.

As a more practical alternative, we may associate a new variable Use[i,j] with each route from i to j, as follows: Use[i,j] takes the value 1 if

```
    sum {p in PROD} Trans[i,j,p]
```

is positive, and is 0 otherwise. Then the fixed cost associated with the route from i to j is fcost[i,j] * Use[i,j], a linear term. To declare these new variables in AMPL, we can say that they are integer with bounds >= 0 and <= 1; equivalently we can use the keyword binary:

```
    var Use {ORIG,DEST} binary;
```

The objective function can then be written as a linear expression:

```
minimize total_cost:
    sum {i in ORIG, j in DEST, p in PROD}
        vcost[i,j,p] * Trans[i,j,p]
  + sum {i in ORIG, j in DEST} fcost[i,j] * Use[i,j];
```

Since the model has a combination of continuous (non-integer) and integer variables, it yields what is known as a mixed-integer program.

To complete the model, we need to add constraints to assure that Trans and Use are related in the intended way. This is the "clever" part of the formulation; we simply modify the Multi constraints cited above so that they incorporate the Use variables:

```
    subject to Multi {i in ORIG, j in DEST}:
        sum {p in PROD} Trans[i,j,p] <= limit[i,j] * Use[i,j];
```

If Use[i,j] is 0, this constraint says that

```
    sum {p in PROD} Trans[i,j,p] <= 0
```

Since it is a sum of nonnegative variables, it must equal 0. On the other hand, when Use[i,j] is 1, the constraint reduces to

```
    sum {p in PROD} Trans[i,j,p] <= limit[i,j]
```

which is the multicommodity limit as before. Although there is nothing in the constraint to directly prevent sum {p in PROD} Trans[i,j,p] from being 0 when Use[i,j] is 1, so long as fcost[i,j] is positive this combination can never occur in an optimal solution. Thus Use[i,j] will be 1 if and only if sum {p in PROD} Trans[i,j,p] is positive, which is what we want. The complete model is shown in Figure 15-1a.

```
set ORIG;    # origins
set DEST;    # destinations
set PROD;    # products

param supply {ORIG,PROD} >= 0;   # amounts available at origins
param demand {DEST,PROD} >= 0;   # amounts required at destinations

    check {p in PROD}:
        sum {i in ORIG} supply[i,p] = sum {j in DEST} demand[j,p];

param limit {ORIG,DEST} >= 0;     # maximum shipments on routes

param vcost {ORIG,DEST,PROD} >= 0; # variable shipment cost on routes
var Trans {ORIG,DEST,PROD} >= 0;    # units to be shipped

param fcost {ORIG,DEST} >= 0;        # fixed cost for using a route
var Use {ORIG,DEST} binary;          # = 1 only for routes used

minimize total_cost:
   sum {i in ORIG, j in DEST, p in PROD} vcost[i,j,p] * Trans[i,j,p]
 + sum {i in ORIG, j in DEST} fcost[i,j] * Use[i,j];

subject to Supply {i in ORIG, p in PROD}:
   sum {j in DEST} Trans[i,j,p] = supply[i,p];

subject to Demand {j in DEST, p in PROD}:
   sum {i in ORIG} Trans[i,j,p] = demand[j,p];

subject to Multi {i in ORIG, j in DEST}:
   sum {p in PROD} Trans[i,j,p] <= limit[i,j] * Use[i,j];
```

Figure 15-1a: Multicommodity model with fixed costs (`multmip1.mod`).

To show how this model might be solved, we add a table of fixed costs to the sample data (Figure 15-1b):

```
param fcost:    FRA  DET  LAN  WIN  STL  FRE  LAF :=
        GARY   3000 1200 1200 1200 2500 3500 2500
        CLEV   2000 1000 1500 1200 2500 3000 2200
        PITT   2000 1200 1500 1500 2500 3500 2200 ;
```

If we apply the same solver as before, the integrality restrictions on the Use variables are ignored:

```
ampl: model multmip1.mod;
ampl: data multmip1.dat;
ampl: solve;
MINOS 5.4: ignoring integrality of 21 variables
MINOS 5.4: optimal solution found.
115 iterations, objective 223504

ampl: option display_eps .000001;
ampl: option display_transpose 10;
```

```
set ORIG := GARY CLEV PITT ;
set DEST := FRA DET LAN WIN STL FRE LAF ;
set PROD := bands coils plate ;

param supply (tr):    GARY    CLEV    PITT :=
              bands    400     700     800
              coils    800    1600    1800
              plate    200     300     300 ;

param demand (tr):
                FRA   DET   LAN   WIN   STL   FRE   LAF :=
        bands   300   300   100    75   650   225   250
        coils   500   750   400   250   950   850   500
        plate   100   100     0    50   200   100   250 ;

param limit default 625 ;

param vcost :=

 [*,*,bands]:  FRA   DET   LAN   WIN   STL   FRE   LAF :=
       GARY    30    10     8    10    11    71     6
       CLEV    22     7    10     7    21    82    13
       PITT    19    11    12    10    25    83    15

 [*,*,coils]:  FRA   DET   LAN   WIN   STL   FRE   LAF :=
       GARY    39    14    11    14    16    82     8
       CLEV    27     9    12     9    26    95    17
       PITT    24    14    17    13    28    99    20

 [*,*,plate]:  FRA   DET   LAN   WIN   STL   FRE   LAF :=
       GARY    41    15    12    16    17    86     8
       CLEV    29     9    13     9    28    99    18
       PITT    26    14    17    13    31   104    20 ;

param fcost:   FRA   DET   LAN   WIN   STL   FRE   LAF :=
       GARY   3000  1200  1200  1200  2500  3500  2500
       CLEV   2000  1000  1500  1200  2500  3000  2200
       PITT   2000  1200  1500  1500  2500  3500  2200 ;
```

Figure 15-1b: Data for Figure 15-1a (`multmip1.dat`).

```
ampl: display sum {i in ORIG, j in DEST, p in PROD}
ampl?     vcost[i,j,p] * Trans[i,j,p];
sum{i in ORIG, j in DEST, p in PROD}
     vcost[i,j,p]*Trans[i,j,p] = 199500

ampl: display Use;
Use [*,*]
:      DET    FRA    FRE    LAF    LAN    STL    WIN     :=
CLEV   1      0.44   0.52   0.36   0.64   0.88   0.32
GARY   0      0      1      0.24   0      1      0
PITT   0.84   1      0.36   1      0.16   1      0.28
;
```

As you can see, the total variable cost is the same as before, and Use assumes a variety of fractional values. This solution tells us nothing new, and there is no simple way to convert it into a good integer solution. An integer programming solver is essential to get any practical results in this situation.

Switching to an appropriate solver, we find that the true optimum with fixed costs is as follows:

```
ampl: option solver cplex; solve;
CPLEX 2.0: optimal integer solution within mipgap; objective 229850
0 simplex iterations
786 integer iterations

ampl: display {i in ORIG, j in DEST} sum {p in PROD} Trans[i,j,p];
sum{p in PROD} Trans[i,j,p] [*,*]
:       DET    FRA    FRE    LAF    LAN    STL    WIN      :=
CLEV    625    275      0    425    350    550    375
GARY      0      0    625      0    150    625      0
PITT    525    625    550    575      0    625      0
;

ampl: display Use;
Use [*,*]
:    DET FRA FRE LAF LAN STL WIN      :=
CLEV   1   1   0   1   1   1   1
GARY   0   0   1   0   1   1   0
PITT   1   1   1   1   0   1   0
;
```

Imposing the integer constraints has increased the total cost from $223,504 to $229,850; but the number of unused routes has increased, to seven, as we had hoped.

Zero-or-minimum restrictions

Although the fixed-cost solution uses fewer routes, there are still some on which the amounts shipped are relatively low. As a practical matter, it may be that even the variable costs are not applicable unless some minimum number of tons is shipped. Suppose, therefore, that we declare a parameter minload to represent the minimum number of tons that may be shipped on a route. We could add a constraint to say that the shipments on each route, summed over all products, must be at least minload:

```
subject to Min_Ship {i in ORIG, j in DEST}:    # WRONG
    sum {p in PROD} Trans[i,j,p] >= minload;
```

But this would force the shipments on every route to be at least minload, which is not what we have in mind. We want the tons shipped to be either zero, or at least minload. To say this directly, we might write:

```
subject to Min_Ship {i in ORIG, j in DEST}:    # NOT ALLOWED
    sum {p in PROD} Trans[i,j,p] = 0 or
    sum {p in PROD} Trans[i,j,p] >= minload;
```

But the current version of AMPL does not accept logical operators in constraints.

The desired zero-or-minimum restrictions can be imposed by employing the variables Use[i,j], much as in the previous example:

```
subject to Min_Ship {i in ORIG, j in DEST}:
    sum {p in PROD} Trans[i,j,p] >= minload * Use[i,j];
```

When total shipments from i to j are positive, Use[i,j] is 1, and this becomes the desired minimum-shipment constraint. On the other hand, when there are no shipments from i to j, Use[i,j] is zero; the constraint reduces to 0 >= 0 and has no effect.

With these new restrictions and a minload of 375, the solution is found to be as follows:

```
ampl: model multmip2.mod;
ampl: data multmip2.dat;

ampl: solve;
CPLEX 2.0: optimal integer solution; objective 233150
0 simplex iterations
1101 integer iterations

ampl: display {i in ORIG, j in DEST} sum {p in PROD} Trans[i,j,p];
sum{p in PROD} Trans[i,j,p] [*,*]
:      DET   FRA   FRE   LAF   LAN   STL   WIN     :=
CLEV   625   425   425     0   500   625     0
GARY     0     0   375   425     0   600     0
PITT   525   475   375   575     0   575   375
;
```

Comparing this to the previous solution, we see that although there are still seven unused routes, they are not the same ones; a substantial rearrangement of the solution has been necessary to meet the minimum-shipment requirement. The total cost has gone up by about 1.4% as a result.

Cardinality restrictions

Despite the constraints we have added so far, origin PITT still serves 6 destinations, while CLEV serves 5 and GARY serves 3. We would like to explicitly add a further restriction that each origin can ship to at most maxserve destinations, where maxserve is a parameter to the model. This can be viewed as a restriction on the size, or cardinality, of a certain set. Indeed, it could in principle be written in the form of an AMPL constraint as follows:

```
subject to Max_Serve {i in ORIG}:    # NOT ALLOWED
    card {j in DEST:
        sum {p in PROD} Trans[i,j,p] > 0} <= maxserve;
```

Such a declaration will be rejected, however, because AMPL does not allow constraints to use sets that are defined in terms of variables.

```
set ORIG;    # origins
set DEST;    # destinations
set PROD;    # products

param supply {ORIG,PROD} >= 0;   # amounts available at origins
param demand {DEST,PROD} >= 0;   # amounts required at destinations

    check {p in PROD}:
        sum {i in ORIG} supply[i,p] = sum {j in DEST} demand[j,p];

param limit {ORIG,DEST} >= 0;    # maximum shipments on routes
param minload >= 0;              # minimum nonzero shipment
param maxserve integer > 0;      # maximum destinations served

param vcost {ORIG,DEST,PROD} >= 0; # variable shipment cost on routes
var Trans {ORIG,DEST,PROD} >= 0;   # units to be shipped

param fcost {ORIG,DEST} >= 0;        # fixed cost for using a route
var Use {ORIG,DEST} binary;          # = 1 only for routes used

minimize total_cost:
   sum {i in ORIG, j in DEST, p in PROD} vcost[i,j,p] * Trans[i,j,p]
 + sum {i in ORIG, j in DEST} fcost[i,j] * Use[i,j];

subject to Supply {i in ORIG, p in PROD}:
   sum {j in DEST} Trans[i,j,p] = supply[i,p];

subject to Max_Serve {i in ORIG}:
   sum {j in DEST} Use[i,j] <= maxserve;

subject to Demand {j in DEST, p in PROD}:
   sum {i in ORIG} Trans[i,j,p] = demand[j,p];

subject to Multi {i in ORIG, j in DEST}:
   sum {p in PROD} Trans[i,j,p] <= limit[i,j] * Use[i,j];

subject to Min_Ship {i in ORIG, j in DEST}:
   sum {p in PROD} Trans[i,j,p] >= minload * Use[i,j];
```

Figure 15-2a: Multicommodity model with further restrictions (multmip3.mod).

Zero-one variables again offer a convenient alternative. Since the variables Use[i,j] are 1 precisely for those destinations j served by origin i, and are zero otherwise, we can write sum {j in DEST} Use[i,j] for the number of destinations served by i. The desired constraint becomes:

```
subject to Max_Serve {i in ORIG}:
    sum {j in DEST} Use[i,j] <= maxserve;
```

Adding this constraint to the previous model, and setting maxserve to 5, we arrive at the mixed integer model shown in Figure 15-2a, with data shown in Figure 15-2b. It is optimized as follows:

```
set ORIG := GARY CLEV PITT ;
set DEST := FRA DET LAN WIN STL FRE LAF ;
set PROD := bands coils plate ;

param supply (tr):    GARY    CLEV    PITT :=
              bands    400     700     800
              coils    800    1600    1800
              plate    200     300     300 ;

param demand (tr):
                 FRA  DET  LAN  WIN  STL  FRE  LAF :=
         bands   300  300  100   75  650  225  250
         coils   500  750  400  250  950  850  500
         plate   100  100    0   50  200  100  250 ;

param limit default 625 ;

param vcost :=

  [*,*,bands]:  FRA  DET  LAN  WIN  STL  FRE  LAF :=
         GARY    30   10    8   10   11   71    6
         CLEV    22    7   10    7   21   82   13
         PITT    19   11   12   10   25   83   15

  [*,*,coils]:  FRA  DET  LAN  WIN  STL  FRE  LAF :=
         GARY    39   14   11   14   16   82    8
         CLEV    27    9   12    9   26   95   17
         PITT    24   14   17   13   28   99   20

  [*,*,plate]:  FRA  DET  LAN  WIN  STL  FRE  LAF :=
         GARY    41   15   12   16   17   86    8
         CLEV    29    9   13    9   28   99   18
         PITT    26   14   17   13   31  104   20 ;

param fcost:   FRA   DET   LAN   WIN   STL   FRE   LAF :=
        GARY  3000  1200  1200  1200  2500  3500  2500
        CLEV  2000  1000  1500  1200  2500  3000  2200
        PITT  2000  1200  1500  1500  2500  3500  2200 ;

param minload := 375 ;

param maxserve := 5 ;
```

Figure 15-2b: Data for Figure 15-2a (`multmip3.dat`).

```
ampl: model multmip3.mod;
ampl: data multmip3.dat;

ampl: solve;
CPLEX 2.0: optimal integer solution; objective 235625
0 simplex iterations
609 integer iterations
```

```
ampl: display {i in ORIG, j in DEST} sum {p in PROD} Trans[i,j,p];
sum{p in PROD} Trans[i,j,p] [*,*]
:       DET   FRA   FRE   LAF   LAN   STL   WIN      :=
CLEV    625   375   550     0   500   550     0
GARY      0     0     0   400     0   625   375
PITT    525   525   625   600     0   625     0
;
```

At the cost of a further 1.1% increase in the objective, rearrangements have been made so that GARY can serve WIN, bringing the number of destinations served by PITT down to five.

Notice that this section's three integer solutions have served WIN from each of the three different origins — a good example of how solutions to integer programs can jump around in response to small changes in the restrictions.

15.3 Practical considerations in integer programming

As a rule, any integer program is much harder to solve than a linear program of the same size and structure. A rough idea of the difference in the previous examples is given by the number of iterations reported by the solvers; it is 45 for solving the original linear multicommodity transportation problem, but ranges from about 600 to 1100 for the mixed integer versions. The computation times vary similarly; the linear programs are solved almost instantly, while the mixed integer ones are noticeably slower.

As the size of the problem increases, the difficulty of an integer program also grows more quickly than the difficulty of a comparable linear program. Thus the practical limits to the size of a solvable integer program will be much more restrictive. Indeed, AMPL can easily generate integer programs that are too difficult for your computer to solve in a reasonable amount of time or memory. Before you make a commitment to use a model with integer variables, therefore, you should consider whether an alternative continuous linear or network formulation might give adequate answers. If you must use an integer formulation, try it first on small collections of data to get an idea of the computer resources required.

If you do encounter an integer program that is difficult to solve, you should consider reformulations that can make it easier. An integer programming solver attempts to investigate all of the different possible combinations of values of the integer variables; although it employs a sophisticated strategy that rules out the vast majority of combinations as infeasible or suboptimal, the number of combinations remaining to be checked can still be huge. Thus you should try to use as few integer variables as possible. For those variables that are not zero-one, lower and upper bounds should be made as tight as possible to reduce the number of combinations that might be investigated.

A solver can derive valuable information about the solutions to an integer program by fixing or restricting some of the integer variables, then solving the linear programming "relaxation" that results when the remaining integer restrictions are dropped. You may be able to assist in this strategy by reformulating the model so that the solution of a relaxation will be closer to the solution of the associated integer problem. In the multicommodity transportation model, for example, if `Use[i,j]` is 0 then each of the individual variables `Trans[i,j,p]` must be 0, while if `Use[i,j]` is 1 then `Trans[i,j,p]` cannot be larger than either `supply[i,p]` or `demand[j,p]`. This suggests that we add the following constraints:

```
subject to Avail {i in ORIG, j in DEST, p in PROD}:
    Trans[i,j,p] <= min(supply[i,p],demand[j,p]) * Use[i,j];
```

Although these constraints do not rule out any previously admissible integer solutions, they tend to force `Use[i,j]` to be closer to 1 for any solution of the relaxation that uses the route from `i` to `j`. As a result, the relaxation is more accurate, and may help the solver to find the optimum integer solution more quickly; this advantage may outweigh the extra cost of handling more constraints.

As this example suggests, the choice of a formulation is much more critical in integer than in linear programming. For large problems, solution times can change dramatically in response to simple reformulations. The effect of a reformulation can be hard to predict, however; it depends on the structure of your model and data, and on the details of the strategy used by your solver. Generally you will need to do some experimentation to see what works best.

You can also try to help a solver by changing some of the default settings that determine how it initially processes a problem and how it searches for integer solutions. Many of these settings can be manipulated from within AMPL, as explained in the separate instructions for using particular solvers.

Finally, a solver may provide options for stopping prematurely, returning an integer solution that has been determined to bring the objective value to within a small percentage of optimality. In some cases, such a solution is found relatively early in the solution process; you may save a great deal of computation time by not insisting that the solver go on to find a provably optimal integer solution.

In summary, experimentation is usually necessary to solve integer or mixed-integer linear programs. Your chances of success are greatest if you approach the use of integer variables with caution, and if you are willing to keep trying alternative formulations, settings and strategies until you get an acceptable solution.

Bibliography

Ellis L. Johnson, Michael M. Kostreva and Uwe H. Suhl, ''Solving 0-1 Integer Programming Problems Arising from Large-Scale Planning Models.'' Operations Research **33** (1985) pp. 803–819. A case study in which preprocessing, reformulation, and algorithmic strategies were brought to bear on the solution of a difficult class of integer linear programs.

George L. Nemhauser and Laurence A. Wolsey, *Integer and Combinatorial Optimization*. John Wiley & Sons (New York, NY, 1988). A survey of integer programming problems, theory and algorithms.

Alexander Schrijver, *Theory of Linear and Integer Programming*. John Wiley & Sons (New York, NY, 1986). A guide to the fundamentals of the subject, with a particularly thorough collection of references.

Exercises

15-1. Exercise 1-1 optimizes an advertising campaign that permits arbitrary combinations of various media. Suppose that instead you must allocate a $1 million advertising campaign among several media, but for each, your only choice is whether to use that medium or not. The following table shows, for each medium, the cost and audience in thousands if you use that medium, and the creative time in three categories that will have to be committed if you use that medium. The final column shows your limits on cost and on hours of creative time.

	TV	Magazine	Radio	Newspaper	Mail	Phone	Limits
Audience	300	200	100	150	100	50	
Cost	600	250	100	120	200	250	1000
Writers	120	50	20	40	30	5	200
Artists	120	80	0	60	40	0	300
Others	20	20	20	20	20	200	200

Your goal is to maximize the audience, subject to the limits.

(a) Formulate an AMPL model for this situation, using a zero-one integer variable for each medium.

(b) Use an integer programming solver to find the optimal solution.

Also solve the problem with the integrality restrictions relaxed and compare the resulting solution. Could you have guessed the integer optimum from looking at the non-integer one?

15-2. Return to the multicommodity transportation problem of Figures 4-1 and 4-2. Use appropriate integer variables to impose each of the restrictions described below. (Treat each part separately; don't try to put all the restrictions in one model.) Solve the resulting integer program, and comment in general terms on how the solution changes as a result of the restrictions. Also solve the corresponding linear program with the integrality restrictions relaxed, and compare the LP solution to the integer one.

(a) Require the amount shipped on each origin-destination link to be a multiple of 100 tons. To accommodate this restriction, allow demand to be satisfied only to the nearest 100 — for example, since demand for bands at FRE is 225 tons, allow either 200 or 300 tons to be shipped to FRE.

(b) Require each destination except STL to be served by at most two origins.

(c) Require the number of origin-destination links used to be as small as possible, regardless of cost.

(d) Require each origin that supplies product p to destination j to ship either nothing, or at least the lesser of demand[j,p] and 150 tons.

15-3. Employee scheduling problems are a common source of integer programs, because it doesn't make sense to schedule a fraction of a person.

(a) Solve the problem of Exercise 4-4(b) with the requirement that the variables Y_{st}, representing the numbers of crews employed on shift s in period t, must all be integer. Confirm that the optimal integer solution just "rounds up" the linear programming solution to the next highest integer.

(b) Similarly attempt to find an integer solution for the problem of Exercise 4-4(c), where inventories have been added to the formulation. Compare the difficulty of this problem to the one in (a). Is the optimal integer solution the same as the rounded-up linear programming solution in this case?

(c) Solve the scheduling problem of Figures 12-4 and 12-5 with the requirement that an integer number of employees be assigned to each shift. Show that this solution is better than the one obtained by rounding up.

(d) Suppose that the number of supervisors required is 1/10 the number of employees, rounded to the nearest whole number. Solve the scheduling problem again for supervisors. Does the integer program appear to be harder or easier to solve in this case? Is the improvement over the rounded-up solution more or less pronounced?

15-4. The so-called knapsack problem arises in many contexts. In its simplest form, you start with a set of objects, each having a known weight and a known value. The problem is to decide which items to put into your knapsack. You want these items to have as great a total value as possible, but their weight cannot exceed a certain preset limit.

(a) The data for the simple knapsack problem could be written in AMPL as follows:

```
set OBJECTS;
param weight {OBJECTS} > 0;
param value {OBJECTS} > 0;
```

Using these declarations, formulate an AMPL model for the knapsack problem.

Use your model to solve the knapsack problem for objects of the following weights and values, subject to a weight limit of 100:

object	a	b	c	d	e	f	g	h	i	j
weight	55	50	40	35	30	30	15	15	10	5
value	1000	800	750	700	600	550	250	200	200	150

(b) Suppose instead that you want to fill several identical knapsacks, as specified by this extra parameter:

```
param knapsacks > 0 integer;   # number of knapsacks
```

Formulate an AMPL model for this situation. Don't forget to add a constraint that each object can only go into one knapsack!

Using the data from (a), solve for 2 knapsacks of weight limit 50. How does the solution differ from your previous one?

(c) Superficially, the preceding knapsack problem resembles an assignment problem; we have a collection of objects and a collection of knapsacks, and we want to make an optimal assignment from members of the former to members of the latter. What is the essential difference between the kinds of assignment problems described in Section 11.2, and the knapsack problem described in (b)?

(d) Modify the formulation from (a) so that it accommodates a volume limit for the knapsack as well as a weight limit. Solve again using the following volumes:

```
object    a   b   c   d   e   f   g   h   i   j
volume    3   3   3   2   2   2   2   1   1   1
```

How do the total weight, volume and value of this solution compare to those of the solution you found in (a)?

(e) How can the media selection problem of Exercise 15-1 be viewed as a knapsack problem like the one in (d)?

(f) Suppose that you can put up to 3 of each object in the knapsack, instead of just 1. Revise the model of (a) to accommodate this possibility, and re-solve with the same data. How does the optimal solution change?

(g) Review the roll-cutting problem described in Exercise 2-6. Given a supply of wide rolls, orders for narrower rolls, and a collection of cutting patterns, we seek a combination of patterns that fills all orders using the least material.

When this problem is solved, the algorithm returns a "dual value" corresponding to each ordered roll width. It is possible to interpret the dual value as the saving in wide rolls that you might achieve for each extra narrow roll that you obtain; for example, a value of 0.5 for a 50" roll would indicate that you might save 5 wide rolls if you could obtain 10 extra 50" rolls.

It is natural to ask: Of all the patterns of narrow rolls that fit within one wide roll, which has the greatest total (dual) value? Explain how this can be regarded as a knapsack problem.

For the problem of Exercise 2-6(a), the wide rolls are 110"; in the solution using the six patterns given, the dual values for the ordered widths are:

    20"   0.0
    40"   0.5
    50"   1.0
    55"   0.0
    75"   1.0

What is the maximum-value pattern? Show that it is not one of the ones given, and that adding it allows you to get a better solution.

15-5. Recall the multiperiod model of production that was introduced in Section 4.2. Add zero-one variables and appropriate constraints to the formulation from Figure 4-4, in order to impose each of the restrictions described below. (Treat each part separately.) Solve with the data of Figure 4-5, and confirm that the solution properly obeys the restrictions.

(a) Require for each product p and week t, that either none of the product is made that week, or at least 2500 tons are made.

(b) Require that only one product be made in any one week.

(c) Require that each product be made in at most two weeks out of any three-week period. Assume that only bands were made in "week −1" and that both bands and coils were made in "week 0".

A

AMPL Reference Manual

AMPL is a language for algebraic modeling and mathematical programming: a computer-readable language for expressing in algebraic notation optimization problems such as linear programming. This appendix summarizes the features of AMPL, with particular emphasis on technical details not fully covered in the preceding chapters.

The following notational conventions are followed. Literal text is printed in `constant width font`, while syntactic categories are printed in *italic font*. Phrases or subphrases enclosed in slanted square brackets *[* and *]* are optional, as are constructs with the subscript *opt*.

A.1 Lexical rules

AMPL models involve variables, constraints, and objectives, expressed with the help of sets and parameters. These are called *model entities*. Each model entity has an alphanumeric name: a string of one or more letters, digits, and underscores, in a pattern that cannot be mistaken for a numeric constant. Upper-case letters are distinct from lower-case letters.

Numeric constants are as in Fortran: an optional sign, a sequence of digits that may contain a decimal point, and an optional exponent field that begins with one of the letters d, D, e or E, as in 1.23D-45. All arithmetic in AMPL is in the same precision (double precision on most machines), so all exponent notations are synonymous.

Literals are strings delimited either by single quotes ' or by double quotes "; the delimiting character must be doubled if it appears within the literal, as in 'x''y', which is a literal containing the three characters x'y. Newline characters may appear within a literal only if preceded by \. The choice of delimiter is arbitrary; 'abc' and "abc" denote the same literal.

Literals are distinct from numeric constants: 1 and '1' are unrelated.

Input is free form; white space (any sequence of space, tab or newline characters) may appear between any tokens. Each statement ends with a semicolon.

Comments begin with # and extend to the end of the current line, or are delimited by /* and */, in which case they may extend across several lines and do not nest. Comments may appear anywhere in declarations, commands, and data.

A.2 Set members

A set contains zero or more elements or members, each of which is an ordered list of one or more components. Each member of a set must be distinct. All members must have the same number of components; this common number is called the set's dimension.

A literal set is written as a comma-separated list of members, between braces { and }. If the set is one-dimensional, the members are simply numeric constants or literal strings, or any expressions that evaluate to numbers or strings:

```
{"a","b","c"}
{1,2,3,4,5,6,7,8,9}
{t,t+1,t+2}
```

For a multidimensional set, each member must be written as a parenthesized comma-separated list of the above:

```
{("a",2),("a",3),("b",5)}
{(1,2,3),(1,2,4),(1,2,5),(1,3,7),(1,4,6)}
```

The value of a numeric member is the result of rounding its decimal representation to a floating-point number. Numeric members that appear different but round to the same floating-point number, such as 1 and 0.01E2, are considered the same.

A.3 Indexing expressions and subscripts

Most entities in AMPL can be defined in collections indexed over a set; individual items are selected by appending a "subscript" to the name of the entity. The range of possible subscripts is indicated by an *indexing expression* in the entity's declaration. Reduction operators, such as sum, also use indexing expressions to specify sets over which operations are iterated.

A subscript is a list of symbolic or numeric expressions, separated by commas and enclosed in square brackets, as in supply[i] and cost[j,p[k]+1,"O+"]. Each subscripting expression must evaluate to a number or a literal. The resulting value or sequence of values must give a member of a relevant one-dimensional or multidimensional indexing set.

An indexing expression is a list of comma-separated set expressions, followed optionally by a colon and a logical "such that" expression, all enclosed in braces:

indexing:
　　{ *sexpr-list* }
　　{ *sexpr-list* : *lexpr* }

sexpr-list:
　　sexpr
　　dummy-member in *sexpr*
　　sexpr-list , *sexpr-list*

Each set expression may be preceded by a dummy member and the keyword in. A dummy member for a one-dimensional set is an unbound name, that is, a name not currently defined. A dummy member for a multidimensional set is a comma-separated list, enclosed in parentheses, of expressions or unbound names; the list must include at least one unbound name.

A dummy member introduces one or more dummy indices (the unbound names in its components), whose *scopes*, or ranges of definition, begin just after the following *sexpr*; an index's scope runs through the rest of the indexing expression, to the end of the declaration using the indexing expression, or to the end of the operand that uses the indexing expression. When a dummy member has one or more expression components, the dummy indices in the dummy member range over a *slice* of the set, i.e., they assume all values for which the dummy member is in the set.

```
{A}                      # a set
{A, B}                   # all pairs, one from A, one from B
{i in A, j in B}         # the same
{i in A, B}              # the same
{i in A, C[i]}           # all pairs, one from A, one from C[i]
{i in A, (j,k) in D}     # 1 from A and 1 (itself a pair) from D
{i in A: p[i] > 0}       # all i in A such that p[i] is positive
{i in A, j in C[i]: i <= j}  # i and j must be numeric
{i in A, (i,j) in D: i <= j} # all pairs with i in A and i,j in D
                         # (same value of i) and i <= j
```

The optional : *lexpr* in an indexing expression selects only the members that satisfy the logical expression and excludes the others. The *lexpr* typically involves one or more dummy indices of the indexing expression.

A.4 Expressions

Various items can be combined in AMPL's arithmetic and logical expressions. In the following summary, an expression that may contain variables is denoted *vexpr*. An expression that may not contain variables is denoted *expr* and is sometimes called a "constant expression", even though it may involve dummy indices. A logical expression, denoted *lexpr*, may contain variables only when it is part of an if expression that yields a *vexpr*. Set expressions are denoted *sexpr*.

Table A-1 summarizes the arithmetic, logical and set operators; the type column indicates whether the operator produces an arithmetic value (A), a logical value (L), or a set value (S).

Arithmetic expressions are formed from the usual arithmetic operators, built-in functions, and arithmetic reduction operators like sum:

expr:
    number
    variable
    expr arith-op expr   *arith-op* is + − less * / mod div ^ **
    unary expr   *unary* is + −
    built-in (*exprlist*)
    if *lexpr* then *vexpr* [else *vexpr*]
    reduction-op indexing expr   *reduction-op* is sum prod max min
    (*expr*)

Built-in functions are listed in Table A-2.

The arithmetic reduction operators are used in expressions like

```
sum {i in Prod} cost[i] * Make[i]
```

The scope of the *indexing* expression extends to the end of the *expr*. If the *expr* contains variables, the resulting expression is a *vexpr*. If the operation is over an empty set, the result is the identity value for the operation: 0 for sum, 1 for prod, Infinity for min, and −Infinity for max.

Logical expressions appear where a value of "true" or "false" is required: in check statements, the "such that" parts of indexing expressions (following the :), and in if *lexpr* then ...

Precedence	Name	Type	Remarks
1	if then else	A, S	A: if no else, then "else 0" assumed S: "else *sexpr*" required
2	or ||	L	
3	exists forall	L	logical reduction operators
4	and &&	L	
5	< <= = == <> != >= >	L	
6	in not in	L	membership in set
6	within not within	L	S within T means set S ⊆ set T
7	not !	L	logical negation
8	union diff symdiff	S	symdiff ≡ symmetric difference
9	inter	S	intersection
10	cross	S	cross or Cartesian product
11	setof .. by	S	set constructors
12	+ - less	A	a less $b \equiv \max(a - b, 0)$
13	sum prod min max	A	arithmetic reduction operators
14	* / div mod	A	div ≡ truncated quotient of integers
15	+ -	A	unary plus, unary minus
16	^ **	A	exponentiation

Operators are listed in increasing precedence order. Exponentiation and if-then-else are right-associative; the other operators are left-associative. The 'Type' column indicates result types: A for arithmetic, L for logical, S for set.

Table A-1: Arithmetic, logical and set operators.

else ... expressions. Numeric values that appear in any of these contexts are implicitly coerced to logical values: 0 is interpreted as false, and all other numeric values as true.

```
lexpr:
    expr
    expr compare-op expr        compare-op is < <= = == != <> > >=
    lexpr logic-op lexpr        logic-op is or || and &&
    not lexpr
    member in sexpr
    member not in sexpr
    sexpr within sexpr
    sexpr not within sexpr
    opname indexing lexpr       opname is exists or forall
    ( lexpr )
```

The in operator tests set membership. Its left operand is a potential set member, i.e., an expression or comma-separated list of expressions enclosed in parentheses, with the number of expressions equal to the dimension of the right operand, which must be a set expression. The within operator tests whether one set is contained in another. The two set operands must have the same dimension.

The logical reduction operators exists and forall are the iterated counterparts of or and and. When applied over an empty set, exists returns false and forall returns true.

Set expressions yield sets.

sexpr:
    { [*member* [, *member* ...]] }
    sexpr set-op sexpr   *set-op* is union diff symdiff inter cross
    opname indexing sexpr   *opname* is union or inter
    expr .. *expr* [by *expr*]
    setof *indexing member*
    if *lexpr* then *sexpr* else *sexpr*
    (*sexpr*)
    interval
    predefined-set
    indexing

Components of members can be arbitrary constant expressions. Section A.6.3 describes *interval*s and *predefined-set*s.

When used as binary operators, union and inter denote the binary set operations of union and intersection. These keywords may also be used as reduction operators.

The .. operator constructs sets. The default by clause is by 1. In general, e_1 .. e_2 by e_3 means the numbers

$$e_1, e_1 + e_3, e_1 + 2e_3, \ldots, e_1 + \left\lfloor \frac{e_2 - e_1}{e_3} \right\rfloor e_3$$

rounded to set members. (The notation $\lfloor x \rfloor$ denotes the floor of x, that is, the largest integer $\leq x$.)

The setof operator is a set construction operator; *member* is either an expression or a comma-separated list of expressions enclosed in parentheses. The resulting set consists of all the *member*s obtained by iterating over the *indexing* expression; the dimension of the resulting expression is the number of components in *member*.

```
ampl: set y := setof {i in 1..5} (i,i^2);
ampl: display y;
set y := (1,1) (2,4) (3,9) (4,16) (5,25);
```

A.4.1 Built-in functions

The built-in functions are listed in Table A-2. The function alias takes as its argument the name of a model entity and returns its *alias*, a literal value described in Section A.5. The functions round(x, n) and trunc(x, n) convert x to a decimal string and round or truncate it to n places past the decimal point (or to $-n$ places before the decimal point if $n < 0$); similarly, precision(x, n) rounds x to n significant decimal digits. For round and trunc, a missing n is taken as 0, thus providing the usual rounding or truncation to an integer.

In addition, several built-in random number generation functions are available, as listed in Table A-3. All are based on a uniform random number generator with a very long period. An initial seed n can be specified with the -sn command-line argument or option randseed; -s or option randseed ' ' instructs AMPL to choose and print a seed. Giving no -s argument is the same as specifying -s1.

Irand224 () returns an integer in the range $[0, 2^{24})$. Given the same seed, an expression of the form floor $(m*$Irand224 $()/n)$ will yield the same value on most computers when m and n are integer expressions of reasonable magnitude (i.e., $|n| < 2^{k-24}$ and $|m| < 2^k$, for machines that correctly compute k-bit floating-point integer products and quotients; $k \geq 47$ for most machines of current interest).

abs (x)	absolute value $\|x\|$
acos (x)	inverse cosine, $\cos^{-1}(x)$
acosh (x)	inverse hyperbolic cosine, $\cosh^{-1}(x)$
alias (v)	alias of model entity v
asin (x)	inverse sine, $\sin^{-1}(x)$
asinh (x)	inverse hyperbolic sine, $\sinh^{-1}(x)$
atan (x)	inverse tangent, $\tan^{-1}(x)$
atan2 (y, x)	inverse tangent, $\tan^{-1}(y/x)$
atanh (x)	inverse hyperbolic tangent, $\tanh^{-1}(x)$
ceil (x)	ceiling of x (next higher integer)
cos (x)	cosine
exp (x)	e^x
floor (x)	floor of x (next lower integer)
log (x)	$\log_e(x)$
log10 (x)	$\log_{10}(x)$
max $(x, y, ...)$	maximum (2 or more arguments)
min $(x, y, ...)$	minimum (2 or more arguments)
precision (x, n)	x rounded to n significant decimal digits
round (x, n)	x rounded to n digits past decimal point
round (x)	x rounded to an integer
sin (x)	sine
sinh (x)	hyperbolic sine
sqrt (x)	square root
tan (x)	tangent
tanh (x)	hyperbolic tangent
trunc (x, n)	x truncated to n digits past decimal point
trunc (x)	x truncated to an integer

Table A-2: Built-in arithmetic and string functions.

Finally, there are functions that operate on sets, for instance, card and next; these are described in Section A.6.

A.4.2 Piecewise-linear terms

Piecewise-linear terms have one of the following forms:

```
<< brkpts ; slopes >> var
<< brkpts ; slopes >> (expr)
<< brkpts ; slopes >> (var, expr)
```

where *brkpts* is a list of breakpoints and *slopes* a list of slopes. Each such list is a comma-separated list of *expr*'s, each optionally preceded by an indexing expression (whose scope covers just the *expr*). The indexing expression must specify a set that is manifestly ordered (see A.6.2), or it can be of the special form

```
{if lexpr}
```

which causes the *expr* to be omitted if the *lexpr* is false.

After the lists of slopes and breakpoints are extended (by indexing over any indexing expressions), the number of slopes must be one more than the number of breakpoints, and the breakpoints

Beta (a, b)	$density(x) = x^{a-1}(1-x)^{b-1}/(\Gamma(a)\Gamma(b)/\Gamma(a+b))$, x in $[0, 1]$
Cauchy()	$density(x) = 1/(\pi(1+x^2))$
Exponential()	$density(x) = e^{-x}$, $x > 0$
Gamma (a)	$density(x) = x^{a-1}e^{-x} / \Gamma(a)$, $x \geq 0$, $a > 0$
Irand224()	integer uniform on $[0, 2^{24})$
Normal (μ, σ)	normal distribution with mean μ, variance σ
Normal01()	normal distribution with mean 0, variance 1
Poisson (μ)	$probability(k) = e^{-\mu}\mu^k/k!$, $k = 0, 1, ...$
Uniform (m, n)	uniform on $[m, n)$
Uniform01()	uniform on $[0, 1)$

Table A-3: Built-in random number generation functions.

must be in non-decreasing numeric order. (There is no requirement on the order of the slopes.) AMPL interprets the result as the piecewise-linear function $f(x)$ defined as follows. Let σ_j, $1 \leq j \leq n$, and b_i, $1 \leq i \leq n - 1$, denote the slopes and breakpoints, respectively, and let $b_0 = -\infty$ and $b_n = +\infty$. Then $f(0) = 0$, and for $b_{i-1} \leq x \leq b_i$, f has slope s_i, i.e., $f'(x) = s_i$. For the forms having just one argument (either a variable *var* or a constant expression *expr*), the result is $f(var)$ or $f(expr)$. The form with two operands is interpreted as $f(var) - f(expr)$. This adds a constant that makes the result vanish when the *var* equals the *expr*.

When piecewise-linear terms appear in an otherwise linear constraint or objective, AMPL collects two or more piecewise-linear terms involving the same variable into a single term.

A.5 Declarations of model entities

Declarations of model entities have the following common form:

> *entity   name   alias*_{opt}   *indexing*_{opt}   *body*_{opt}   ;

where *name* is an alphanumeric name that has not previously been assigned to an entity by a declaration, *alias* is an optional literal, *indexing* is an optional indexing expression, and *entity* is one of the keywords

```
set
param
var
arc
minimize
maximize
subject to
node
```

The *entity* may be omitted, in which case `subject to` is assumed. The *body* of various declarations consists of other, mostly optional, phrases that follow the initial part. Each declaration ends with a semicolon.

Declarations may appear in any order, except that each name must be declared before it is used.

As with piecewise-linear terms, a special form of indexing expression is allowed for variable, constraint, and objective declarations:

    {if *lexpr*}

If the logical expression *lexpr* is true, then a simple (unsubscripted) entity results; otherwise the entity is excluded from the model, and subsequent attempts to reference it cause an error message. For example, this declaration includes the variable Test in the model if the parameter testing has been given a value greater than 100:

```
param testing;
var Test {if testing > 100} >= 0;
```

A.6 Set declarations

A set declaration has the form

> *set declaration*:
>   set *name alias*_{opt} *indexing*_{opt} *attribs*_{opt} ;

in which *attribs* is a list of attributes optionally separated by commas:

> *attrib*:
>   dimen *n*
>   within *sexpr*
>   := *sexpr*
>   default *sexpr*

The dimension of the set is either the constant positive integer *n*, or the dimension of *sexpr*, or 1 by default. The phrase within *sexpr* requires the set being declared to be a subset of *sexpr*. Several within phrases may be given. The := phrase specifies a value for the set; it implies that the set will not be later given a value in a data statement (A.12.1). The default phrase specifies a default value for the set, to be used if no value is given in a data statement. The := and default phrases are mutually exclusive. If neither is given and the set is not defined by a data statement, references to the set during model generation cause an error message.

The *sexpr* in a := or default phrase can be { }, the empty set, which then has the dimension implied by any dimen or within phrases in the declaration, or 1 if none is present. In other contexts, { } denotes the empty set of dimension 1.

Recursive definitions of indexed sets are allowed, so long as the assigned values can be computed in a sequence that only references previously computed values. For example,

```
set nodes;
set arcs within nodes cross nodes;

param max_iter := card(nodes)-1; # card(s) = number of elements in s

set step {s in 1..max_iter} dimen 2 :=
    if s == 1
        then arcs
        else step[s-1] union setof {k in nodes,
                (i,k) in step[s-1], (k,j) in step[s-1]} (i,j);

set reach := step[max_iter];
```

computes in set reach the transitive closure of the graph represented by nodes and arcs.

A.6.1 Cardinality function

The function `card` operates on any finite set: `card` (*sexpr*) returns the number of members in *sexpr*. If *sexpr* is an indexing expression, the parentheses may be omitted. For example,

```
card({i in A: x[i] >= 4})
```

may also be written

```
card{i in A: x[i] >= 4}
```

A.6.2 Ordered sets

A named one-dimensional set may have an order associated with it. Declarations for ordered sets include one of the phrases

```
ordered [ by [ reversed ] sexpr ]
circular [ by [ reversed ] sexpr ]
```

The keyword `circular` indicates that the set is ordered and wraps around, i.e., its first member is the successor of its last, and its last member is the predecessor of its first.

Sets of dimension two or higher may not be `ordered` or `circular`.

If set S is `ordered` by T or `circular` by T, then set T must itself be an ordered set that contains S, and S inherits the order of T. If the ordering phrase is by `reversed` T, then S still inherits its order from T, but in reverse order.

If S is `ordered` or `circular` and no ordering by *sexpr* is specified, then one of two cases applies. If the members of S are explicitly specified by a `:= {` *member-list* `}` expression in the model or by a list of members in a data statement, S adopts the ordering of this list. If S is otherwise computed from an assigned or default *sexpr* in the model, AMPL will retain the order of manifestly ordered sets (explained below) and is otherwise free to pick an arbitrary order.

Functions of ordered sets are summarized in Table A-4.

If S is an expression for an ordered set of n members, e is the jth member of S, and k is an integer expression, then next (e, S, k) is member $j + k$ of S if $1 \le j + k \le n$, and an error otherwise. If S is `circular`, then next (e, S, k) is member $1 + ((j + k - 1) \bmod n)$ of S. The function `nextw` (`next` with wraparound) is the same as `next`, except that it treats all ordered sets as `circular`; prev $(e, S, k) \equiv$ next $(e, S, -k)$, and prevw $(e, S, k) \equiv$ nextw $(e, S, -k)$.

Several abbreviations are allowed. If k is not given, it is taken as 1. If both k and S are omitted, then e must be a dummy index in the scope of an indexing expression that runs over S, as in $\{e$ in $S\}$.

Five other functions apply to ordered sets: first (S) returns the first member of S, last (S) the last, member (j, S) returns the jth member of S, ord (e, S) and ord0 (e, S) the ordinal position of member e in S. If e is not in S, then ord0 returns 0, whereas ord complains of an error. Again, ord(e) = ord(e, S) and ord0(e) = ord0(e, S) if e is a dummy index in the scope of an indexing expression for S.

Some sets are manifestly ordered, such as arithmetic progressions, intervals, subsets of ordered sets, `if` expressions whose `then` and `else` clauses are both ordered sets, and set differences (but not symmetric differences) whose left operands are ordered.

`card` (S)	number of members in set S
`next` (e, S, k)	member k positions after member e in set S
`next` (e, S)	same, with $k = 1$
`next` (e)	next member of set for which e is dummy index
`nextw` (e, S, k)	member k positions after member e in set S, wrapping around
`nextw` (e, S)	wrapping version of `next`(e, S)
`nextw` (e)	wrapping version of `next`(e)
`prev` (e, S, k)	member k positions before member e in set S
`prev` (e, S)	same, with $k = 1$
`prev` (e)	previous member of set for which e is dummy index
`prevw` (e, S, k)	member k positions before member e in set S, wrapping around
`prevw` (e, S)	wrapping version of `prev`(e, S)
`prevw` (e)	wrapping version of `prev`(e)
`first` (S)	first member of S
`last` (S)	last member of S
`member` (j, S)	jth member of S; $1 \leq j \leq$ `card`(S), j integer
`ord` (e, S)	ordinal position of member e in S
`ord` (e)	ordinal position of member e in set for which it is dummy index
`ord0` (e, S)	ordinal position of member e in S; 0 if not present
`ord0` (e)	same as `ord`(e)

Table A-4: Functions of sets.

```
set U;                          # not ordered
set A := 1..n;                  # ordered (arithmetic progression)
set B := i..j by k;             # ordered (arithmetic progression)
set C ordered;                  # data order retained
set D := {i in C: x[i] in U};   # ordered subset
set E := D diff U;              # ordered
set T := U union D;             # not ordered
set V := U diff D;             # not ordered
set W := D symdiff U;          # not ordered
```

A.6.3 Intervals and other infinite sets

For convenience in specifying ordered sets and prototypes for imported functions (Section A.14), there are several kinds of infinite sets. Iterating over infinite sets is forbidden, but one can check membership in them and use them to specify set orderings. The most natural infinite set is the interval, which may be closed, open, or half-open and follows standard mathematical notation. There are intervals of real (floating-point) numbers and of integers, introduced by the keywords `interval` and `integer` respectively:

interval:
```
interval [a, b] ≡ {x: a ≤ x ≤ b},
interval (a, b] ≡ {x: a < x ≤ b},
interval [a, b) ≡ {x: a ≤ x < b},
interval (a, b) ≡ {x: a < x < b},
integer [a, b] ≡ {x: a ≤ x ≤ b and x ∈ ℑ},
integer (a, b] ≡ {x: a < x ≤ b and x ∈ ℑ},
integer [a, b) ≡ {x: a ≤ x < b and x ∈ ℑ},
integer (a, b) ≡ {x: a < x < b and x ∈ ℑ}
```

where a and b denote arbitrary arithmetic expressions, and $\Im$ denotes the set of integers. In function prototypes (A.14) and the declaration phrases

```
in interval
within interval
ordered by [ reversed ] interval
circular by [ reversed ] interval
```

the keyword `interval` may be omitted.

The predefined infinite sets `Reals` and `Integers` are the sets of all floating-point numbers and integers, respectively, in numeric order. The predefined infinite sets `ASCII`, `EBCDIC`, and `Display` all represent the universal set of strings and numbers from which members of any one-dimensional set are drawn; `ASCII` is ordered by the ASCII collating sequence, `EBCDIC` by the EBCDIC collating sequence, and `Display` by the ordering used in the `display` command (Section A.13). Numbers precede literals in `Display` and are ordered numerically; literals are sorted by the ASCII collating sequence. For example:

```
ampl: set s := {'c', 'b', 'a', 10, 1, 1.5, 2};
ampl: set t ordered by Display := s;
ampl: display t;
set t := 1 1.5 2 10 a b c;

ampl: set u ordered by ASCII := s;
ampl: display u;
set u := 1 1.5 10 2 a b c;
```

A.7 Parameter declarations

Parameter declarations have a list of optional attributes, optionally separated by commas:

parameter declaration:
       `param` *name* *alias*$_{opt}$ *indexing*$_{opt}$ *attributes*$_{opt}$ `;`

attribute:
    `binary`
    `integer`
    `symbolic`
    relop expr
    `in` *sexpr*
    `:=` *expr*
    `default` *expr*

relop:
    `<   <=   =   ==   !=   <>   >   >=`

The keyword `integer` restricts the parameter to be an integer; `binary` restricts it to 0 or 1. If `symbolic` is specified, then the parameter may assume any literal or numeric value (rounded as for set membership), and the attributes involving <, <=, >= and > are disallowed; otherwise the parameter is numeric and can only assume a numeric value.

The attributes involving comparison operators specify that the parameter must obey the given relation. The `in` attribute specifies a check that the parameter lies in the given set.

The `:=` and `default` attributes are analogous to the corresponding ones in set declarations, and are mutually exclusive.

Recursive definitions of indexed parameters are allowed, so long as the assigned values can be computed in a sequence that only references previously computed values. For example,

```
param comb 'n choose k' {n in 0..N, k in 0..n}
    := if k = 0 or k = n then 1 else comb[n-1,k-1] + comb[n-1,k];
```

computes the number of ways of choosing n things k at a time.

In a recursive definition of a symbolic parameter, the keyword `symbolic` must precede all references to the parameter.

A.7.1 Check statements

Check statements supply assertions to help verify that correct data have been read or generated; they have the syntax

check *[indexing_{opt} :] lexpr* ;

Each `check` is evaluated after all data values have been read.

A.7.2 Infinity

`Infinity` is a predefined parameter; it is the threshold above which upper bounds are considered absent (i.e., infinite), and `-Infinity` is similarly the threshold below which lower bounds are considered absent. Thus given

```
set A;
param Ub{A} default  Infinity;
param Lb{A} default -Infinity;
var x {i in A} >= Lb[i], <= Ub[i];
```

components of `x` for which no `Lb` value is given in a data section are unbounded below and components for which no `Ub` value is given are unbounded above. One can similarly arrange for optional lower and upper constraint bounds.

A.8 Variable declarations

Variable declarations begin with the keyword `var`:

variable declaration:
        `var` *name alias_{opt} indexing_{opt} attributes_{opt}* ;

Optional attributes of variable declarations may be separated by commas; these attributes include

attribute:
```
    binary
    integer
    >= expr
    <= expr
    := expr
    default expr
    = vexpr
    coeff indexing_opt  constraint vexpr
    cover indexing_opt  constraint
    obj indexing_opt  objective vexpr
```

As with parameters, `integer` restricts the variable to integer values, and `binary` restricts it to 0 or 1. The `>=` and `<=` phrases specify bounds, and the `:=` phrase an initial value. The `default` phrase specifies a default for initial values that may be provided in a data statement (A.12.2); `default` and `:=` are mutually exclusive. The `= vexpr` phrase is allowed only if none of the previous attributes appears; it makes the variable a *defined variable* (A.15).

The `coeff` and `obj` phrases are for columnwise coefficient generation; they specify coefficients to be placed in the named constraint or objective, which must have been previously declared using the placeholder `to_come` (see A.9 and A.10). The scope of *indexing* is limited to the phrase, and may have the special form

```
    {if lexpr}
```

which contributes the coefficient only if the *lexpr* is true. A `cover` phrase is equivalent to a `coeff` phrase in which the *vexpr* is 1.

Arcs are special network variables, declared with the keyword `arc` instead of `var`. They may contribute coefficients to special `node` constraints (A.9) via optional attribute phrases of the form

```
    from indexing_opt  node vexpr_opt
    to indexing_opt  node vexpr_opt
```

These phrases are analogous in syntax to the `coeff` phrase, except that the final *vexpr* is optional; omitting it is the same as specifying 1.

A.9 Constraint declarations

The form of a constraint declaration is

constraint declaration:
```
    [ subject to ] name alias_opt indexing_opt
        [ := initial_dual ] [ default initial_dual ] [ : constraint expression ] ;
```

The keyword `subject to` in a constraint declaration may be abbreviated to `subj to` or `s.t.` or may be omitted entirely; all are synonyms. The optional `:=` *initial_dual* specifies an initial guess for the dual variable (Lagrange multiplier) associated with the constraint. Again, `default` and `:=` clauses are mutually exclusive, and `default` is for initial values not given in a data statement. Constraint declarations must specify a constraint in one of the following forms:

constraint expression:
     vexpr <= *vexpr*
     vexpr =   *vexpr*
     vexpr >= *vexpr*
     expr <= *vexpr* <= *expr*
     expr >= *vexpr* >= *expr*

To enable columnwise coefficient generation for the constraint, one of the *vexpr*s may have one of the following forms:

```
to_come + vexpr
vexpr + to_come
to_come
```

Terms for this constraint that are specified in a `var` declaration (A.8) are placed at the location of `to_come`.

Nodes are special constraints that may send flow to or receive flow from arcs. Their declarations begin with the keyword `node` instead of `subject to`. Pure transshipment nodes do not have a constraint body; they must have "flow in" equal to "flow out". Other nodes are sources or sinks; they specify constraints in one of the forms above, except that they may not mention `to_come`, and exactly one *vexpr* must have one of the following forms:

```
net_in + vexpr
net_out + vexpr
vexpr + net_in
vexpr + net_out
net_in
net_out
```

The keyword `net_in` is replaced by an expression representing the net flow into the node: the terms contributed to the node constraint by `to` phrases of `arc` declarations (A.8), minus the terms contributed by `from` phrases. The treatment of `net_out` is analogous; it is the negative of `net_in`.

A.10 Objective declarations

The declaration of an objective is one of

objective declaration:
     `maximize` *name alias*$_{opt}$ *indexing*$_{opt}$ [: *expression*] ;
     `minimize` *name alias*$_{opt}$ *indexing*$_{opt}$ [: *expression*] ;

and may specify an expression in one of the following forms:

expression:
     vexpr
     `to_come` + *vexpr*
     vexpr + `to_come`
     `to_come`

The `to_come` forms permit columnwise coefficient generation, as for constraints (A.9). Specifying none of the above expressions is the same as specifying ": `to_come`".

A.11 Suffix notation for auxiliary values

Variables, constraints, and objectives have a variety of associated auxiliary values. For example, variables have bounds and reduced costs, and constraints have dual values and slacks. These values are accessed as *name* . *suffix*, where *name* is a simple or subscripted variable, constraint, or objective name, and . *suffix* is one of the possibilities listed in Tables A-5, A-6, and A-7.

For a constraint, the .body, .lb, and .ub values correspond to a modified form of the constraint. If the constraint involves a single inequality, subtract the right-hand side from the left, then move any constant terms to the right-hand side; if the constraint involves a double inequality, similarly subtract any constant terms in the middle from all three expressions (left, middle, right). Then the constraint has the form $lb \le body \le ub$, where lb and ub are (possibly infinite) constants.

The following rules determine lower and upper dual values (c .ldual and c .udual) for a constraint c. The solver returns a single dual value, c .dual, which might apply either to $body \ge lb$ or to $body \le ub$. For an equality constraint ($lb = ub$), AMPL uses the sign of c .dual to decide. For a minimization problem, c .dual > 0 implies that the same optimum would be found if the constraint were $body \ge lb$, so AMPL sets c .ldual = c .dual and c .udual = 0; similarly, c .dual < 0 implies that c .ldual = 0 and c .udual = c .dual. For a maximization problem, the inequalities are reversed.

For inequality constraints ($lb < ub$), AMPL uses nearness to bound to decide whether c .ldual or c .udual equals c .dual. If $body - lb < ub - body$, then c .ldual = c .dual and c .udual = 0; otherwise, c .ldual = 0 and c .udual = c .dual.

Model declarations may reference any of the suffixed values described in Tables A-5, A-6 and A-7. This is most often useful in new declarations that are introduced after one model has already been translated and solved. In particular, the suffixes .val and .dual are provided so that new constraints can refer to current optimal values of the primal and dual variables and of the objective.

A.12 Standard data format

AMPL supports a standard format to describe the sets and parameter values that it combines with a model to yield a specific optimization problem.

A data section consists of a sequence of tokens (literals and strings of printing characters) separated by white space (spaces, tabs, and newlines). Tokens include keywords, literals, numbers, and the delimiters ( ) [ ] : , ; := *. A statement is a sequence of tokens terminated by a semicolon. Comments may appear as in model mode. In all cases, arrangement of data into neat rows and columns is for human readability; AMPL ignores formatting.

A data section begins with a data command and ends with end-of-input or with a command that returns to model mode (A.13).

In a data section, model entities may be assigned values in any convenient order, independent of their order of declaration.

A.12.1 Set data

Statements defining sets consist of the keyword set, the set name, an optional :=, and the members. A one-dimensional set is most easily specified by listing its members, optionally separated by commas. The single or double quotes of a literal string may be omitted if the string is alphanumeric but does not specify a number:

`.init`	current initial guess
`.init0`	initial initial guess (set by `:=`, data, or `default`)
`.lb`	current lower bound
`.lb0`	initial lower bound
`.lb1`	weaker lower bound from presolve
`.lb2`	stronger lower bound from presolve
`.lrc`	lower reduced cost (for `var >= lb`)
`.lslack`	lower slack (`val - lb`)
`.rc`	reduced cost
`.slack`	`min(lslack, uslack)`
`.ub`	current upper bound
`.ub0`	initial upper bound
`.ub1`	weaker upper bound from presolve
`.ub2`	stronger upper bound from presolve
`.urc`	upper reduced cost (for `var <= ub`)
`.uslack`	upper slack (`ub - val`)
`.val`	current value

Table A-5: Dot suffixes for variables.

`.body`	current value of constraint *body*
`.dinit`	current initial guess for dual variable
`.dinit0`	initial initial guess for dual variable (set by `:=`, data, or `default`)
`.dual`	current dual variable
`.lb`	lower bound
`.lbs`	lb for solver (adjusted for fixed variables)
`.ldual`	lower dual value (for `body >= lb`)
`.lslack`	lower slack (`body - lb`)
`.slack`	`min(lslack, uslack)`
`.ub`	upper bound
`.ubs`	ub for solver (adjusted for fixed variables)
`.udual`	upper dual value (for `body <= ub`)
`.uslack`	upper slack (`ub - body`)

Table A-6: Dot suffixes for constraints.

`.val`	current value

Table A-7: Dot suffix for objectives.

```
data;
set S := 'a' "b" c;              # equivalent to set S := a b c;
set T := 2B2, 3C3, '4D4';        # quotes needed; 4D4 is numeric
set U := 11 "11";                # two different members
set Q := '''' """";              # equivalent to set Q := "'" '"'
set P := DDT '2,4-D' '2,4,5-T';  # quotes needed; comma not alphabetic
set Spread := 'Lotus 1-2-3' 'Microsoft Excel 4.0';  # quotes needed
```

To define a compound set, one can list all members. Each member is a parenthesized, comma-separated list of components, and successive members have an optional comma between them:

```
set A := (b,y) (c,x) (c,z), (d,y), (d,z);
```

Alternatively, one can describe the members of a two-dimensional set by a table, as in

```
set A :=
        :    x    y    z    :=
        b    -    +    -
        c    +    -    +
        d    -    +    +    ;
```

In such a table, the row labels are for the first subscript, the columns for the second; "+" stands for a pair that occurs in the set, and "-" for a pair that does not. The colon introduces a table, and is mandatory in this context. If (tr) precedes the colon, the table is transposed, interchanging the roles of rows and columns.

For sets of dimension greater than two, one can give tables along slices. For instance, the definitions

```
set ABC := (a,p,x) (a,p,z) (a,q,x) (a,q,y) (b,p,y) (b,p,z) (b,q,z);
```

and

```
set ABC :=
     (a,*,*)    :    x    y    z :=
                p    +    -    +
                q    +    +    -
     (b,*,*)    :    y    z :=
                p    +    +
                q    -    +  ;
```

both specify the same set. It is not necessary to give a column labeled x in the slice (b,*,*), since there are no members of the form (b,*,x) in ABC.

The examples above illustrate two-dimensional (2D) tables, in which each row describes several items. A *template* for such a table, for example (a,*,*), contains two *'s, of which the first is ordinarily replaced by an entry's row label, the second by the entry's column label; the roles of the row and column labels are interchanged when (tr) immediately precedes (or replaces) the : at the beginning of the table. The default template for the first 2D table in a set statement is (*,*).

In a one-dimensional (1D) table each row describes a single member. Templates for 1D tables may contain any number of *'s; in a 1D table, the first entry in a row replaces the template's first *, the second entry the second *, and so on. Using 1D tables, we can express the set ABC in many equivalent ways, such as

```
set ABC :=
     (a,*,x) p q
     (*,*,y) a q  b p
     (a,p,z)
     (b,*,z) p q ;
set ABC :=
     (a,p,*) x z (a,q,*) x y
     (b,p,*) y z (b,q,z) ;
```

```
set ABC :=
    (*,*,*)          # default for set of dimen 3: could be omitted
     a p x   a p z   a q x   a q y
     b p y   b p z   b q z ;
```

One can also mix 1D and 2D tables:

```
set ABC :=
    (a,*,*)   :    x    y    z :=
                p    +    -    +
                q    +    +    -
    (b,p,y) (b,*,z) p q ;
```

In general, a `set` statement involves a sequence of 1D and 2D set tables. 1D tables start with either a new template (after which a `:=` is optional) or with `:=` alone, in which case the previous template is retained. The default (initial) template is `(*,...,*)`, that is, as many `*`'s as the set's dimension. 2D tables start with an optional new template, followed by `:` or `(tr)` (and an optional `:`), followed by a list of column labels and a `:=`. Templates containing no `*`'s stand for themselves.

For indexed sets, each component set must be given in a separate data statement:

```
set PROD;
set AREA {PROD};

set PROD := bands coils ;
set AREA[bands] := east north ;
set AREA[coils] := east west export ;
```

It is not necessary to specify subset members in the same order as in their parent set.

A.12.2 Parameter data

Statements defining parameters are analogous in most respects to statements defining sets. They begin with the keyword `param`, followed by the name of the parameter. For simple parameters, an optional `:=` and the parameter value complete the statement, as in

```
param T := 4;
```

Parameters indexed by sets have values presented in a sequence of 1D and 2D tables similar to those for `set` statements. The simplest case is a parameter indexed by a simple set:

```
param init_stock := iron 7.32 nickel 35.8;
```

where the data are a sequence of name-value pairs. This can be punctuated with commas and white space (any sequence of space, tab and newline characters) for readability.

Several parameters indexed by the same set may be given in a single table:

```
param:    init_stock  cost    value   :=
   iron        7.32   .025    -.01
   nickel      35.8   .03      .02 ;
```

In this form of the `param` statement, the parameter names are the column labels that follow the colon; the row names are the subscripts. Parameters with two or more subscripts may also be given values by such a table; all of the parameters must have the same number of subscripts, and each line of the table must start with that number of subscripts.

A variation on this form allows defining a set while defining parameters indexed over the set; the name of the set comes between colons immediately after `param`:

```
param    :raw:    init_stock    cost    value  :=
         iron        7.32       .025     -.01
         nickel     35.8        .03      .02 ;
```

The set named between the colons may have dimension n greater than 1, in which case the first n items in each row of the table specify the set member and subscript for the remaining items.

Two-dimensional parameters are defined analogously to two-dimensional sets: a name, `:=`, a list of columns, a colon, and a sequence of rows. For example, this defines a 2D set of parameters `profit`:

```
param profit :=
        :        1       2       3       4      :=
        nuts    1.73    1.8     1.6     2.2
        bolts   1.82    1.9     1.7     2.5
        washers 1.05    1.1     .95     1.33 ;
```

Such a table may be input in transposed form; for instance, the definition of `profit` is equivalent to

```
param profit :=
        (tr)     nuts    bolts   washers :=
        1        1.73    1.82    1.05
        2        1.8     1.9     1.1
        3        1.6     1.7     .95
        4        2.2     2.5     1.33    ;
```

As in `set` statements, the notation `(tr)` means *transpose*; it implies a 2D table, and a `:` after it is optional.

A table may be given in several chunks of columns. For example, a data set might include

```
param cf75 :=
      :          CAN_260    CAN_310   CAN_335   AMM_SULF    UREA :=
ALEXANDRIA         0           0        5.0        3.0       1.0
BEHERA            1.0          0       25.0       90.0      35.0
GHARBIA            0           0       17.0       60.0      28.0

      :          SSP_155    C_250_55  C_300_100    DAP :=
ALEXANDRIA        8.0          0         0          0
BEHERA           64.0         1.0        .1         .1
GHARBIA          57.0         1.0        .2         .1 ;
```

In this context, a `:` by itself signifies the start of another series of columns (i.e., of another 2D table with the current template), which is followed by another `:=`.

A data item may be be specified as ``.'' rather than an explicit value. This is normally taken as a missing value, and a reference to it in the model would cause an error message. If there is a `default` value, however, each missing value is determined from that default. A default value may be specified either through a default phrase in a parameter's declaration in the model (A.7), or from an optional phrase

```
default r
```

that follows the parameter's name in the data statement. In the latter case, r must be a numeric constant:

```
param cf75  default 0.0   :=
     :          CAN_260    CAN_335    AMM_SULF     UREA :=
ALEXANDRIA         .          5.0        3.0        1.0
BEHERA           1.0         25.0       90.0       35.0
GHARBIA            .         17.0       60.0       28.0

     :          SSP_155    C_250_55   C_300_100    DAP :=
ALEXANDRIA       8.0          .          .          .
BEHERA          64.0         1.0        .1         .1
GHARBIA         57.0         1.0        .2         .1 ;
```

Default-value symbols may appear in both 1D and 2D tables. The `defaultsym` statement (which is recognized only in a data section) changes the default-value symbol to an arbitrary symbol or quoted string. For instance,

```
defaultsym --;
```

changes the default-value symbol to ``--''. The

```
nodefaultsym;
```

statement arranges that no default-value symbol is recognized. A stack of default-value symbols is maintained, with the current symbol at the top. The `defaultsym` statement pushes a new symbol onto the stack, and `nodefaultsym` pushes a "no symbol" indicator onto the stack. The statement

```
defaultsym;
```

(without a symbol) pops the stack.

Parameters having three or more indices may be given values by a sequence of 1D and 2D tables, one or more for each slice of nondefault values. Templates for `param` statements are indicated with square brackets and `*`'s, as in `[a,*,*]` or `[*,MIC_C,*]`. The default initial 1D template is `[*,…,*]` (as many `*`'s as the dimen of the parameter's indexing set). In 2D tables, the default that the first `*` corresponds to rows of the table can be overridden by a `(tr)` before (or instead of) the : before the column names. Suppose, for example, that parameter `sales` is declared by the AMPL statement

```
param sales{1..N, color, flavor};
```

A corresponding data set might include

```
param N := 3;

set color := red green blue yellow;
set flavor := vanilla chocolate;

param sales :=

[*,*,vanilla]:
                red     green   blue    yellow  :=
        1       1.33    2.42    9.87    7.2
        2       4.3     3.8     4.27    6.3
        3       -.2     2       1.5     2.1

[*,*,chocolate]:
                red     green   blue    yellow  :=
        1       3.7     -5.2    6.8     7.6
        2       2.9     4       4.1     3.6
        3       7.7     8.8     9.9     0.1     ;
```

In summary, a `param` statement defining one indexed parameter starts with the keyword `param` and the name of the parameter, followed by an optional default value and an optional `:=`. Then comes a sequence of 1D and 2D `param` tables, which are similar to 1D and 2D `set` tables, except that templates involve square brackets rather than parentheses, 2D tables contain numbers (or, for a symbolic parameter, literals) rather than +'s and -'s, and 1D tables corresponding to a template of k *'s contain $k + 1$ rather than k columns, the last being a column of numbers or default symbols (. 's). A special form, the keyword `param`, an optional default value, and a single (untransposed) 2D table, defines several parameters indexed by a common set, and another special form, `param` followed by the parameter name, an optional `:=`, and a numeric value, defines a scalar parameter.

Variable and constraint names may appear in data statements anywhere that a parameter name may appear, to specify initial values for variables and for the dual variables associated with constraints. The rules for default values are the same as for parameters. The keyword `var` is a synonym for `param` in data statements.

A.13 Command language

The invocation of AMPL normally causes it to enter a command environment, where the commands described below can be entered interactively. The model declarations and data statements described above are also accepted as commands. Depending on the operating system under which AMPL is run, the invocation may be accompanied by one or more *command-line arguments* that affect AMPL's initial state; the details are explained in the system-specific documentation.

The AMPL command environment recognizes two modes. In *model* mode, where it starts upon invocation, it recognizes model declarations (A.5) and all of the commands described below. Upon seeing a `data` statement, it switches to *data* mode, where only data statements (A.12) are recognized. It returns to model mode upon encountering a keyword that cannot begin a data statement.

A phrase of the form

        include *filename*

causes the indicated file to be interpolated. Here, and in subsequent contexts where a *filename* appears, if *filename* involves semicolons, quotes, or non-printing characters, it must be given as a literal, i.e., '*filename*' or "*filename*". `include` commands may be nested; they are recognized in both model and data mode. The sequences

        model; include *filename*
        data; include *filename*

may be abbreviated

        model *filename*
        data *filename*

AMPL recognizes the commands listed in Table A-8. Commands are not part of a model, but cause AMPL to act as described below. Commands other than `data`, `end`, `include`, `quit` and `shell` cause AMPL to enter model mode.

close	close a file
data	switch to data mode; optionally include file contents
display	print model entities and expressions
drop	drop a constraint or objective
end	end input from current input file
fix	freeze a variable at its current value
include	include file contents
let	change data values
model	switch to model mode; optionally include file contents
objective	select an objective to be optimized
option	set or display option values
print	print model entities unformatted
printf	print model entities formatted
quit	terminate AMPL
reset	reset specified entities to their initial state
restore	undo a drop command
shell	temporary escape to operating system to run commands
solution	import variable values as if from a solver
solve	send current instance to a solver and retrieve solution
update	allow updating data
unfix	undo a fix command
write	write out a problem instance

Table A-8: Commands.

A.13.1 Printing commands: display, print and printf

The display, print, and printf commands print arbitrary expressions. They have the forms

```
display  [ indexing: ] disparglist redirection opt ;
print    [ indexing: ] arglist redirection opt ;
printf   [ indexing: ] fmt , arglist redirection opt ;
```

If *indexing* is present, its scope extends to the end of the command, and it causes one execution of the command for each member in the *indexing* set. The format string *fmt* is like a printf format string in the C programming language and is explained more fully below.

An *arglist* is a (possibly empty, comma-separated) list of expressions and *iterated-arglist*s; an *iterated-arglist* has one of the forms

indexing expr
indexing (*nonempty-arglist*)

where *expr* is an arbitrary expression. The *expr*s can also involve simple or subscripted variable, constraint, and objective names; a constraint name represents the constraint's current dual value. *Disparglist* is described below.

The optional *redirection* has one of the forms

>filename
>>filename

%d	signed decimal notation
%u	unsigned decimal notation
%o	unsigned octal notation, without leading 0
%x	unsigned hexadecimal, using abcdef, without leading 0x
%X	unsigned hexadecimal, using ABCDEF, without leading 0X
%c	single character
%s	string
%f	double-precision floating-point
%e	double-precision floating-point, exponential notation using e
%E	double-precision floating-point, exponential notation using E
%g	double-precision floating-point, using %f or %e
%G	double-precision floating-point, using %f or %E
%%	literal %

Table A-9: Conversion specifications in printf formats.

The file is opened the first time a printing command specifies *filename* in a *redirection*; the first form of *redirection* causes the file to be overwritten, while the second form causes output to be appended to the current contents. Once open, *filename* remains open until a reset or unless explicitly closed by a close command:

    close *filename* ;

As long as *filename* remains open, either form of *redirection* causes output to be appended to the file's current contents.

The print command prints the items in its *arglist* on one line, separated by spaces; the separator may be changed with option print_separator. Literals are quoted only if they would have to be quoted in data mode. By default, numeric expressions are printed to full precision, but this can be changed with option print_precision or option print_round, as described below.

The printf command prints the items in its *arglist* as dictated by its format string *fmt*. It behaves like the printf function in C or AWK. Most characters in the format string are printed verbatim. Conversion specifications are an exception. They start with a % and end with a format letter, as summarized in Table A-9. Between the % and the format letter there may be any of −, for left-justification; +, which forces a sign; 0, to pad with leading zeros; a minimum field width; a period; and a precision giving the maximum number of characters to be printed from a string or digits to be printed after the decimal point for %f and %e or significant digits for %g or minimum number of digits for %d. Field widths and precisions are either decimal numbers or a *, which is replaced by the value of the next item in the *arglist*. Each conversion specification consumes one or (when *'s are involved) more items from the *arglist* and formats the last item it consumes. With %g, a precision of 0 (%.0g) specifies the shortest decimal string that rounds to the value being formatted. The standard C escape sequences are allowed: \a (*alert* or *bell*), \b (*backspace*), \f (*formfeed*), \n (*newline*), \r (*carriage return*), \t (*horizontal tab*), \v (*vertical tab*), and \x*d* and \x*dd*, where *d* denotes a hexadecimal digit.

The display command formats various entities in tables or lists, as appropriate. Its *disparglist* is similar to an *arglist* for print or printf, except that an item to be displayed can also be a set expression or the unsubscripted name of an indexed parameter, variable, constraint, or set; furthermore iterated *arglist*s cannot be nested, i.e., they are restricted to the forms

indexing expr
indexing (exprlist)

where *exprlist* is a nonempty, comma-separated list of expressions. The `display` command prints scalar expressions and sets individually, and partitions indexed entities into groups having the same number of subscripts, then prints each group in its own table or sequence of tables.

By default, the `display` command rounds numeric expressions to six significant figures, but this can be changed with the options `display_precision` or `display_round`, as described below.

A.13.2 Options and environment variables

AMPL maintains the values of a variety of *options* that influence the behavior of commands and solvers. Options resemble the "environment variables" of the DOS and UNIX operating systems; in fact AMPL inherits its initial options from the environments of these systems, as explained in the system-specific documentation. AMPL supplies its own defaults for many options, however, if they are not inherited in this way.

The `option` command provides a way to examine and change options. It has one of the forms

    option *redirection*$_{opt}$;
    option *envname* [*evalue*] [, *envname* [*evalue*] ...] *redirection*$_{opt}$;

The first form prints all options that have been changed or whose default may be provided by AMPL. In the second form, if an *evalue* is present, it is assigned to *envname*; otherwise the value (a character string) currently associated with *envname* is printed. An *envname* is either a name or a name-pattern, which is a name containing one or more `*`'s. In a name-pattern, a `*` stands for an arbitrary sequence, possibly empty, of name characters, and thus may match multiple names; for example

    option *col*;

lists all options whose names contain the string "`col`".

An *evalue* is a white-space separated sequence of one or more literals, numbers, and references to options of the form `$`*envname* or `$$`*envname*, in which *envname* contains no `*`'s; `$`*envname* means the current value of *envname*, and `$$`*envname* means the default value, i.e., the value inherited from the operating system, if any, or provided by AMPL. The quotes around a literal may be omitted if what remains is a name or number. The displayed option values are in a format that could be read as an `option` command; the redirection operations for printing commands can be used with `option` as well.

The initialization of some options may be determined by command-line arguments. The `-?` command-line argument produces a listing of these options and their command-line equivalents.

Several options whose names end with `_precision` control the precision with which floating-point numbers are converted to printable values; positive values imply rounding to that many significant figures, and 0 or other values imply rounding to the shortest decimal string that, when properly rounded to a machine number, would yield the number in question. If set to integral values, `$display_round` and `$print_round` override `$display_precision` and `$print_precision`, respectively. For example, `$display_round` *n* causes the `display` command to round numeric values to *n* places past the decimal point (or to $-n$ places before the decimal point if $n < 0$). Options that affect printing include those shown in Table A-10.

display_1col	maximum elements for a 1D table to be displayed one element per line
display_eps	display absolute numeric values < $display_eps as zero
display_max_2d_cols	if > 0, maximum data columns in a 2D display
display_precision	precision for display command when $display_round is not numeric
display_round	places past decimal for display command to round
display_transpose	transpose tables if rows − columns < $display_transpose
display_width	maximum line length for print and display commands
gutter_width	separation between columns for display command
objective_precision	precision for objective value displayed by solver
omit_zero_cols	if nonzero, omit all-zero columns from displays
omit_zero_rows	if nonzero, omit all-zero rows from displays
output_precision	precision used in nonlinear expression (.nl) files
print_precision	precision for print command when $print_round is not numeric
print_round	places past decimal for print command to round
print_separator	separator for values printed by print command

Table A-10: Options that control printing.

Sometimes it is convenient to have option settings remembered across AMPL sessions. Under operating systems from which AMPL can inherit environment variables as described above, the options OPTIONS_IN, OPTIONS_INOUT, and OPTIONS_OUT provide one way to do this. If OPTIONS_IN is nonnull in the inherited environment, it names a file (meant to contain option commands) that AMPL reads before processing command-line arguments or entering its command environment. OPTIONS_INOUT is similar to OPTIONS_IN; AMPL reads file OPTIONS_INOUT (if nonnull) after OPTIONS_IN. At the end of execution, if OPTIONS_INOUT is nonnull, AMPL writes the current option settings to file OPTIONS_INOUT. If nonnull, OPTIONS_OUT is treated like OPTIONS_INOUT at the end of execution.

A.13.3 The solve command

The solve command has the form

    solve redirection_opt ;

It causes AMPL to write the current translated problem to temporary files in directory $TMPDIR (unless the current optimization problem has not changed since a previous write command), to invoke a solver, and to attempt to read optimal primal and dual variables from the solver. If this works, the optimal variable values become available for printing with display, print, and printf commands. The optional *redirection* is for the solver's standard output.

The current value of the solver option determines the solver that is invoked. Appending '_oopt' to $solver gives the name of an option which, if defined with a nonnull string, determines (by the first letter of the string) the style of temporary problem files written; otherwise, AMPL uses its generic binary output format (style b). For example, if $solver is supersol, then $supersol_oopt, if nonnull, determines the output style. The command-line option −o? shows a summary of the currently supported styles.

AMPL passes two command-line arguments to the solver: the *stub* of the temporary files followed by −AMPL. AMPL expects the solver to write dual and primal variable values to file *stub* .sol, preceded by commentary that, if appropriate, reports the objective value to $objective_precision significant digits. In reading the solution, AMPL rounds the primal variables to solution_round places past the decimal point if solution_round is an integer, or to solution_precision significant figures if solution_precision is a positive integer; the defaults for solution_round and solution_precision imply no rounding.

A variable always has a current value. A variable declaration or data section can specify the initial value, which is otherwise 0. The option reset_initial_guesses controls the initial guess conveyed to the solver. If reset_initial_guesses has its default value of 0, then the current variable values are conveyed as the initial guess. Setting reset_initial_guesses to 1 causes the original initial values to be conveyed. Thus reset_initial_guesses affects the starting guess for a second solve command, as well as for an initial solve command that follows a solution command (described below).

A constraint always has an associated current dual variable value (Lagrange multiplier). The initial dual value is 0 unless otherwise given in a data section or specified in the constraint's declaration by a := *initial_dual* or a default *initial_dual* phrase. Whether a dual initial guess is conveyed to solvers is governed by the option dual_initial_guesses. Its default value of 1 causes AMPL to pass the current dual variables (if reset_initial_guesses is 0) or the original initial dual variables to the solver; if dual_initial_guesses is set to 0, AMPL will omit initial values for the dual variables.

AMPL's presolve phase computes two sets of bounds on variables. The first set reflects any sharpening of the declared bounds implied by eliminated constraints. The other set incorporates sharpenings of the first set that presolve deduces from constraints it cannot eliminate from the problem. The problem has the same solutions with either set of bounds, but solvers that use active-set strategies (such as the simplex method) may have more trouble with degeneracies with the sharpened bounds. Solvers often run faster with the first set, but sometimes run faster with the second. By default, AMPL passes the first set of bounds, but if option var_bounds is 2, AMPL passes the second set. The .lb and .ub suffixes for variables always reflect the current setting of $var_bounds; .lb1 and .ub1 are for the first set, .lb2 and .ub2 for the second set.

If the output style is m, AMPL writes a conventional "MPS" format file, and option $integer_markers determines whether integer variables are indicated in the MPS file by 'INTORG' and 'INTEND' 'MARKER' lines. By default, $integer_markers is 1, causing these lines to be written; specifying

        option integer_markers 0;

causes AMPL to omit the 'MARKER' lines.

A.13.4 *The* solution *command*

The solution command has the form

        solution *filename* ;

This causes AMPL to read primal and dual variable values from *filename*, as though written by a solver during execution of a solve command.

A.13.5 *The* write *command*

The write command has the form

Key	File	Description
a	*stub*.adj	constant added to objective values
c	*stub*.col	AMPL names of variables the solver sees
e	*stub*.env	environment (written by a solve command)
f	*stub*.fix	variables eliminated from the problem because their values are known
p	*stub*.spc	MINOS ''specs'' file for output style m
r	*stub*.row	AMPL names of constraints and objectives the solver sees
s	*stub*.slc	constraints eliminated from the problem
u	*stub*.unv	unused variables

Table A-11: Auxiliary files.

```
write  [ outopt-value ]  ;
```

in which the optional *outopt-value* must adhere to the quoting rules for a *filename*. If *outopt-value* is present, write sets $outopt to *outopt-value*. Whether or not *outopt-value* is present, write then writes the translated problem as $outopt dictates: the first letter of $outopt gives the output style (A.13.3), and the rest is used as a ''stub'' to form the names of the files that are created. For example, if $outopt is "m/tmp/diet", the write command will create an MPS file in /tmp/diet.mps, and auxiliary files /tmp/diet.row, /tmp/diet.col, and so forth. The solve command's rules for initial guesses, bounds, and integer markers apply.

A.13.6 Auxiliary files

The solve and write commands may cause AMPL to write some auxiliary files. For the solve command, appending _auxfiles to $solver gives the name of an option that governs the auxiliary files written; for the write command, $auxfiles plays this role. The auxiliary files shown in Table A-11 are written only if the governing option's value contains the indicated key letter.

A.13.7 The objective, drop and restore commands

These commands have the forms

```
objective  objective-name  ;
drop  indexing_opt  constr-or-obj-name  ;
restore  indexing_opt  constr-or-obj-name  ;
```

where *constr-or-obj-name* is the possibly subscripted name of a constraint or objective. The drop command instructs AMPL not to transmit the indicated entity (in write and solve commands); the restore command cancels the effect of a corresponding drop command. If *constr-or-obj-name* is not subscripted but is the name of an indexed collection of constraints or objectives, drop and restore affect all members of the collection.

The objective command arranges that only the named objective is passed to the solver. Issuing an objective command is equivalent to dropping all objectives, then restoring the named objective.

A.13.8 The `fix` and `unfix` commands

These commands have the forms

```
fix indexing_opt var-name ;
unfix indexing_opt var-name ;
```

where *var-name* is the possibly subscripted name of a variable. The `fix` command instructs AMPL to treat the indicated variable (in `write` and `solve` commands) as though fixed at its current value, i.e., as a constant; the `unfix` command cancels the effect of a corresponding `fix` command. If *var-name* is not subscripted but is the name of an indexed collection of variables, `fix` and `unfix` affect all members of the collection.

A.13.9 The `shell` command

The `shell` command provides a temporary escape to the operating system, if such is permitted, to run commands.

```
shell 'command-line' ;
shell ;
```

The first version runs *command-line*, which is contained in a literal string. In the second version, AMPL invokes an operating-system shell, and control returns to AMPL when that shell terminates. Before invoking the shell, AMPL writes a list of current options and their values to the file (if any) named by option `shell_env_file`. The name of the shell program is determined by option `SHELL`.

A.13.10 The `reset` and `update` commands

The `reset` command has several forms.

```
reset ;
```

causes AMPL to forget all model declarations and data and to close all files opened by redirection, while retaining option settings.

```
reset options ;
```

causes AMPL to restore all options to their initial state. It ignores the current `OPTIONS_IN` and `OPTIONS_INOUT`; the files they name can be included manually, if desired.

```
reset data ;
```

causes AMPL to forget all assignments read in data mode and allows reading data for a different problem instance.

```
reset data name-list ;
```

causes AMPL to forget any data read for the entities in *name-list*; commas between names are optional.

```
update data ;
```

permits all data to be updated once in subsequent data sections: current values may be overwritten, but no values are discarded.

```
        update data name-list ;
```

grants update permission only to the the entities in *name-list*.

A.13.11 The `let` command

The `let` command

```
        let indexing_opt name := expr ;
```

changes the value of the (possibly indexed) set, parameter or variable *name* to the value of the expression. If *name* is a set or parameter, its declaration must not specify a `:=` phrase.

A.13.12 The `quit` and `end` commands

The `quit` command causes AMPL to stop without writing any files implied by `$outopt`, and the `end` command causes AMPL to behave as though it has reached the end of the current input file, without reverting to model mode. At the top level of command interpretation, either command terminates an AMPL session.

A.14 Imported functions

Sometimes it is convenient to express models with the help of functions that are not built into AMPL. AMPL has facilities for importing functions and optionally checking the consistency of their argument lists. **Note:** The practical details of using imported functions are highly system-dependent. This section is concerned only with syntax; specific information will be found in system-specific documentation.

An imported function may need to be evaluated to translate the problem; for instance, if it plays a role in determining the contents of a set, AMPL must be able to evaluate the function. In this case the function must be linked, perhaps dynamically, with AMPL. On the other hand, if an imported function's only role is in computing the value of a constraint or objective, AMPL never needs to evaluate the function and can simply pass references to it on to a (nonlinear) solver.

Imported functions must be declared in a `function` declaration before they are referenced. This statement has the form

```
        function name alias_opt (domain-spec)_opt type_opt [ pipe litseq_opt [ format fmt ] ] ;
```

in which *name* is the name of the function, and *domain-spec* amounts to a function prototype:

> *domain-spec*:
>   *domain-list*
>   . . .
>   *nonempty-domain-list* , . . .

A *domain-list* is a (possibly empty, comma-separated) list of set expressions, asterisks (*'s), and *iterated-domain-list*s:

> *iterated-domain-list*:
>   *indexing* (*nonempty domain-list*)

An *iterated-domain-list* is equivalent to one repetition of its *domain-list* for each member in the *indexing* set, and the domain of dummy variables appearing in the *indexing* extends over that *domain-list*.

Omitting the optional (*domain-spec*) in the `function` declaration is the same as specifying (`...`). The function must be invoked with at least or exactly as many arguments as there are sets in the *domain-spec* (after *iterated-domain-list*s have been expanded), depending on whether or not the *domain-spec* ends with `...`. AMPL checks that each argument corresponding to a set in the *domain-list* lies in that set. A `*` by itself in a *domain-list* signifies no domain checking for the corresponding argument.

Type can be `symbolic` or `random` or both; `symbolic` means the function returns a literal (string) value rather than a numeric value, and `random` indicates that the "function" may return different values for the same arguments, i.e., AMPL should assume that each invocation of the function returns a different value.

The keyword `pipe` indicates that this is a *pipe function*, which means AMPL should start a separate process for evaluating the function. Each time a function value is needed, AMPL writes a line of arguments to the function process, then reads a line containing the function value from the process. (Of course, this is only possible on systems that allow multiple processes.) A *litseq* is a sequence of one or more adjacent literals, which AMPL concatenates and passes to the operating system (i.e., to $SHELL) as the description of the process to be invoked. In the absence of a *litseq*, AMPL passes a single literal, whose value is the name of the function. If the optional `format` *fmt* is present, *fmt* must be a format, suitable for `printf`, that tells AMPL how to format each line it sends to the function process. If no *fmt* is specified, AMPL uses spaces to separate the arguments it passes to the pipe function.

For example:

```
ampl: function mean2 pipe "awk '{print ($1+$2)/2}'";
ampl: display mean2(1,2) + 1;
mean2(1, 2) + 1 = 2.5
```

The function `mean2` is expected (by default) to return numeric values; AMPL will complain if it returns a string that does not represent a number.

The following functions are `symbolic`, to illustrate formatting and the passing of arguments.

```
ampl: function f1 symbolic pipe "awk '" '{printf "x%s\n", $1}' "'";
ampl: function g1 symbolic pipe 'awk ''{printf "XX%s\n", $1}''';
ampl: function cat symbolic pipe format ">>%s<<\n";

ampl: display f1(2/3);
f1(2/3) = x0.66666666666666667

ampl: display g1('abc');
g1('abc') = XXabc

ampl: display cat('some words');
cat('some words') = ">>'some words'<<"
```

The declaration of `f1` specifies a *litseq* of 3 literals, while `g1` specifies one literal; `cat`, having an empty *litseq*, is treated as though its *litseq* were `'cat'`. The literals in each *litseq* are stripped of the quotes that enclose them, have one of each adjacent pair of these quotes removed, and have (*backslash*, *newline*) pairs changed to a single newline character; the results are concatenated to produce the string passed to the operating system as the description of the process to be started. Thus for the four pipe functions above, the system sees the commands

```
awk '{print ($1+$2)/2}'
awk '{printf "x%s\n", $1}'
awk '{printf "XX%s\n", $1}'
cat
```

respectively. Function cat illustrates the optional format *fmt* phrase. If *fmt* results in a string
that does not end in a newline, AMPL appends a newline character. If no *fmt* is given, each
numeric argument is converted to the shortest decimal string that rounds to the argument value.

Caution: The line returned by a pipe function must be a complete line, i.e., must end with a
newline character, and the pipe function process must flush its buffers to prevent deadlock. (Pipe
functions do not work with most UNIX programs, because they don't flush output at the end of each
line.)

Imported functions may be invoked with conventional functional notation, as illustrated above.
In addition, iterated arguments are allowed. More precisely, if f is an imported function, an invocation of f has the form f (*arglist*) in which *arglist* is as for the print and printf commands
— a (possibly empty, comma-separated) list of expressions and *iterated-arglists*:

```
ampl: function mean pipe 'awk ''{x = 0\
        for(i = 1; i <= NF; i++) x += $i\
        printf "%.17g\n", x/NF}''';
ampl: display mean({i in 1..100} i);
mean({i in 1 .. 100} (i)) = 50.5

ampl: display mean({i in 1..50}(i,i+50));
mean({i in 1 .. 50} (i, i + 50)) = 50.5

ampl: display mean({i in 0..90 by 10}({j in 1..10} i + j));
mean({i in 0 .. 90 by 10} ({j in 1 .. 10} (i + j))) = 50.5
```

A.15 Defined variables

In some nonlinear models, it is convenient to define named values that contribute somehow to
the constraints or objectives. For example,

```
set A;
var v {A};
var w {A};
subject to C {i in A}: w[i] = vexpr;
```

where v*expr* is an expression involving the variables v.

As things stand, the members of C are constraints, and we have turned an unconstrained problem into a constrained one, which may not be a good idea. Setting option substout to 1
instructs AMPL to eliminate the collection of constraints C. It does so by telling solvers that the
constraints define the variables on their left-hand sides, so that, in effect, these defined variables
become named common subexpressions.

When $substout is 1, a constraint such as C that provides a definition for a defined variable
is called a *defining constraint*. AMPL decides which variables are defined variables by scanning the
constraints once, in order of their appearance in the model. A variable is eligible to become a
defined variable only if its declaration imposes no constraints on it, such as integrality or bounds.
Once a variable has appeared on the right-hand side of a defining constraint, it is no longer eligible
to be a defined variable — without this restriction, AMPL might have to solve implicit equations.
Once a variable has been recognized as a defined variable, its subsequent appearance on the left-

hand side of what would otherwise be a defining constraint results in the constraint being treated as an ordinary constraint.

Some solvers give special treatment to linear variables because their higher derivatives vanish. For such solvers, it may be helpful to treat linear defined variables specially. Ordinarily, variables involved in the right-hand side of the equation for a defined variable appear to solvers as nonlinear variables, even if they are used only linearly in the right-hand side. By doing Gaussian elimination rather than conveying linear variable definitions explicitly, AMPL can arrange for solvers to see such right-hand variables as linear variables. This often causes fill-in, i.e., makes the problem less sparse, but it may give the solvers a more accurate view of the problem. Setting option `linelim` to 1 instructs AMPL to treat linear defined variables in this special way; by default, it treats all defined variables alike.

A variable declaration may have a phrase of the form = *vexpr*, where *vexpr* is an expression involving variables. Such a phrase indicates that the variable is to be defined with the value *vexpr*. Such defining declarations allow some models to be written more compactly.

Recognizing defined variables is not always a good idea — it leads to a problem in fewer variables, but one that may be more nonlinear and that may be more expensive to solve because of loss of sparsity. By using defining constraints (instead of using defining variable declarations) and translating and solving a problem both with `$substout = 0` and with `$substout = 1`, one can see whether recognizing defined variables is worthwhile. On the other hand, if recognizing a defined variable seems clearly worthwhile, defining it in its declaration is often more convenient than providing a separate defining constraint; in particular, if all defined variables are defined in their declarations, one need not worry about `$substout`.

One restriction on defining declarations is that subscripted variables must be defined before they are used. For example,

```
var x;
var y{i in 1..2} = if i == 1 then x*x else x*y[i-1];
```

is fine, but

```
var x;
var y{i in 2..1 by -1} = if i == 1 then x*x else x*y[i-1];
```

elicits an error message.

A.16 Reserved and predefined words

These words are reserved, and may not be used in other contexts:

```
Infinity   data      else      if        less      setof     then
binary     default   exists    in        max       sum       union
by         dimen     forall    integer   min       symbolic  within
```

The other keywords, function names, etc., are predefined, but their meanings can be redefined. For example, the word `prod` is predefined as a product operator analogous to `sum`, but it can be redefined in a declaration, such as

```
set prod;   # products
```

Once a word has been redefined, the original meaning is inaccessible.

A.17 Synonyms

AMPL provides synonyms for several keywords and operators; the preferred forms are on the left:

```
^            **
=            ==
<>           !=
and          &&
coeff        coef
diff         difference
dimen        dimension
floor        int
inter        intersect        intersection
not          !
or           ||
prod         product
subject to   subj to          s.t.
```

Index

1. variables can be given initial values by

   var unknown : = initial value;

2. the model can include several objective function,
   and optimized by one of them, which is selected as:
   model name.mod;
   data name.dat;
   objective name; Path
   solve;

3. two sided constraints const-expr $\leq \ell(c, x) \leq$ const expr.
   One sided constraint   cost-expr (or $\ell(c, x)$) $\lesseqgtr \ell(c, x)$